AF581727

In-Depth on the Fouling and Antifouling of Ion-Exchange Membranes

In-Depth on the Fouling and Antifouling of Ion-Exchange Membranes

Editors

Lasâad Dammak
Natalia Pismenskaya

MDPI • Basel • Beijing • Wuhan • Barcelona • Belgrade • Manchester • Tokyo • Cluj • Tianjin

Editors
Lasâad Dammak
Université Paris-Est
France

Natalia Pismenskaya
Kuban State University
Russia

Editorial Office
MDPI
St. Alban-Anlage 66
4052 Basel, Switzerland

This is a reprint of articles from the Special Issue published online in the open access journal *Membranes* (ISSN 2077-0375) (available at: https://www.mdpi.com/journal/membranes/special_issues/Ion_Exchange_Membranes).

For citation purposes, cite each article independently as indicated on the article page online and as indicated below:

LastName, A.A.; LastName, B.B.; LastName, C.C. Article Title. *Journal Name* **Year**, *Volume Number*, Page Range.

ISBN 978-3-0365-2724-6 (Hbk)
ISBN 978-3-0365-2725-3 (PDF)

Contents

About the Editors

Lasâad Dammak is a professor specialized in Materials Science and in Industrial Process Engineering, at the Institut de Chimie et des Matériaux Paris-Est, which is a joint research unit between the Centre National de le Recherche Scientifique (CNRS, France) and the Université Paris-Est Créteil (UPEC, France). His research activities are mainly focused on ion-exchange materials and their applications in membrane processes in various fields, such as water treatment and purification, food industry, and energy production. These activities concern the synthesis, physicochemical, mechanical, and structural characterization of membranes, the modeling of their microstructures, and the study of their fouling throughout their life cycle, whereby antifouling solutions are provided to extend the life of these materials. He has participated in many national and international projects in this field.

Natalia Pismenskaya is a professor at the Department of Physical Chemistry and Head of the Laboratory of Electromembrane Phenomena at the Institute of Membranes, Kuban State University (Russia). Her research interests are mainly focused on the fundamental aspects of mass transfer in systems with ion-exchange membranes. These activities concern the development of experimental methods for investigating the phenomena of concentration polarization and studying the transfer of substances capable of participating in protonation–deprotonation reactions toward revealing the mechanisms of fouling during electrodialysis purification and concentration of various solutions, including the processing of liquid media in the food industry. She leads and participates in many national and international scientific projects in addition to teaching postgraduate and undergraduate students in this field.

Preface to "In-Depth on the Fouling and Antifouling of Ion-Exchange Membranes"

The use of ion-exchange membranes (IEMs) has accelerated over the past two decades in a wide variety of industrial processes (electrodialysis, electro-electrodialysis, electrolysis, dialysis, etc.) for applications related to the chemical, pharmaceutical, and food industries, energy production, and water treatments, among others. Organic and mineral fouling (or scaling) phenomena are two major factors limiting the efficiencies of IEM processes and performances (which reduce the selectivity and stability of IEMs, increase their electrical resistance, decrease the energy efficiency of the process, etc.) leading to significant economic losses. While the current washing, cleaning, and sterilization processes (antifouling treatments) make it possible to restore some of the loss in IEM performance, they frequently result in the degradation of membrane material.

Another essential point in fouling studies is to choose the best analysis and diagnostic technique to evaluate any degree of magnitude or to observe any type of object on the surface or within the membrane mass.

This book is a collection of the scientific contributions to the Special Issue "Fouling and Antifouling of Ion-Exchange Membranes" of the journal *Membranes*. It focuses on recent advancements in techniques for diagnosing and characterizing the fouling effects on membranes, the mechanisms governing this complex phenomenon, and the various innovative and economically viable solutions for reducing fouling.

Lasâad Dammak, Natalia Pismenskaya
Editors

 *membranes*

Editorial

In-Depth on the Fouling and Antifouling of Ion-Exchange Membranes

Lasâad Dammak [1,*] and Natalia Pismenskaya [2]

1 Institut de Chimie et des Matériaux Paris-Est (ICMPE), Université Paris-Est Créteil, CNRS, ICMPE, UMR 7182, 2 Rue Henri Dunant, 94320 Thiais, France
2 Department of Physical Chemistry, Kuban State University, 149 Stavropol'skaya Str., 350040 Krasnodar, Russia; n_pismen@mail.ru
* Correspondence: dammak@u-pec.fr; Tel.: +33-145171786

Citation: Dammak, L.; Pismenskaya, N. In-Depth on the Fouling and Antifouling of Ion-Exchange Membranes. *Membranes* **2021**, *11*, 962. https://doi.org/10.3390/membranes11120962

Received: 1 December 2021
Accepted: 3 December 2021
Published: 7 December 2021

This work is a synthesis of several in-depth studies on fouling and antifouling phenomena of ion-exchange membranes (IEMs). It is motivated by the increasing interest of the scientific and industrial communities in the diagnosis, quantification, and understanding of the different fouling aspects, and also in applications for its reduction. The main aim of such a synthesis is to help users of processes associated with ion-exchange membranes to find the best solution in order to extend the duration of membrane life and to reduce the operating costs of industrial processes.

Separation processes associated with ion-exchange membranes (IEM-processes), and with their two main families of dialysis (absence of electric current) and electrodialysis (application of an electric current) membranes, were first established for water treatment and desalination, and are still highly recommended in this field for their high water recovery, long life, and acceptable electricity consumption. Today, thanks to technological progress, these IEM-processes, especially electrodialysis, and the emerging new IEMs, have been extended to many other applications (chemical, food, pharmaceutical, and energy industries, etc.). This expansion of uses has also generated several problems, such as limitation of IEMs' lifetimes due to different ageing phenomena (because of organic and/or mineral compounds). If these aspects are not sufficiently controlled and mastered, the use and the efficiency of IEM-processes will be limited since they will no longer be competitive or profitable compared to other separation methods. The current commercial IEMs have excellent performances in IEM-processes; however, organic foulants such as proteins, surfactants, polyphenols, or other natural organic matter, can adhere to their surface (especially when using anion-exchange membranes (AEMs)) forming a colloid layer or they can infiltrate the membrane matrix, which leads to an increase in ion-transport resistance, resulting in higher energy consumption, lower water recovery, loss of membrane permselectivity and current efficiency as well as lifetime limitation. It has therefore become necessary and very useful to develop research areas to understand these fouling phenomena in order to act appropriately to limit their negative effects. Significant developments have been achieved in recent years and, currently, research is very active on this topic.

In this Special Issue, we have collected contributions on recent advancements in techniques for diagnosis and characterization of the fouling effects on ion-exchange membranes, the mechanisms governing this complex phenomenon, and the various innovative and economically viable solutions for reducing fouling. The Special Issue contains nine articles. Two reviews on ion-exchange membrane fouling during the electrodialysis process in the food industry are included; three studies are focused on the formation of fouling both on the surface and inside the IEMs, one of which involves a mathematical modeling of this phenomenon; three other studies are devoted to the antifouling processes, and the ninth study is dedicated to scaling (mineral fouling).

Dammak et al. [1] and Pismenskaya et al. [2] reviewed a significant number (about 380) of recent and relevant high-quality scientific papers, research articles and reviews studying the phenomena of IEM fouling during the ED process in the food industry with a special focus on the last decade. They first classified the different types of fouling according to the most commonly used classifications. Then, the fouling effects and the characterization methods and techniques as well as the different fouling mechanisms and interactions where presented, analyzed, discussed, and illustrated in a very in-depth manner. The relationships between the nature of the foulant and the structural, physicochemical, and transport properties and behaviors of ion-exchange membranes in an electric field were analyzed using experimental data and modern mathematical models. The implications of traditional chemical cleaning are taken into account in this analysis and modern "mild" membrane cleaning methods are discussed. Therefore, conclusions and outlook are provided, highlighting the main technical challenges, the current status of the fouling and antifouling phenomena of ion-exchange membranes in the food industry, and the key points for the future R&D. Finally, many challenges for the near future are identified and realistic suggestions are made.

Kozmai et al. [3] show the broad possibilities of electrochemical impedance spectroscopy for assessing the capacitance of interphase boundaries and the resistance and thickness of the foulant layer through the example of an AMX-Sb homogeneous membrane separating a red-wine solution from a 0.02 M sodium chloride solution. This enabled them to determine to what extent foulants affect the electrical resistance of ion-exchange membranes, the ohmic resistance and the thickness of diffusion layers, the intensity of water splitting, and the electroconvection in under- and over-limiting current modes. The authors established that short-term (10 h) contact of the AMX-Sb membrane with wine reduces the water splitting due to the screening of fixed groups on the membrane surface by wine components. On the contrary, biofouling, which develops upon a longer membrane operation, enhances water splitting, due to the formation of a bipolar structure on the AMX-Sb surface. This bipolar structure is composed of a positively charged surface of the anion-exchange membrane and negatively charged outer membranes of microorganisms. The island-like distribution of foulants on the membrane surface contributes to the development of electroconvection. Using optical microscopy and microbiological analysis, Kozmai et al. found that more intense biofouling was observed on the AMX-Sb surface that had not been in contact with wine.

Todeschini et al. [4] were interested in the herring milt hydrolysate (HMH) which, like many fish products, has the drawback of being associated with off-flavors. As odor is an important criterion, an effective deodorization method targeting the volatile compounds responsible for off-flavors was developed. First, the authors identified 15 volatile compounds known for their main contribution to the odor of HMH. Then, they evaluated the performances of the electrodialysis (ED) process to remove these 15 volatile compounds as well as trimethylamine, dimethylamine, and trimethylamine oxide, by testing the impact of both hydrolysate pH (4 and 7) and pulsed electric field (PEF) current modes (duration of a DC pulse and pause). The ED performance was compared with that of a deaerator by assessing three hydrolysate pH values (4, 7, and 10). The initial pH of HMH had a huge impact on the targeted compounds, while ED had no effect. Finally, the authors showed that the fouling formation, resulting from electrostatic and hydrophobic interactions between HMH constituents and ion-exchange membranes (IEM), the occurrence of water splitting at IEM interfaces, due to the reaching of the limiting current density, and the presence of water dissociation catalyzers were considered as the major limiting ED-process conditions.

Skolotneva et al. [5] focused on the electrochemical methods utilizing reactive electrochemical membranes (REMs) as a promising technology for efficient degradation and mineralization of organic compounds in natural, industrial, and municipal wastewaters. They proposed a two-dimensional (2D) model considering the transport and reaction of organic species with hydroxyl radicals generated at a TiO_x REM operated in flow-through mode, which takes into account convection, diffusion, and chemical reaction constants. This model allows the determination of unknown parameters of the system by treatment of experimental data and predicts the behavior of the electrolysis setup. The authors obtained a good agreement in the calculated and experimental degradation rate of a model pollutant (foulants molecules) at different permeate fluxes and current densities. The model also provides an understanding of the current density distribution over an electrically heterogeneous surface and its effect on the distribution profile of hydroxyl radicals and diluted species. Skolotneva et al. proved that the removal percentage of paracetamol increases with decreasing the pore radius by foulant molecules and/or increasing the porosity of the membrane. This effect becomes more pronounced as the current density increases. The model highlights how convection, diffusion, and reaction limitations have to be taken into consideration for understanding the effectiveness of the process.

Andreeva et al. [6] investigated scaling on the surface of a heterogeneous MK-40 cation-exchange membrane and on the same membrane, the surface layer of which contained polyaniline. The studies were carried out using a solution of a mixture of Na_2CO_3, KCl, $CaCl_2$, and $MgCl_2$ salts. Voltammetry and chronopotentiometry with the simultaneous control of the desalted solution pH provided information on the effect of current density on the concentration polarization of the membrane systems under study. A mineral scale on the surfaces of the studied membranes was found by scanning electron microscopy. Two limiting currents, which were observed on the current–voltage curve of the modified membrane, indicated the formation of a bipolar interface between the polyaniline layer with positively charged amino groups and the pristine membrane with negatively charged sulfone fixed groups. Enhanced water splitting due to the formation of this bipolar interface caused more significant scaling on the modified membrane compared to the commercial MK-40.

Nichka et al. [7] compared the effect of five different solution flow rates, corresponding to Reynolds numbers of 162, 242, 323, 404, and 485, combined with the use of pulsed electric field current mode to avoid protein fouling of the bipolar membrane (BPM) during electrodialysis of skim milk. They proved that application of PEF almost completely prevented fouling formation by proteins on the cationic interface of the BPM, regardless of the flow rate or Reynolds number. Indeed, under the PEF mode of current, the weight of protein fouling was negligible in comparison with the continuous current (CC) mode (0.07 ± 0.08 mg/cm^2 vs. 5.56 ± 2.40 mg/cm^2). When CC mode was applied, a Reynolds number equal to or higher than 323 corresponded to a minimal value of protein fouling of the BPM. The authors explain that the positive effect of both increasing the flow rate and using a PEF is due to the fact that during pauses, the solution flow flushes the accumulated protein from the membrane while at the same time there is a decrease in concentration polarization (CP) and consequently a decrease in H^+ flux at the feed solution/cationic interface of the BPM, minimizing fouling formation and accumulation.

Gil et al. [8] focused on modifying the surface of the MK-40 cation-exchange membrane to reduce fouling in wastewater ED by enhancing electroconvection. The surface of this membrane was covered with the perfluorosulfonic acid (PFSA) polymer film doped with TiO_2 nanoparticles. It was found that changes in surface characteristics conditioned by such modification lead to an increase in the limiting current density due to the stimulation of electroconvection, which develops according to the mechanism of electroosmosis of the first kind. The greatest increase in the current compared to the pristine membrane was obtained by modification with the film being 20 μm thick and containing 3 wt% of TiO_2. The sample containing 6 wt% of TiO_2 provided higher mass transfer in over-limiting current modes due to the development of nonequilibrium electroconvection. A 1.5-fold increase in the thickness of the modifying film reduced the positive effect of introducing TiO_2 nanoparticles due to (1) partial shielding of the nanoparticles on the surface of the modified membrane and (2) a decrease in the tangential component of the electric force, which affects the development of electroconvection.

Bdiri et al. [9] used enzymatic agents as biological solutions for cleaning ion-exchange membranes fouled by organic compounds during ED treatments in the food industry. They tested the cleaning efficiency of three enzyme classes (β-glucanase, protease, and polyphenol oxidase) chosen for their specific actions on polysaccharides, proteins, and phenolic compounds, respectively, fouled on a homogeneous cation-exchange membrane (referred to as CMX-Sb) and used for tartaric stabilization of red wine by ED in industry. First, enzymatic cleaning tests were performed using each enzyme solution separately with two different concentrations (0.1 and 1.0 g/L) at different incubation temperatures (30, 35, 40, 45, and 50 °C). The evolution of membrane parameters, such as electrical conductivity, ion-exchange capacity, and contact angle was determined to estimate the efficiency of the membrane′s principal action as well as its side activities. Based on these tests, these authors determined the optimal operating conditions for optimal recovery of the studied characteristics. Then, they tested two strategies combining these enzymes. The first one consisted of cleaning the fouled membranes with three successive enzyme solutions. The second one combined the use of two enzymes simultaneously in an enzyme mixture. In each case they took into account the optimal conditions of the enzymatic activity (concentration, temperatures, and pH). This study led to significant results, indicating effective external and internal cleaning by the studied enzymes. For example, they obtained a recovery rate of at least 25% of the electrical conductivity, 14% of the ion-exchange capacity, and 12% of the contact angle, and demonstrated the presence of possible enzyme combinations for the enhancement of the global cleaning efficiency or reducing cleaning durations. These important results prove, for the first time, the applicability of enzymatic cleanings to membranes and the inertia of their action towards the polymer matrix to the extent that the choice of enzymes is specific to the fouling substrates.

Funding: This research received no external funding.

Conflicts of Interest: The authors declare no conflict of interest.

References

1. Dammak, L.; Fouilloux, J.; Bdiri, M.; Larchet, C.; Renard, E.; Baklouti, L.; Sarapulova, V.; Kozmai, A.; Pismenskaya, N. A Review on Ion-Exchange Membrane Fouling during the Electrodialysis Process in the Food Industry, Part 1: Types, Effects, Characterization Methods, Fouling Mechanisms and Interactions. *Membranes* **2021**, *11*, 789. [CrossRef]
2. Pismenskaya, N.; Bdiri, M.; Sarapulova, V.; Kozmai, A.; Fouilloux, J.; Baklouti, L.; Larchet, C.; Renard, E.; Dammak, L. A Review on Ion-Exchange Membranes Fouling during Electrodialysis Process in Food Industry, Part 2: Influence on Transport Properties and Electrochemical Characteristics, Cleaning and Its Consequences. *Membranes* **2021**, *11*, 811. [CrossRef]
3. Kozmai, A.; Sarapulova, V.; Sharafan, M.; Melkonian, K.; Rusinova, T.; Kozmai, Y.; Pismenskaya, N.; Dammak, L.; Nikonenko, V. Electrochemical Impedance Spectroscopy of Anion-Exchange Membrane AMX-Sb Fouled by Red Wine Components. *Membranes* **2021**, *11*, 2. [CrossRef] [PubMed]
4. Todeschini, S.; Perreault, V.; Goulet, C.; Bouchard, M.; Dubé, P.; Boutin, Y.; Bazinet, L. Assessment of the Performance of Electrodialysis in the Removal of the Most Potent Odor-Active Compounds of Herring Milt Hydrolysate: Focus on Ion-Exchange Membrane Fouling and Water Dissociation as Limiting Process Conditions. *Membranes* **2020**, *10*, 127. [CrossRef] [PubMed]
5. Skolotneva, E.; Trellu, C.; Cretin, M.; Mareev, S. A 2D Convection-Diffusion Model of Anodic Oxidation of Organic Compounds Mediated by Hydroxyl Radicals Using Porous Reactive Electrochemical Membrane. *Membranes* **2020**, *10*, 102. [CrossRef] [PubMed]
6. Andreeva, M.A.; Loza, N.V.; Pis'menskaya, N.D.; Dammak, L.; Larchet, C. Influence of Surface Modification of MK-40 Membrane with Polyaniline on Scale Formation under Electrodialysis. *Membranes* **2020**, *10*, 145. [CrossRef] [PubMed]
7. Nichka, V.S.; Geoffroy, T.R.; Nikonenko, V.; Bazinet, L. Impacts of Flow Rate and Pulsed Electric Field Current Mode on Protein Fouling Formation during Bipolar Membrane Electroacidification of Skim Milk. *Membranes* **2020**, *10*, 200. [CrossRef] [PubMed]
8. Gil, V.; Porozhnyy, M.; Rybalkina, O.; Butylskii, D.; Pismenskaya, N. The Development of Electroconvection at the Surface of a Heterogeneous Cation-Exchange Membrane Modified with Perfluorosulfonic Acid Polymer Film Containing Titanium Oxide. *Membranes* **2020**, *10*, 125. [CrossRef] [PubMed]
9. Bdiri, M.; Bensghaier, A.; Chaabane, L.; Kozmai, A.; Baklouti, L.; Larchet, C. Preliminary Study on Enzymatic-Based Cleaning of Cation-Exchange Membranes Used in Electrodialysis System in Red Wine Production. *Membranes* **2019**, *9*, 114. [CrossRef] [PubMed]

Review

A Review on Ion-Exchange Membrane Fouling during the Electrodialysis Process in the Food Industry, Part 1: Types, Effects, Characterization Methods, Fouling Mechanisms and Interactions

Lasâad Dammak [1,*], Julie Fouilloux [1], Myriam Bdiri [1], Christian Larchet [1], Estelle Renard [1], Lassaad Baklouti [2], Veronika Sarapulova [3], Anton Kozmai [3] and Natalia Pismenskaya [3]

1 Institut de Chimie et des Matériaux Paris-Est (ICMPE), Université Paris-Est Créteil, CNRS, ICMPE, UMR 7182, 2 Rue Henri Dunant, 94320 Thiais, France; julie.fouilloux@u-pec.fr (J.F.); myriam.bdiri@u-pec.fr (M.B.); larchet@u-pec.fr (C.L.); e.renard@u-pec.fr (E.R.)

2 Department of Chemistry, College of Sciences and Arts at Al Rass, Qassim University, Ar Rass 51921, Saudi Arabia; bakloutilassaad@yahoo.fr

3 Department of Physical Chemistry, Kuban State University, 149, Stavropol'skaya Str., 350040 Krasnodar, Russia; vsarapulova@gmail.com (V.S.); kozmay@yandex.ru (A.K.); n_pismen@mail.ru (N.P.)

* Correspondence: dammak@u-pec.fr; Tel.: +33-145171786

Citation: Dammak, L.; Fouilloux, J.; Bdiri, M.; Larchet, C.; Renard, E.; Baklouti, L.; Sarapulova, V.; Kozmai, A.; Pismenskaya, N. A Review on Ion-Exchange Membrane Fouling during the Electrodialysis Process in the Food Industry, Part 1: Types, Effects, Characterization Methods, Fouling Mechanisms and Interactions. *Membranes* **2021**, *11*, 789. https://doi.org/10.3390/membranes11100789

Academic Editor: Marek Gryta

Received: 5 September 2021
Accepted: 11 October 2021
Published: 16 October 2021

Abstract: Electrodialysis (ED) was first established for water desalination and is still highly recommended in this field for its high water recovery, long lifetime and acceptable electricity consumption. Today, thanks to technological progress in ED processes and the emergence of new ion-exchange membranes (IEMs), ED has been extended to many other applications in the food industry. This expansion of uses has also generated several problems such as IEMs' lifetime limitation due to different ageing phenomena (because of organic and/or mineral compounds). The current commercial IEMs show excellent performance in ED processes; however, organic foulants such as proteins, surfactants, polyphenols or other natural organic matters can adhere on their surface (especially when using anion-exchange membranes: AEMs) forming a colloid layer or can infiltrate the membrane matrix, which leads to the increase in electrical resistance, resulting in higher energy consumption, lower water recovery, loss of membrane permselectivity and current efficiency as well as lifetime limitation. If these aspects are not sufficiently controlled and mastered, the use and the efficiency of ED processes will be limited since, it will no longer be competitive or profitable compared to other separation methods. In this work we reviewed a significant amount of recent scientific publications, research and reviews studying the phenomena of IEM fouling during the ED process in food industry with a special focus on the last decade. We first classified the different types of fouling according to the most commonly used classifications. Then, the fouling effects, the characterization methods and techniques as well as the different fouling mechanisms and interactions as well as their influence on IEM matrix and fixed groups were presented, analyzed, discussed and illustrated.

Keywords: ion-exchange membrane; electrodialysis; food industry; foulant identification; fouling mechanisms

1. Introduction

Electrodialysis (ED) was first established for water treatment applications, principal-ly for water desalination [1–4] and is still highly used and recommended in this field for its high water recovery, long lifetime compared to other usual technologies and acceptable electricity consumption [5]. In addition to that, the ED process is very flexible and is basically regulated to be an efficient desalination method that could be controlled by fixing the current (or potential drop) in the process. Today, thanks to technological progress related to the development of ED processes and the emergence of new ion-exchange membranes (IEMs) [6–10], this technique has been extended to other applications, for example, in the food industry for milk demineralization, deacidification and demineralization of beverages such as sugarcane and cranberries juices, tartaric stabilization of wine, desalination of cheese whey, treatment of glucose syrup, [11–15] etc.

This expansion of uses has increased the interest in ED techniques but has also generated several problems such as limitation of IEM membrane lifetime due to different age-ing mechanisms (prolonged use, cleaning operations, fouling phenomena, etc.) [16–18]. If these aspects are not sufficiently controlled and mastered, the use and the efficiency of ED processes will be limited, since it will no longer be competitive or profitable compared to other treatments and separation methods.

In our present work, we focused on the issues of IEM fouling in the ED industry and it should be noted that this is a subject of common interest to all types of membranes used in all baromembrane [2,19,20] and electromembrane [21–24] methods. As was defined and explained in our previous review on IEM cleanings and strategies of fouling prevention during ED in the food industry [25], the terms "fouling" and "scaling" designate organic matters and mineral matters, respectively, and the same definition has been adopted in this review.

Commercial IEMs show excellent performance in the ED processes, especially with new progress in membrane synthesis and manufacturing. However, the organic foulants such as proteins, surfactants, polyphenols or other natural organic matters [26–30] can adhere to the membrane surface forming a colloid layer or can infiltrate the membrane matrix, which leads to the increase in electrical resistance, resulting in higher energy con-sumption, lower water recovery and lifetime limitation [22,24,31,32]. Until now, membrane fouling has still represented a severe problem in the ED processes, especially when using anion exchange membranes (AEMs) [33,34]. IEM fouling not only can increase the electrical resistance but also decreases the lifetime of these membranes [5,35].

The interest in fouling, scaling and ageing phenomena has generated an increase in the number of studies and publications that deal with the different aspects of these issues. In this context, we propose to present the results of a statistical study on the evolution of the amount of published scientific work and research carried out on the themes of fouling and/or scaling of different membrane types during the last decade. Figure 1 presents the evolution of the number of publications on the topic of fouling and scaling of IEMs com-paring to other industrial filtration membranes from 2010 to 2020.

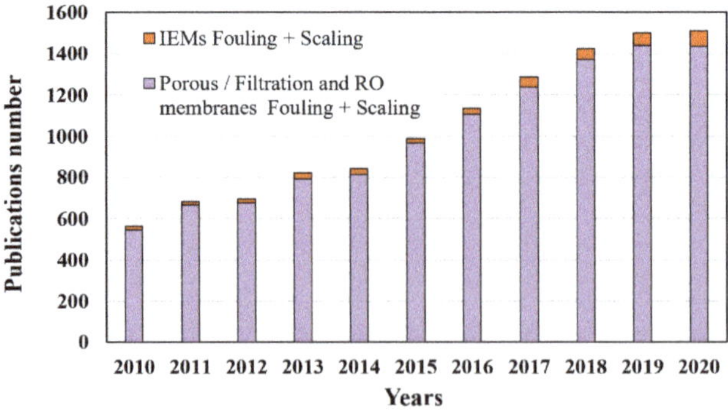

Figure 1. Evolution of the number of publications on the topic of fouling and scaling of ion-exchange membranes vs. other industrial filtration membranes from 2010 to 2020. Source of statistics: Web of Science. Keywords: "ion-exchange membrane" AND "fouling" OR "ion-exchange membrane" AND "scaling" OR "porous membrane" AND "fouling" OR "porous membrane" AND "scaling" OR "filtration membrane" AND "fouling" OR "filtration membrane" AND "scaling" OR "reverse osmosis membrane" AND "fouling" OR "reverse osmosis membrane" AND "scaling".

A continuous increase in the number of publications can be seen for both electromembrane and baromembrane technologies during the last decade, which confirms the growing interest for membrane fouling phenomena. It should also be noted that studies on IEMs do not exceed 5–6% of the published works on these themes each year. The slowdown in the rate of increase in the number of publications in 2020 is probably related to a decrease in scientific activity following the global health crisis. The large number of publications on porous, filtration and reverse osmosis (RO) membranes is due on the one hand to the older development of these techniques, and on the other hand to a greater diversity of their applications.

Figure 2 presents the evolution of the number of reviews on the topic of IEM fouling comparing to other industrial filtration membranes from 2010 to 2020. An overall increase in the number of reviews can be observed over the last 10 years despite some fluctuations from year to year. The ratio of the works carried out on the IEMs to that carried out on the filtration membranes varies between 0% and 22% in the last decade. In all cases, the number of reviews on IEM fouling is always lower or even negligible. Therefore, it is essential to further develop the studies and the research carried out on the fouling phenomena of IEMs.

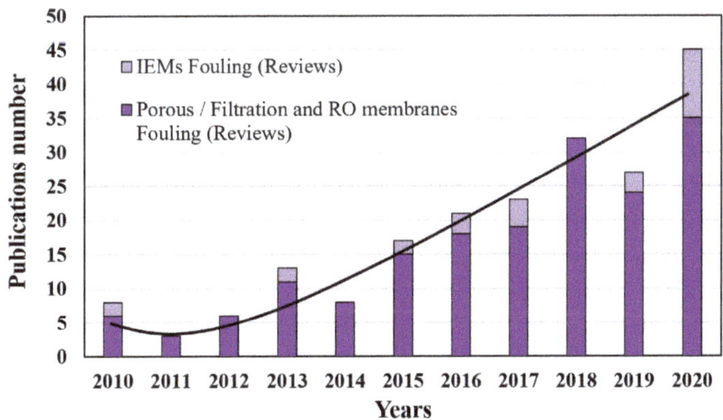

Figure 2. Evolution of the reviews number on the topic of ion-exchange membrane fouling vs. other industrial filtration membranes from 2010 to 2020. Source of statistic: Web of Science.

The present work reviewed a significant amount of scientific publications, research and reviews studying the phenomena of IEM fouling during ED process in food industry with a special focus on the last decade. It first classified the different types of fouling ac-cording to the most commonly used classifications. Then, the fouling effects, the characterization methods and techniques and the different fouling mechanisms and interactions are presented and illustrated.

2. The Nature of Foulants and Methods of Their Identification

The word "fouling" was used as a generic term to talk about all types of fouling in-cluding organic, inorganic, biological and colloidal matters in old publication [21,36]. Grossman et al. [21] used the term "fouling" to refer to the formation of a layer of impurities on the surface of the ion exchange membrane (IEM) and "poisoning" to refer to their penetration inside the membrane. In some cases, there is a confusion between the terms "clogging" and "fouling". The latter term can be used for both IEMs and filtration membranes [37] but the term "clogging" is more appropriate for porous membranes when talking about cake formation on the filtration membrane surface due to pore clogging and pore blocking or clogging of membrane defects by a layer of foulants [38,39].

It would be more judicious to use a common vocabulary to designate these phenomena for the community that works on these issues. In this review as in some recent studies, the word "fouling" refers to organic matters specifically but can be used to designate all fouling phenomena independently of the foulants' nature; the word "scaling" refers to mineral compounds and the terms "external" and "internal" designate the fouling and scaling of the membrane surface or bulk of the matrix, respectively.

The fouling phenomenon is defined as the deposition of undesirable molecules or substances of different nature (particles, colloids, macromolecules or salts) on the surface of the membrane or their adsorption inside the matrix due to different interactions and mechanisms depending on the nature of the foulants and the membrane material. All foulants are classified by their size [40], nature [41], as well as by the strength of binding to the membrane material [41]. Substances that can enter liquid media of the food industry differ in size as follows [42]:

- Dissolved substances (size less than 1 nm), including ions of inorganic substances, as well as ions and molecules of organic acids, saccharides, amino acids, proteins, phe-nolic compounds, etc.
- Colloidal particles (size from 1 nm to 1 μm) formed by inorganic, organic substances or their mixtures, the surface of which has a positive or negative charge;
- Suspended particles (larger than 1 μm) include biological objects (viruses, bacteria, and fungi), fragments of biological cells, colloidal aggregates and salt crystals.

Classification of substances by their nature includes [41,43–45]: inorganic (mineral) substances; organic compounds; colloidal particles; biological objects. Such division is rather conditional, since some organic dissolved compounds under certain conditions become colloidal particles, and not only salts of inorganic (gypsum, $CaSO_4\ 2H_2O$), but also those of organic (potassium hydrogen tartrate, KHT) acids can be poorly soluble [46].

In recent years, great strides have been made in the identification of foulants and in the study of their morphology in systems with IEMs. Some of them are examined below.

2.1. *Identification of Foulants*

2.1.1. Mineral Foulants

Mineral precipitation and crystal growth on the surface and in the pores of IEMs are characteristic of electrodialysis (ED) water conditioning or whey demineralization. Tradi-tionally, this type of fouling has been studied using scanning electron microscopy (SEM) combined with energy dispersive X-ray spectrometry (EDS) [47–50]. SEM allows one to determine the localization of mineral deposits and to identify the formed crystals by their shape. The EDS in turn gives an idea of the chemical composition of crystals and amorphous formations. For example, gypsum crystals have a characteristic acicular structure, as shown in Figure 3. X-ray diffraction (XRD) or X-ray absorption fine structure (EXAFS) methods [47,51] also seem to be very informative for identifying crystallographic inorgan-ic foulants. It should be noted that drying of the samples precedes the use of these methods. This preparation can lead to the destruction of membranes and crystals. In addition, the listed methods can only be applied when removing the IEM from the dialyzer, electrodialyzer, or membrane reactor. Therefore, high-resolution optical microscopy is becoming more and more popular. This method allows one to study swollen membranes, as well as to determine the IEM color depending on the pH and the composition of the bathing solu-tion (Figure 4). Using optical microscopy, successful attempts have been made to study scaling in situ [48] (Figure 5). In recent years, antiscalants that have fluorescent properties have been developed. Embedding in the structure of gypsum crystals [52,53], these an-tiscalants are indicators of sediment on the membrane surface. Thus far, such studies have been carried out in relation to reverse osmosis (RO) membranes (Figure 6). However, this method can become very informative for studying scaling (fouling by mineral components) in IEM systems.

The relatively new methods "Scanning ion conductance microscopy" (SICM) [54] and "Scanning electrochemical microscopy" (SECM) [55] allow scanning of the IEM sur-face and visualization of conductive and non-conductive areas that represent, within some distance, the surface morphology of the test material. For example, in one of the modifications of these methods [56,57], a pair of very fine microcapillaries (of about 5 μm in diameter) measures the potential drop on both sides of the membrane in an ED cell at a given current density. These capillaries can move in three mutually perpendicular direc-tions (X, Y, Z). The processing of chronopotentiogramms measured in situ at the nodes of 2D or 3D lattice under real electrical, chemical and hydraulic conditions of ED provide information on the IEM surface geometry. Moreover, this geometry is in good agreement with SEM images of dry (Figure 7) or swollen membranes obtained using optical micros-copy.

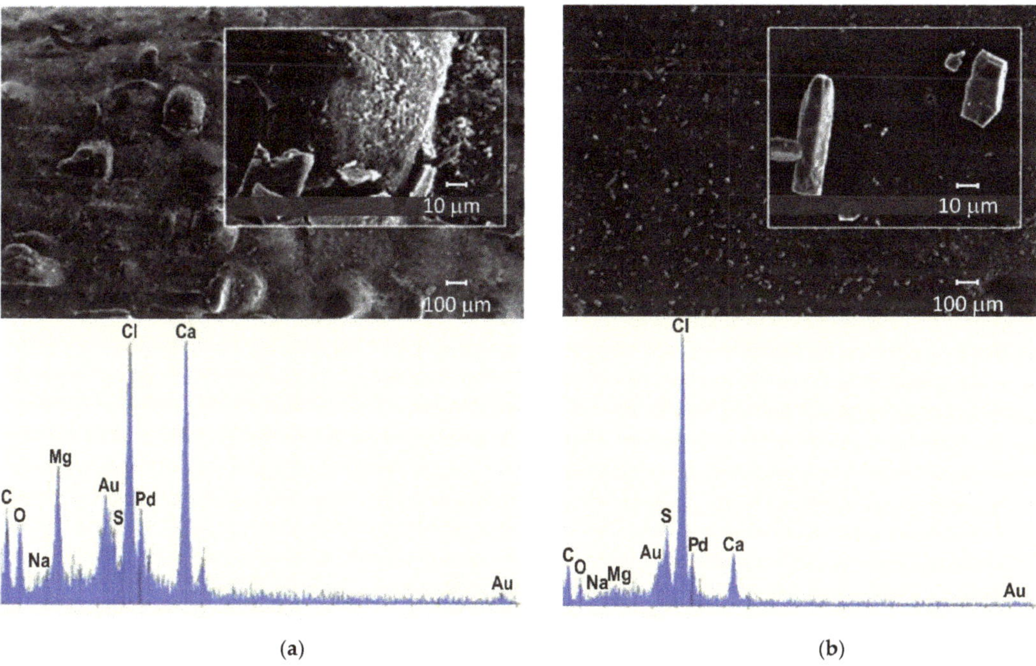

Figure 3. SEM images and EDS spectrograms of Neosepta CMX-Sb (Astom, Shunan, Japan) cation exchange membrane after operation in ED milk processing at pH = 5.4: surface facing to concentrate (**a**) or diluate (**b**). Adapted from [50].

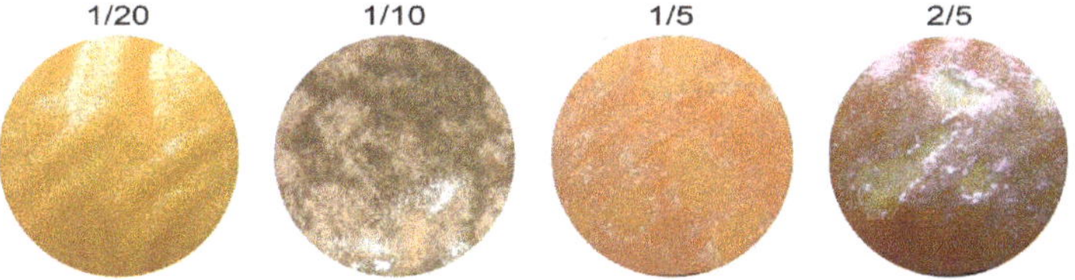

Figure 4. In-situ optic observation of scaled Selemion AMV (Asahi Glass, Tokyo, Japan) anion exchange membrane after 1 (**a**) and 7 (**b**) hour operation in a solution containing Ca^{2+} and SO_4^{2-} ions. The study was carried out using a stereo microscope (Zeiss Discovery V8 Stereosope with total magnitude of X6.3–50.4, Norway) connected to a digital camera (Power-shot A640, Canon). Adapted from [48].

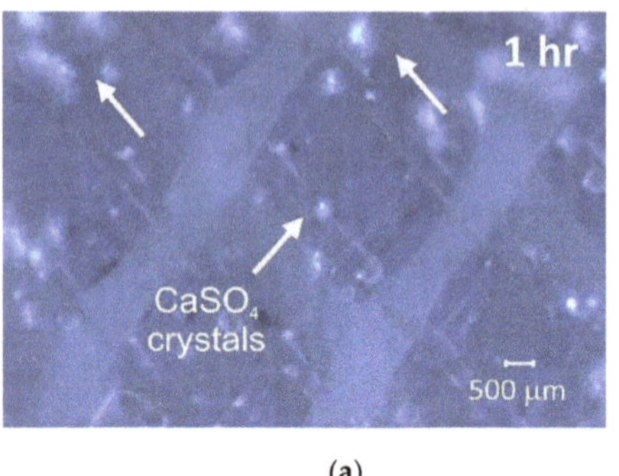

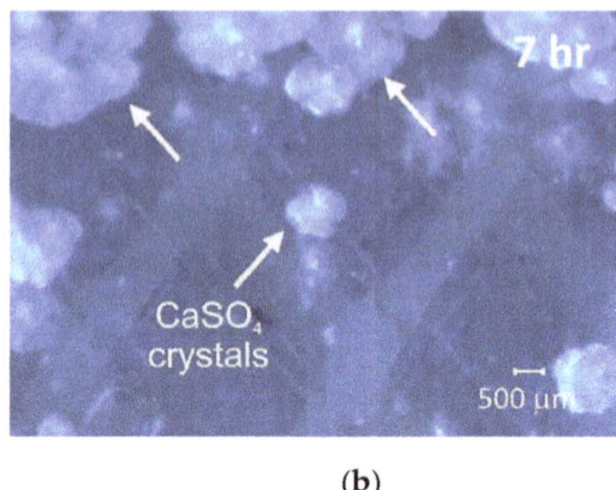

(**a**) (**b**)

Figure 5. SEM (**a**) and 3D fluorescent optic (**b**) images of gypsum deposit on reverse osmosis RE182 (CSM Co., Seoul, Korea) membrane surface (**b**). Adapted from [52].

The relatively new methods "Scanning ion conductance microscopy" (SICM) [54] and "Scanning electrochemical microscopy" (SECM) [55] allow scanning of the IEM sur-face and visualization of conductive and non-conductive areas that represent, within some distance, the surface morphology of the test material. For example, in one of the modifications of these methods [56,57], a pair of very fine microcapillaries (of about 5 µm in diameter) measures the potential drop on both sides of the membrane in an ED cell at a given current density. These capillaries can move in three mutually perpendicular direc-tions (X, Y, Z). The processing of chronopotentiogramms measured in situ at the nodes of 2D or 3D lattice under real electrical, chemical and hydraulic conditions of ED provide information on the IEM surface geometry. Moreover, this geometry is in good agreement with SEM images of dry (Figure 7) or swollen membranes obtained using optical micros-copy.

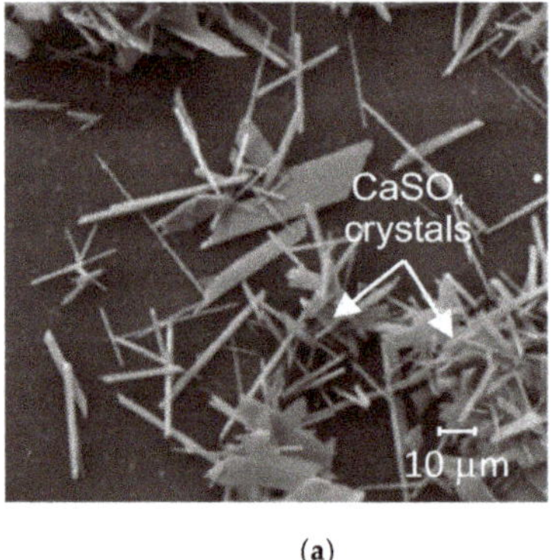

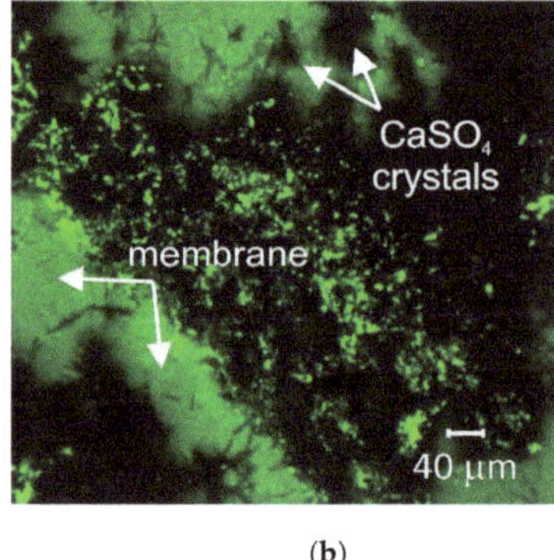

(**a**) (**b**)

Figure 6. Photographs of Neosepta CMX-Sb membrane taken from the side of concentration compartment after ED treatment of solutions imitating whey (pH = 12). Numbers above the photo show magnesium/calcium ratios in the solution. Adapted from [58].

Comparison of the results of blank experiments carried out, for example, in a NaCl solution, with the data obtained in multicomponent solutions prone to precipitation, allows one to monitor the scaling process and its effect on the conductivity of the membrane surface [59].

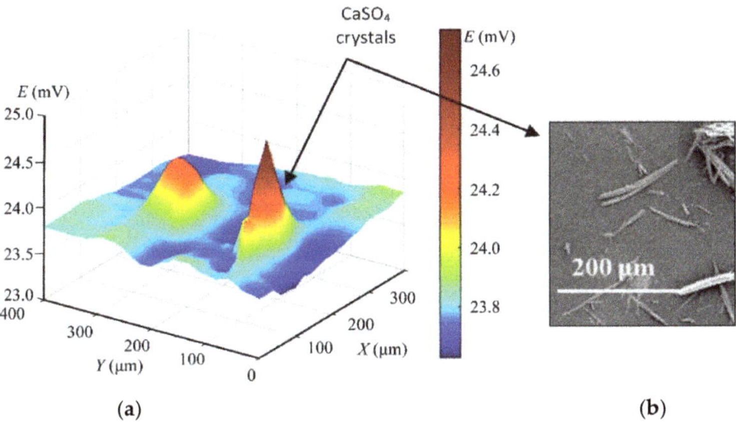

Figure 7. SECM 3D potential drop distribution at the surface of the swollen Neosepta CMX (Astom, Shunan, Japan) membrane with $CaSO_4$ crystals (**a**) and SEM image of the same (dry membrane) surface (**b**). Adapted from [59].

2.1.2. Organic Foulants and Colloidal Particles

Organic molecules of BSA, humate, carbonic acids, amino acids, anionic surfactants such as SDBS, peptides [13,24,34,46,60–69] and more recently phenolic and polyphenolic compounds including anthocyanins [26,31,32,70–75] are the main substances that are found in the industrial environments of the food industry or their imitations. Many organic foulants are known to be negatively charged at neutral pH. The interaction of such substances with Ca^{2+} and Mg^{2+} cations and positively charged fixed groups is the reason for the formation of a thick gel-like film on the anion exchange membrane (AEM) surface [76–78]. At the same time, the presence of oily compounds in the processed solution does not lead to such fouling. The difficulty in identifying organic foulants lies in the fact that the processed solutions, as a rule, contain not one, but several substances close in elemental composition. For example, wine and fruit or berry juices can contain up to 600 components [26,31,32,79,80]. Moreover, the chemical structure of IEM and foulants is often too similar. Inside membranes, foulants can form compounds that were not present in the processed fluids [79,80].

Identification of Typical Chemical Elements

The EDS method is widely used for the identification of organic foulants or other substances on the IEM surface [81,82] (Figure 8). At the same time, drying of IEM, preceding, for example, SEM and EDS, has a destructive effect on the colloidal particles and the three-dimensional structure of most organic foulants. In addition, EDS gives reliable re-sults only if the depth of the analyzed layer is more than 1 μm and is of little use for the analysis of thinner layers [83]. This limits the use of the traditional method of SEM-EDS and requires the employment of other suitable methods of identifying substances.

A wide range of methods are used to identify typical chemical elements characteristic of foulants. For example, Cheesman et al. [84] used ^{31}P nuclear magnetic resonance spec-troscopy, molybdate colorimetry and inductively coupled plasma optical emission spec-trometry to identify phosphorus in organic and condensed inorganic compounds ad-sorbed by AEMs. X-ray photoelectron spectroscopy (XPS) or Rutherford backscattering spectroscopy (RBS) is used to analyze the elemental composition of a thin (up to 10 nm) surface layer [51,83]. Note that XPS and RBS are informative if the chemical composition of foulants differs significantly from that of the IEM [83,85,86]. However, they are not very useful for identifying proteins, amino acids, carboxylic acids and polyphenols, which are composed of the same chemical elements as the IEM.

Figure 8. EDS element mapping of the fouled anion-exchange membrane cross-section. Foulant is sodium dodecyl sulfate (SDS). Adapted from [82].

Total nitrogen content analysis is mostly used for the detection and identification of proteins, peptides, amino acids, and other low molecular weight compounds [13,62–65,87] in IEM. The total nitrogen content can be determined using the Dumas method. In this method, dried membrane samples are combusted in the presence of O_2 at high temperature of about 900 °C to 1000 °C to convert the sample into CO_2, water and nitrogen. Then, the products of combustion are passed through a thermoelectric cooler to remove the water. The nitrogen gases are quantified using a thermal conductivity detector where He gas is used as reference. For the study of IEM fouling, several authors have used LECO nitrogen quantification method which is simple, rapid and automated but leads to mem-brane destruction [64]. This technique requires large sample size of about 150 mg dry membrane to obtain accurate and reproducible results. A pristine AEM, which contains fixed amino groups, is usually used as reference or control [79]. For example, Langevin et al. [63] used nitrogen content analysis to investigate the fouling of Neosepta CMX-Sb and AMX-Sb (Astom, Shunan, Japan) with a soy protein hydrolysate solution (SPHS) and study the effects of the different pre-treatments on fouling formation. The results showed that that cation exchange membrane (CEM) was almost two times more sensitive to peptide fouling than AEM. The nitrogen content observed for the CEM pretreated in HCl solution after soaking in the SPHS would be 12% higher than the ones pretreated by distilled water and NaOH (Figure 9). Persico et al. [65] preconized these methods to characterize peptide fouling on AEMs by a tryptic whey protein hydrolysate where it appeared that peptide charge modifications related to alkalinization, have a severe impact on AEM fouling. The authors also studied the formation of peptide layers and its adsorption mechanisms on AEMs and CEMs [63] depending on the pH of feed solution.

Identification of Characteristic Chemical Groups

Attenuated total reflection infrared spectroscopy (ATR–FTIR) is one of the most commonly used methods for the identification of organic fouling for IEMs and other membranes [56,57,59,67,88]. For example, Suwal et al. [80] used the ATR–FTIR to investigate the membrane surface (facing the anode or cathode) sensitive to peptide or amino acid fouling and deterioration after their use in consecutive ED-UF treatments. It should be noted that, in attenuated total reflection mode, this spectroscopy allows the identification of functional groups present on a thickness of about 1 μm.

Bdiri et al. [26,31,32] investigated the fouling of Neosepta AMX-Sb and CMX-Sb used for the tartaric stabilization of red wine by phenolic compounds. The comparison and the interpretation of the ATR–FTIR spectra of pristine and fouled membranes (Figure 10) showed, principally, the presence of highly hydrated C=O, –COOH due to the accumulation of phenolic acids and phenolic compounds in AEMs which confirmed the affinity of organic acids for AEMs. The results also showed the intensification of C=C stretching bands of polyphenol aromatic rings in both membranes but this was more intense in CEMs.

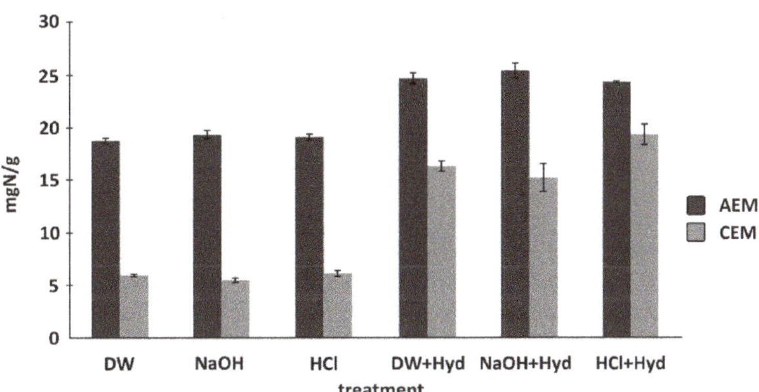

Figure 9. Evaluation of nitrogen content of CMX-Sb and AMX-Sb before (DW, NaOH, HCl) and after soaking in soy protein hydrolysate solution (DW + Hyd, NaOH + Hyd, HCl + Hyd) for 24 h (adapted from [63]). IEMs were pretreated in distilled water (DW), 1 M NaOH or HCl before the experiment.

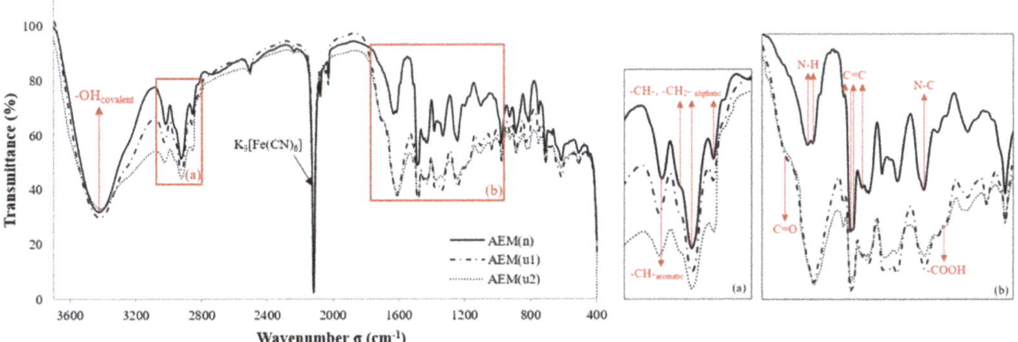

Figure 10. ATR–FTIR spectra of pristine (AEM(n)) and fouled (AEM(u1), AEM(u2)) anion exchange membrane Neosepta AMX-Sb used in the tartaric stabilization of red wine [31].

Xie et al. demonstrated the capabilities of the well-resolved Synchrotron Fourier transform infrared mapping [89] for quantifying organic foulants such as alginate (Figure 11) in membrane distillation. We believe that this relatively new method can be extremely useful for obtaining information on the localization and accumulation of specific organic foulants in the case of IEMs as well.

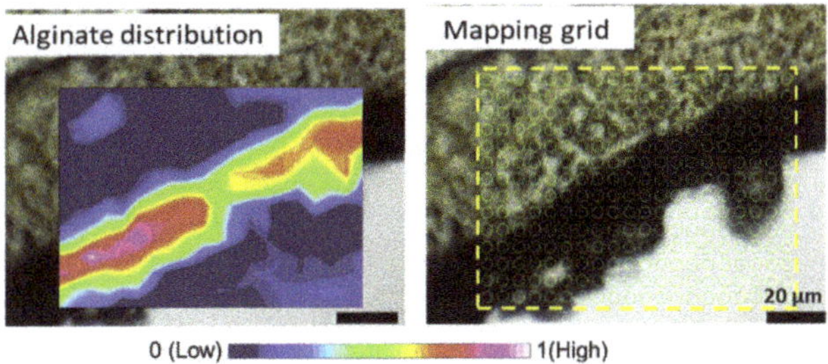

Figure 11. Synchrotron Fourier transform infrared mapping of alginate distribution and abundancy in the cross-section of membrane (in the left) and mapping grid (in the right). Spectrum integration of characteristic wavenumber was carried out to plot the map. Adapted from [89].

It is known that some organic groups are chromophores and fluorophores. These properties allow the use of classical optical microscopy and optical spectroscopy techniques to study the participation of these substances in fouling [90]. For example, the representatives of polyphenols—anthocyanins, are key substances in the fouling of ion exchange materials used for wine stabilization and juice conditioning. These substances contain a chromophore group, the structure of which (and the color of anthocyanin) depends on the pH of the medium [91]: red flavylium cation (pH < 3); colorless carbinol pseudobase (pH = 4~5); purple quinoidal anhydrobase (pH = 6~7); deep blue anhydrobase anion (pH = 7~8); green anhydrobase dianion (pH = 8~10) and yellow chalcone dianion (pH > 11). In natural solutions containing a mixture of anthocyanins, the color palette may undergo some changes, but the general trend remains (Figure 12a). It was shown in [92] that this property of anthocyanins can be used to assess their structure within ion-exchange materials. Indeed, the color of aliphatic ion exchange resins equilibrated with anthocyanin-containing solutions (pH from 3 to 9) is in good agreement with the ATR–FTIR results. However, ATR–FTIR provides little information in the case of IEM, which has an aromatic matrix, due to the similarity of its structure to that of anthocyanins. In this case, changes in the color of the studied samples, for example aromatic anion exchange resin AV-17-2P (KHIMIMPEX LLC, Kiev, Ukraine), (Figure 12b) are easy to estimate using optical microscopy and the color indication scale [33,92]. Sarapulova et al. [33,92], using this method, established that the pH of the internal solution is shifted to acidic (cation exchange resins and CEMs) or alkaline values (anion exchange resins and AEMs) as compared to an equilibrium solution. Therefore, the structure of anthocyanins inside ion-exchange materials differs from their structure in an external solution (Figure 12b). The reason is Donnan exclusion of co-ions [93]: hydroxyl ions or protons, that are products of protonation–deprotonation of polar groups, substances entering the IEM or water.

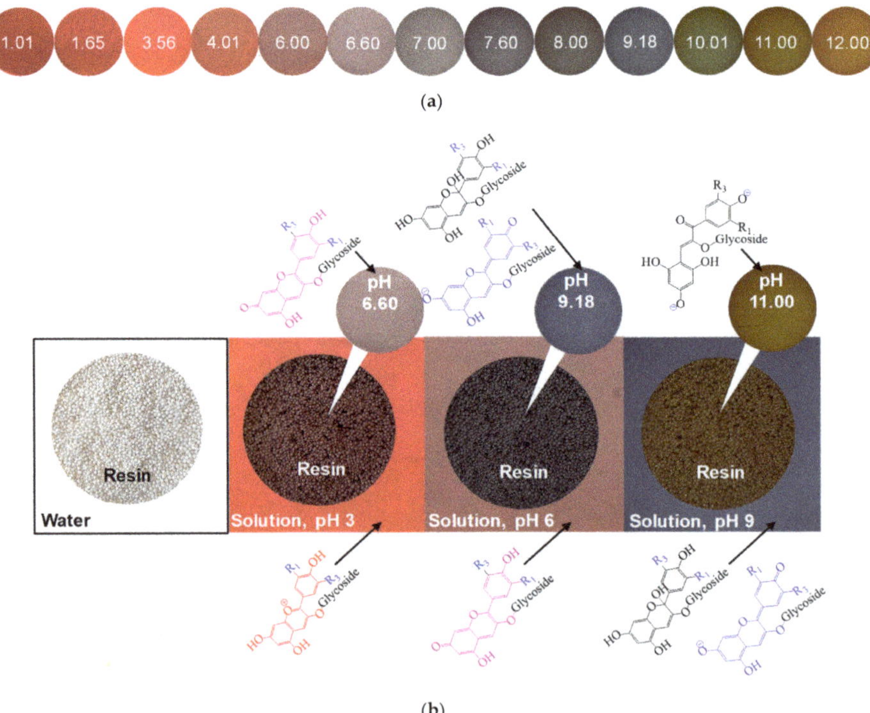

Figure 12. Colors of anthocyanin mixture solution depending on pH (**a**) as well as color of the aromatic anion-exchange resins AV-17-2P (KHIMIMPEX LLC, Kiev, Ukraine) equilibrated with distilled water and anthocyanin solutions of pH 3, 6 or 9 (**b**). Adapted from [92].

Another alternative to ATR–FTIR is Raman spectroscopy, the use of which does not require preliminary drying of the studied membrane samples [51,94]. Surface-enhanced Raman spectroscopy (SERS) and Tip-enhanced Raman spectroscopy (TERS) can be used for real-time monitoring of membrane fouling. For example, Virtanen et al. [95] used Ra-man spectroscopy to study the interactions between the surface of the polyethersulfone membrane and vanillin, which is a foulant. Chen et al. [96] studied the adsorption and aggregation of proteins and polysaccharides in the pores of microfiltration membranes. Figure 13 illustrates an example of SERS application to study the adsorption of myoglobin on the surface of polyvinylidene fluoride (PVDF) membrane. Here, the peak height was determined as 752 cm^{-1}. Note that this technique is limited by the high fluorescence of some polymers constituting the membranes, such as polystyrene-based membranes.

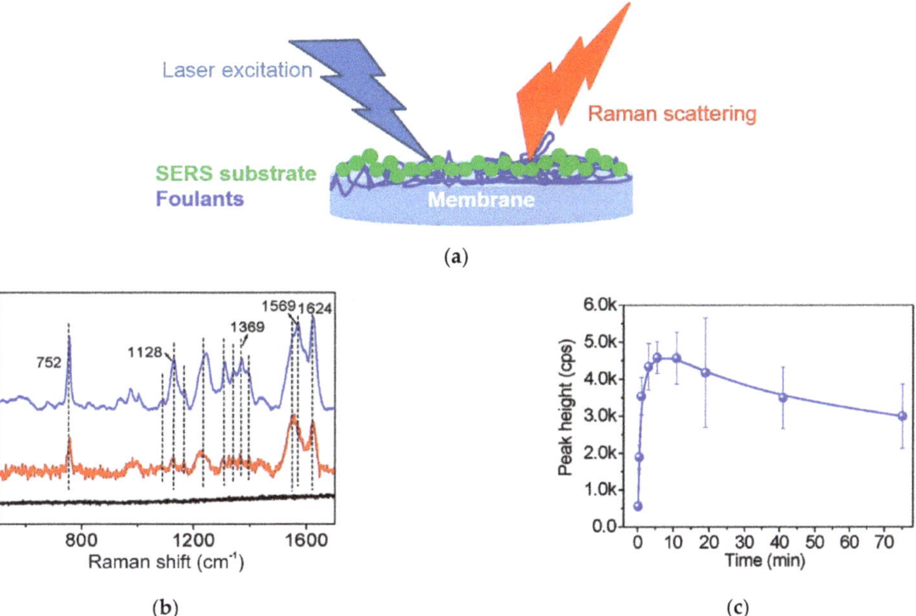

Figure 13. Surface-enhanced Raman Spectroscopy applied to monitor PVDF membranes fouling by proteins. (**a**) is a schematic illustration of SERS measurement; (**b**) is SERS spectrum of myoglobin on a PVDF membrane (blue line) with Ag sol, Raman spectrum of pure myoglobin solid (red line), and myoglobin on a PVDF membrane without Ag sol (black line). (**c**) is obtained at 752 cm^{-1}. Adapted from [97].

Fluorescence-based techniques have become increasingly popular in recent years. Fluorescence spectroscopy, the UV-vis spectra and fluorescence excitation–emission matrix (EEM) analysis allow detection of the UV humic-like substances' adsorption on ion exchange resins [98]; EEM coupled with parallel factor analysis (PARAFAC) can provide some structural information about the metal binding with some organic matter, as well as organic substances (for example humic acid and protein) interactions [98–102]; two di-mensional fluorescence/Fourier transform infrared correlation spectroscopy gives information concerning the localization of functional groups of humic substances in complex metal cations [103]. For example, Peiris et al. [104] applied a fluorescence-based technique in combination with the Surface plasmon resonance (SPR) method to study the adsorption of α-lactalbumin, protein-like matter as well as colloidal particles. They found that inter-particle or inter-molecular physical-level interactions between these substances can be detected by signal attenuation in the SPR measurements.

Identification of Substances Included in the Composition of Foulant

Spectrophotometric techniques, where each of the components are determined at a given wavelength [92,105], are the simplest and most commonly used methods for the quantitative determination of various foulants in solutions. For example, the mass concentration of the sum of anthocyanins in terms of cyanidin-3-glucoside is determined based on the change in the light absorption (wavelength of 510 nm) when the pH of the studied solutions changes from 1 to 4.4 units [92]. Protein concentration is measured at 595 nm using the chromogenic agent Coomassie brilliant [105], etc.

Sodium dodecyl sulphate–polyacrylamide gel electrophoresis (SDS-PAGE) allows separation of proteins based on differences in their molecular weight (Figure 14) [68]. Mo-lecular weight (MW) standards are used for protein identification.

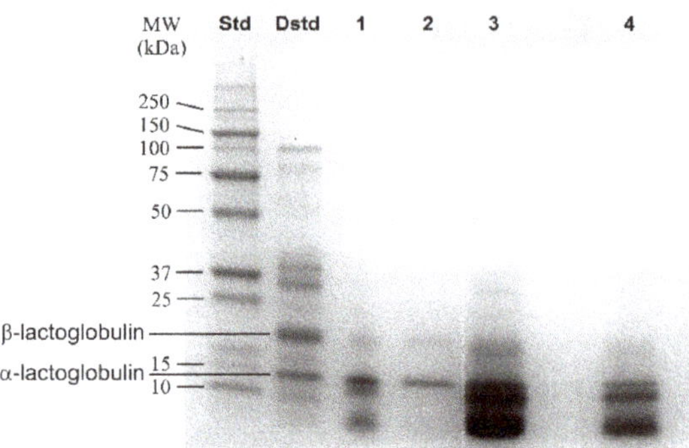

Figure 14. SDS-PAGE profile of molecular weight standard (Std), moleculer weight dairy standard (Dstd), whey protein hydrolysate before centrifugation (1), supernatant (2), precipitate (3) and gel on the AEM surface after electrodialysis (4). Adapted from [68].

The Fourier transform-ion cyclotron resonance-mass spectrometry (FT-ICR-MS) method should also be mentioned. Ray et al. [106] are sure that FT-ICR-MS is the most in-formative tool to detect organic matters at molecular levels.

Chromatographic methods are widely used for IEM fouling investigation [26,46,68, 72,73,105,107–109]. In the case of organic fouling with substances of different molecular weights, Size-exclusion (SEC) High-Liquid Performance Chromatography (HPLC) was found to be an effective method to identify foulants [46,107–109]. For exam-ple, Bdiri et al. [26] developed an extraction method of organic foulants from fouled Ne-osepta AMX-Sb and CMX-Sb with phenolic compounds during tartaric stabilization of red wine by ED. A mixture of four solvents at 25% volume each (acetone, methanol, isopropanol and ultra-pure water) was used. Then, specific Ultra-High-Liquid Performance Chromatography (UPLC) and HPLC methods were performed for the identification of phenolic compounds extracted and it was, for the first time, possible to identify numerous phenolic compounds adsorbed in the membranes (examples in Table 1).

Mass spectrometry (MS) coupled to the UPLC (UPLC-MS) [68] or to HPLC (HPLC-MS) [63] gives important and significant information about organic fouling and leads to a specific identification of the fouling nature by allowing choice of the appropriate chromato-graphic method to analyze each sample. Note that a necessary requirement for the application of this method is the preliminary extraction of foulants from the IEM by extraction methods that have to be previously optimized and have the appropriate standards. Aqueous solutions of mineral salts [110], individual organic substances [101,111] or mixtures of organic solvents [26,73] are used as stripping solutions. The samples are preliminarily subjected to finer grinding [73] to ensure the extraction of high molecular weight compounds, for example, proanthocyanidins, from the IEM. In the course of such pre-treatment, colloidal particles and some chemical compounds are destroyed; the extraction of foulants from the membrane may be incomplete, etc. The macromolecules of food polyphenols and anthocyanins [112,113], protein mixtures [114], etc., adsorbed onto the membrane surface are desorbed with a laser adsorbing solution known as a MALDI matrix and directly analyzed in MS [114].

Table 1. Phenolic compounds identified in IEMs after tartaric stabilization of red wine by ED in industry (μg eq. Gallic acid/g dried membrane) during 6000 (CMX-Sb) and 1738 (AMX-Sb) hours. Adapted from [26].

Phenolic Compaunds	CMX-Sb	AMX-Sb
Quercetin	68.1 ± 9.6	40.4 ± 9.1
Quercetin-3-glucoside	10.9 ± 2.0	2.4 ± 0.9
Quercetin-3-galactoside	1.9 ± 1.1	3.3 ± 0.3
Quercetin-3-rhamnoside	—	7.2 ±0.8
Kaempferol	2.0 ± 1.8	10.0 ± 0.8
Kaempferol-3-glucoside	2.6 ± 0.6	—
Myricetin-3-glucoside	2.3 ± 1.4	—
Isorhamnetin	176.8 ± 8.6	37.8 ± 8.9
4-hydroxylbenzoic acid	8.5 ± 3.3	44.3 ± 21.1
Protocatechic acid	28.0 ± 4.4	26.8 ± 2.8
Vanillic acid	9.1 ± 3.9	18.8 ± 5.5
Piceid	3.7 ± 2.9	1.4 ± 0.6
Resveratrol	5.5 ± 0.8	22.4 ± 4.7

2.1.3. Biofouling

Membrane biofouling starts with the attachment of microbial cells at their surface. The interactions underlying this initial attachment are hydrophobic and electrostatic, leading to cell growth and multiplication thanks to soluble nutrients present in the feed water or organic foulants already present as a conditioning film. The microorganism metabolism will then excrete extracellular polymeric substances (EPS) which will create a three-dimensional matrix. EPS are high molecular weight substances being polysaccharides, proteins, nucleic acids and lipids, and account for 90% of biofilm dry mass [115]. In a nutshell, a biofilm is composed of microorganisms (10%w) living in a self-produced hydrated EPS matrix (90%w) providing mechanical stability, adhesion surface and an in-terconnected network promoting their metabolism.

Biofouling is most often identified by standard microbiological methods [66]. In particular, smear-prints of the membrane surface are applied to degreased glass slides, dried, and then stained by the Gram method using a carbolic solution of gentian violet, Lugol's and fuchsin solution. Gram-negative microorganisms painted in a pink–red color, and Gram-positive microorganisms becomes blue–violet (Figure 15). Using this method, for example, it was found [67] that the side of AEM, which is exposed to the NaCl solution, is more susceptible to biofouling. The opposite side, which is in contact with wine and con-tains ethyl alcohol and polyphenols (that possess an antibacterial activity), hardly undergoes biofouling.

Some of the microorganisms can be detected using SEM [116,117] (Figure 16a) or atomic force microscopy (AFM) [33,116] (Figure 16b).

The major disadvantage of SEM imaging is that samples are subjected to harsh treatments before visualization which can damage the biofilm. To overcome this issue, Environmental Scanning Electron Microscopy (ESEM), which is a form of SEM allowing imaging of hydrated specimens, can be used. This technique allows direct imaging of un-damaged hydrated biofilms at high magnification without prior preparation of the samples [118]. Biofilms are observed in their natural hydrated structure, thus avoiding any shrinkage compared to SEM. This technique has been used, for instance, by Luo et al. [119,120] to characterize IEM used in microbial desalination cells. It allowed them to highlight the formation of a porous biofouling gel-like layer on AEM surface, formed by bacterial growth. This visual observation was coupled and confirmed with EDS, which exhibited a high amount of carbon, oxygen, phosphate, sulfur, zinc, and potassium, as well as the absence of chlorine and fluorine present in AEM polymer. This confirms that the AEM surface was completely biofouled. Pictures are not as sharp as SEM ones, thus decreasing the optical analysis precision [121], but it is necessary when wanting to observe swelled/hydrated membranes. ESEM can be coupled with SEM [122] or AFM [123] to obtain higher resolution information.

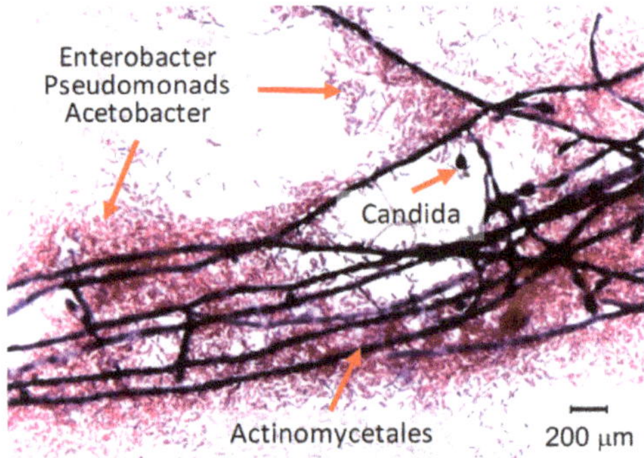

Figure 15. Optical images of the inoculation from touch smears taken from Neosepta AMX-Sb surfaces facing concentration compartment. Data obtained in a laboratory electrodialysis cell at tartrate stabilization of wine. Adapted from [67].

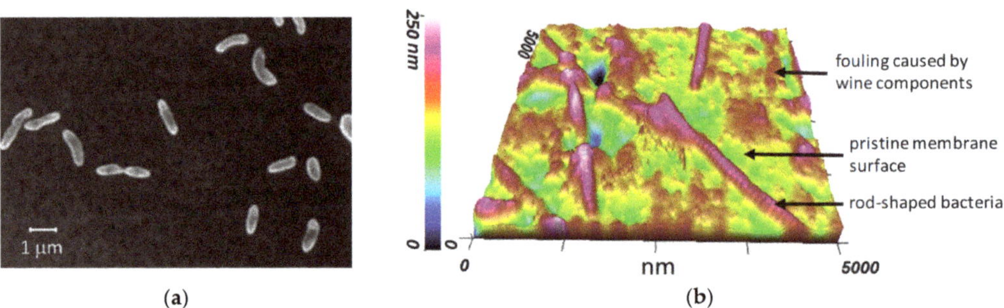

Figure 16. Detection of some microorganisms using (**a**) the emission scanning electron microscope (FE-SEM; image of Neosepta AMX (Astom, Shunan, Japan) membrane after 20 h contact with P. putida bacterial suspension. Adapted from [117]) and (**b**) the atomic force microscopy (AFM image of Neosepta AMX-Sb sample (air-dried) which was soaked in red wine during 72 h. Adapted from [33]).

In order to gain broader understanding of biofilm formation, identifying and quantifying microorganisms, as well as EPS, is crucial. As concerns over biofouling in ED pro-cesses are relatively new, few studies have been carried out to apply methods used in mi-crobiology for biofouled IEMs. However, those methods could be very useful to acquire a better understanding of biofouling, such as knowledge on deposition/adhesion mecha-nisms which could result in appropriate biofouling control strategies.

There are several techniques used to identify microorganisms, most of which start with DNA extraction followed by PCR amplification. In the case of 16S rRNA-gene se-quencing, DNA is then sequenced, and data are analyzed to allow determination of bacte-rial community diversity. The 16S rRNA gene is a commonly used genetic marker because of its sufficient length, presence in almost all bacteria and unaltered gene sequence over time. This technique is widely used to identify microorganisms in biofilm, for instance, of microbial desalination cells [120], as well as reverse osmosis membranes [123,124].

Quantitative real-time PCR (qPCR) is a fast method used to quantify the total number of microorganisms. Unknown DNA samples are compared against a standard curve of 16S rRNA genes [125] leading to an estimation of bacterial cell density [126]. Another way to evaluate bacterial cell density is Heterotrophic Plate Count (HPC) [125,127]. There is a wide variety of HPC methods, but their shared goal is to estimate the number of live and culturable bacteria in water.

After PCR amplification, DNA can also undergo a Denaturing Gradient Gel Electrophoresis (DGGE), which is an electrophoresis technique using a chemical gradient to denature DNA moving across an acrylamide gel. This technique was used by Wu et al. [128] to identify microbial communities' footprints attached on membrane bioreactors surfaces and their evolution with time. It allowed them to highlight that microbial communities shifted with solid retention time, therefore impacting EPS composition. The major advantage of this technique is that it provides an easy and rapid estimation of microbial di-versity with less bias than traditional sequencing [129]; however, it should be noted that it is a destructive method. PCR-DGGE has also been used, coupled with 16S rRNA se-quencing, to characterize biofilm bacterial community composition in a nanofiltration membrane used for wastewater treatment [125,126] and also in microbial desalination cells for domestic wastewater treatment [120].

Terminal Restriction Fragment Length Polymorphism (T-RFLP) is another commu-nity fingerprint technique which uses PCR-amplificated DNA, with fluorescent tags, which is then digested by restriction enzymes leading to terminal restriction fragments of various lengths. Those fragments are then separated according to their size by capillary gel electrophoresis equipped with a fluorescence detector [130]. This method has been used by Chen et al. [131] and Gao et al. [127] to analyze the biofilm microbial community in membrane bioreactors (MBR).

Biofilms are diverse microbial assemblages and Automated Ribosomal Intergenic Spacer Analysis (ARISA) is a powerful technique for determining and estimating microbial richness and diversity. Vanysacker et al. [126] used ARISA to measure species rich-ness in a biofouled microfiltration membrane used in MBRs.

Fluorescence In situ Hybridization (FISH) is a molecular cytogenetic technique where fluorescent probes bind to chromosome parts with a high degree of complementarity. DNA samples first undergo dehydration, then staining with fluorescent probes, followed by observation using Confocal Laser Scanning Microscopy (CLSM) [132]. Genus-specific probes for FISH, such as rRNA-targeted oligonucleotide probes [125,132], have been developed to enable specific and simultaneous identification of multiple microbial species constituting biofilms [133]. CLSM in combination with fluorescent probes can also be used to identify major biomolecules but it often results in nonspecific binding and misidentification [134]. Thus, the FISH technique allows identification, visualization, and quantification of specific bacteria within the biofilm microbial community. However, the procedure is complex and time-consuming. Another major disadvantage is that the dehydration step can destroy the original structure of bio-aggregates and biofilms.

EPS', such as proteins and polysaccharides, identification and quantification can also be achieved with various methods. Sodium Dodecyl Sulphate–Polyacrylamide Gel Electrophoresis (SDS-PAGE) is an easy and low-cost electrophoresis method used to separate proteins, extracted from biofilms, by their molecular mass on a polyacrylamide gel, regardless of their charge due to SDS. This technique is often used to identify proteins in biofouled membrane bioreactors (MBRs) [129,130].

Matrix-Assisted Laser Desorption/Ionization-Time of Flight Mass Spectrometry (MALDI-TOF MS) is a new simple, fast and cost-effective tool for surface characterization. Proteins and peptides present in the biofilm are desorbed with a laser adsorbing solution (MALDI matrix) and analyzed in mass spectrometry (MS) to measure their exact size and signal intensity for quantification [24]. This technique also allows bacterial identification, as peptides and proteins are specific to each bacterial species [124].

EPS protein can also be quantified by the Bradford assay which is a colorimetric quantitative protein determination assay based on a shift of absorbance of a dye when binding with protein. A similar method can be used for polysaccharide content, where sugar groups are reduced by phenol and sulfuric acid and their absorbance is measured (DuBois assay) [135]. Sweity et al. used both this technique to study EPS adherence and viscoelastic properties in MBR [136]. They showed that fouling was linked to EPS adherence as a polysaccharide content increase resulted from a strong EPS adherence leading to a reduced membrane permeability.

All the techniques used to characterize membrane biofilm are shown in the summary Figure 17.

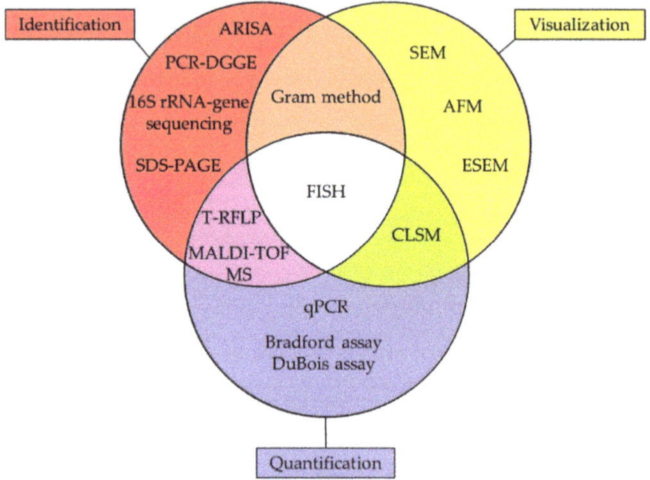

Figure 17. Overview of the different techniques presented to characterize biofilms on IEMs membranes.

2.2. Characterization of the Interaction of Foulants with the Membrane Surface

2.2.1. Localization of Foulants and Surface Roughness Parameters

As already mentioned in Section 2.1.1., SEM or SEM combined with EDS are widely used to determine the localization of mineral substances on the surface and in the volume of IEM [47,48,50,137]. Much less often, these methods are used to determine the localiza-tion of organic foulants [47] and to make comparisons with the surface roughness.

Atomic force microscopy provides some insight into the presence of organic matter on membrane surfaces. Fouling with these substances (wine components, amino acids, proteins, etc.) leads to a change in adhesion forces [74,138,139], as well as surface roughness parameters in comparison with the pristine IEM [33,79,140]. Figure 18 shows an example of such changes in the case of membrane fouling during ED fractionation of pro-tein (snow crab by-products) hydrolysate [79]. It should be noted that the recorded param-eters of the fouled surface relief correspond to the operating conditions of IEM in real membrane stacks only when using an AFM (and SEM) modifications, which allow one to study swollen samples or membranes in water [141] because the size of organic foulants is highly dependent on the degree of hydration.

Modern modifications of optical microscopy make it possible to examine the surface and cross-sections of swollen samples [33,67,73], to determine the localization of foulants (Figure 19) and the thickness of the film formed by them on the IEM surface, based on the difference in color of the pristine membrane and foulant.

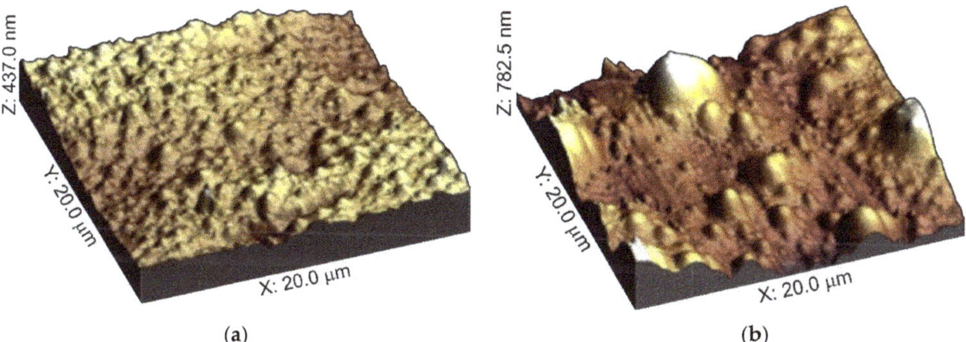

Figure 18. Three-dimensional atomic force microscopy AFM images representing surface of pristine (**a**) and fouled with peptides and amino acids (**b**) Neosepta AMX-Sb membrane. Adapted from [79].

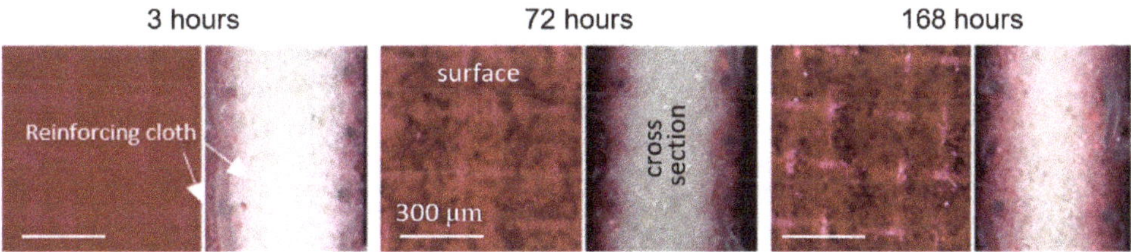

Figure 19. Changes in the color of the surface and cross-section of the MK-40 (Shchekinoazot LTD, Pervomaiskii, Russia) cation-exchange membrane during its fouling with anthocyanins (red color) and proanthocyanins (brownish-violet color), which are components of cranberry juice. The contact time of the membrane with cranberry juice in hours is indicated above the images.

Confocal Laser Scanning Microscopy (CLSM) is a powerful technique to obtain sharp images of a sample that would otherwise appear blurred when viewed under a conventional microscope [97]. It uses the light reflected from the sample (epitransmission or re-flectance) or captures the fluorescent light that is excited in the sample by the incident beam (epifluorescence). One of the main advantages of this method is that potential organic foulants, such as proteins, peptides, polysaccharides or biofoulants, could be stained with different fluorescent probes which facilitates the quantitative characterization and the visualization of fouling and polymer adsorption, and could lead to the de-termination of the possible interactions between proteins and membranes. For example, Reichert et al. [142] studied CLSM images to investigate the adsorption of two model proteins, BSA and lysozyme on commercially available CEM (Sartobind S) and AEM (Sartobind Q) for the protein purification process (Figure 20), Vaselbehagh et al. [143] com-pared CLSM images of PDA-modified AMX (Neosepta—Astom, Shunan, Japan) carried out to improve the biofouling resistance of the membrane during ED treatments and it was possible to clearly observe stained viable bacteria on the surface; Herzberg et al. identified living and dead cells on the surface of various CEMs and AEMs [144]. While interest in the CLSM method is growing, it remains less used than SEM and AFM.

Optical Coherence Tomography (OCT) enables in situ visualization and quantifica-tion of fouling layers dominated by scaling [145–147], monitoring of biofouling develop-ment in membrane [148], quantitative analysis of membrane fouling by oil emulsions [149], etc. An example of biofouling visualization obtained by this method is shown in Figure 21.

Another effective way to determine the thickness of the foulant film (and its electrical resistance) is to analyze the parameters of the high-frequency arc of the electrochemical impedance spectrum (EIS) obtained in the absence of a constant electric field [67,82,150–153]. For such analysis, mathematical models [154], as well as the method of electri-cal equivalent circuits [67,149,150] are used. If the foulant and the membrane material have significantly different values of electrical capacitance, an additional arc that charac-terizes the foulant layer appears in the high-frequency range of the impedance spectrum [67,147,149–151].

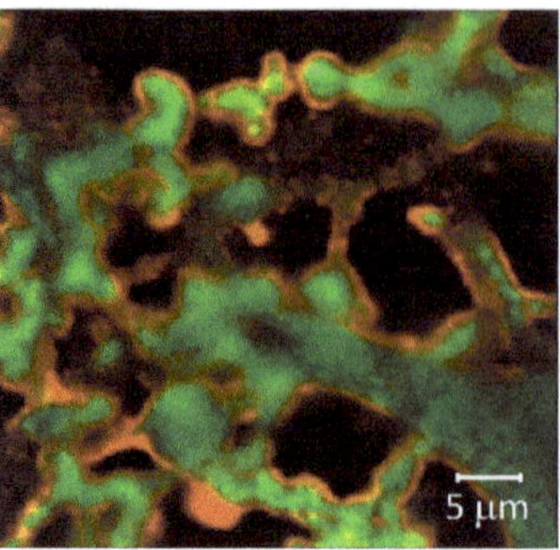

Figure 20. Confocal Laser Scanning Microscopy image of lysozyme protein adsorption on Sartobind S (Sartorius, Göttingen, Germany) cation exchange membrane in a 50 mM potassium phosphate buffer with pH = 7.5 at equilibrium. Membrane backbone was labelled with 6-DTAF (green) and the lysozyme was coupled to Cy3 dye (red) [139].

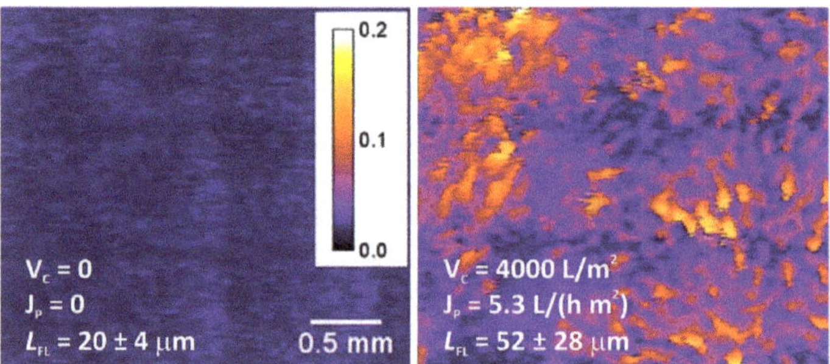

Figure 21. Optical Coherence Tomography 2D topographic image of the membrane/solution interface in distillation process depending on amount of produced condensate (referred to as cumulated volume V_C.) and permeate flux, J_P. The colored bar shows the height of the biofoulant layer above the surface of a flat membrane (in mm). L_{FL} indicates the average height of the foulant layer. Adapted from [142].

2.2.2. Membrane Surface Charge and Hydrophobicity

Other techniques used to characterize the charge and degree of hydrophobicity of the membranes are very useful in determining if the fouling is impregnated or not. These techniques are global (several mm^2 of the surface) and ex-situ.

Contact angle. Organic foulants very often contain highly hydrated polar groups (carboxyl, hydroxyl, phosphoric acid, amino groups, etc.). Quite often these groups are attached to hydrophobic aromatic chains (polyphenols, aromatic amino acids, etc.). Therefore, the hydrophilic/hydrophobic balance of the IEM surface provides important information about foulants and the nature of their interaction with the IEM material. For example, fouling with phenylalanine (aromatic amino acids) [34], polyacrylamide [155] or bovine serum albumin and humic compounds [69] showed a significant increase in the surface hydrophobicity of fouled IEM as compared to the pristine one. In the case of phe-nolic compounds, the hydrophobicity decreases for AEMs while it always increases for CEMs [26,32,33] due to the different orientation of the hydrophilic and hydrophobic com-ponents of the foulant particles, which is determined by the sign of the electric charge of fixed membrane groups [31]. The hydrophilic/hydrophobic balance of a surface is usually determined by the values of the contact angle [26,31–34,69,152,156], which is formed between the tangent drawn to the surface of the liquid–gas phase and the IEM surface with the vertex located at the point of contact of the three phases. The contact angle is al-ways measured inside the liquid phase.

The sessile drop method is the most commonly used. In this method a drop of liquid is placed on the test surface (Figure 22). Despite the apparent simplicity of the method, the results depend on many factors [157–159]: the chemical nature and volume of the test liquid; the height from which the drop falls; the water content of the sample; the position of the sample relative to the gravitational field; the time elapsed since the drop touches the test surface; software used to process the obtained images; etc. That is why the data obtained by different researchers are often very different. Sometimes the error range turns out to be larger than that expected due to the changes in the hydrophilic/hydrophobic balance of the surface caused by fouling.

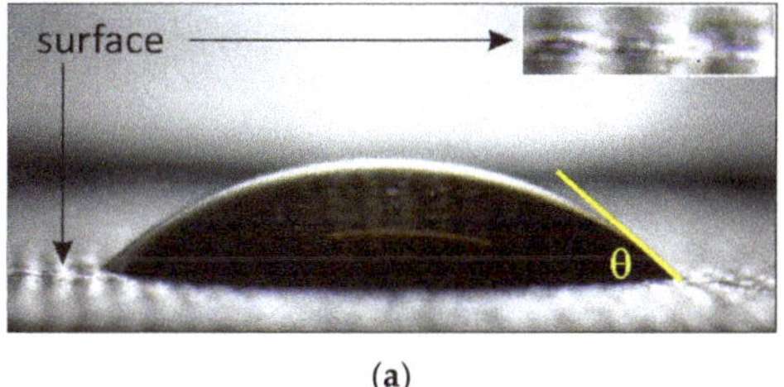

(a)

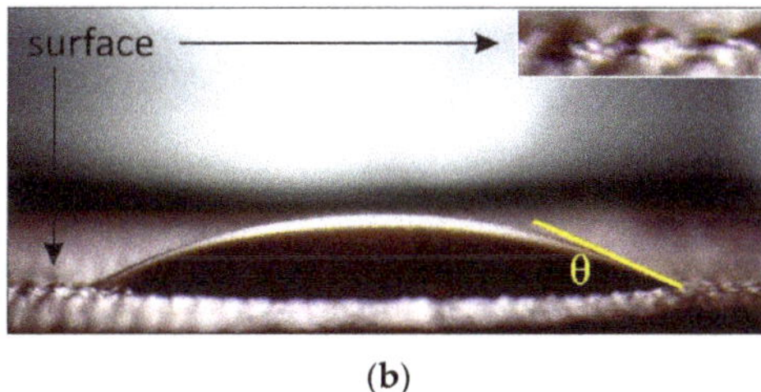

(b)

Figure 22. Contact angles of swollen CJMC-5 membrane (Hefei Chemjoy Polymer Material Co., Hefei, China) surface using sessile drop method: pristine (**a**) (soaked in 0.01 M NaCl solution, $\theta = 42 \pm 10°$) and fouled in red wine (**b**) (soaked for 3 days in a polyphenol extract from the wine pulp, $\theta = 24 \pm 10°$).

In addition, cracks can form on the IEM surface when dry samples are used in measurements. Contact of foulants with hydrophobic air can lead to a change in the orientation of their organic chains and, accordingly, to a change in the hydrophilic/hydrophobic balance of the investigated surface in comparison with the aqueous medium. To break down these factors, contact angles are increasingly measured on swollen IEM samples [33,160]. The captive bubble method [161] also seems to be very promising but is used very rarely so far [162]. Note that all these techniques are relatively simple to implement.

Zeta (the electrokinetic) potential (ζ) is the electric potential measured at the fluid slipping plane along the IEM surface [163]. The slipping plane located at the boundary between the diffuse electric layer and the adsorption (dense) electric layer, or in the diffuse layer near this boundary. The diffuse layer contains mobile counterions, which are at-tracted to the IEM surface due to electrostatic forces. The adsorptive electric layer is directly adjacent to the membrane material and is formed as a result of electrostatic interactions of membrane fixed groups with counterions and specific adsorption of foulants. The magnitude of ζ can provide information about electrostatic or charge repulsion and attraction forces in the specific case of organic particles interactions with IEM [30,46,69,164–166] and plays a vital role in fouling. It is appropriate to measure this parameter before, during and after membrane operation in food industry processes [24]. However, only few studies have focused on ζ potential analysis for characterizing IEM fouling by proteins, peptides and amino acids or other organic foulants such as humate [114], since the current and standard automatic zeta potential measurement equipment is not suitable for IEMs. The point is that IEMs are conductors of the second kind (they participate in the ion transport), and this property significantly distorts the results.

Our view is that more reliable values of ζ, as well as surface charge, can be obtained from tangential streaming potential measurements. For example, a laboratory-made gap cell which described in detail by Sabbatovskii et al. [167] is applied in the paper [33]. This cell is similar to Anton Paar SurPASS 3 [168]. The streaming potential measurements are carried out using two Ag/AgCl electrodes connected to multimeter and an electrolyte solution pumped at the pressure drop between inlet and outlet of the channel in the range from 0.125 bar to 0.625 bar. The channel was formed by two identical IEMs.

These studies showed that the surface charge of the anion-exchange membrane AMX-Sb contacted with wine changes from positive to negative when the pH of the feed solution changes from 3.5 to 6.7 (Figure 23). These data provided additional confirmation of anthocyanins (which change their electric charge depending on pH) adsorption by the IEM surface.

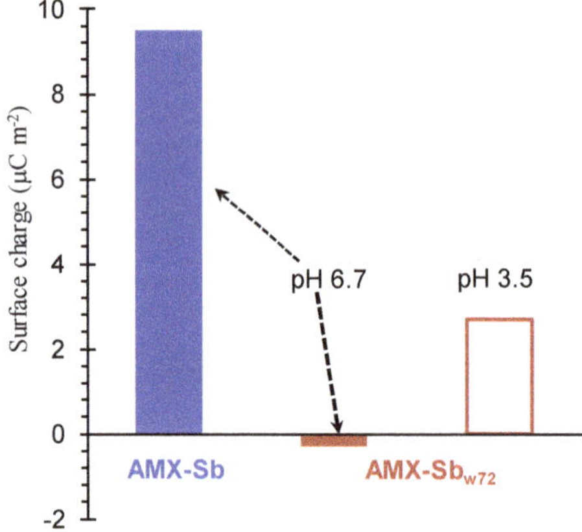

Figure 23. Surface charge of the pristine (AMX-Sb) and fouled in wine (AMX-Sb_W) membranes in a 0.02 M NaCl solution with pH 6.7 and 3.5. Adapted from [33].

Note that the correct determination of the parameters of the hydrophilic/hydrophobic balance, as well as the zeta potential and surface charge, requires normalization to the true area of the investigated surface [169], which can be found, for example, using 3D SEM, AFM or profilometry data.

3. Mechanisms of Foulants Interaction with Ion-Exchange Materials

We should mention that one or more substances present in food, sometimes even in very low concentrations, with a high affinity for the membrane material, are usually re-sponsible for fouling. Some compounds can slowly adsorb on the surface and/or in the membrane bulk and irreversibly change its structure. Phenomenon of organic fouling of IEMs can be quick, cumulative and destructive at the same time during long term contact with the treated media. It depends on the membrane material and the fixed groups charge and the nature of interactions between organic particles and IEM. Let us examine the main mechanisms of foulants interaction with ion-exchange membranes.

3.1. Physicochemical Interactions

Electrostatic, hydrophobic–hydrophobic π–π (stacking) as well as ion–dipole (hydrogen bonds) and dipole–dipole (Van der Waals) interactions underlie the fouling of ion-exchange materials.

Electrostatic interactions between counterions and IEM fixed groups are determined by the value of the membrane exchange capacity and the electric charge that the foulant has in solution (fouling of the IEM surface) or acquires upon penetration into the internal IEM solution (fouling of the IEM bulk). The vast majority of components that make up the liquid media of the food industry (amino acids, polybasic carboxylic and inorganic acids, proteins, anthocyanins, etc.) have amphoteric properties (they are ampholytes). They enter into protonation–deprotonation reactions with water and with each other. Therefore, their electric charge depends on the pH of the medium and the dissociation reaction (and protonation–deprotonation) constants of ampholytes (Ka). Proteins, which are components of milk, whey, and animal blood, exhibit the most complex behavior, because several amino acids with their own dissociation constants of amino groups and carboxyl groups are included in each of them.

Persico et al. [62,65,68] performed 3D mapping of the electrostatic charge of various whey proteins using molecular dynamics simulation and evaluated the effect of electrostatic interactions on the possibility of IEM surface fouling with these high molecular weight substances. They found [62] that the interaction of positively charged fixed AEM groups with proteins that are enriched in carboxyl groups is most likely for the pH of the treated solution exceeding the pKa (≈4) of these groups. A decrease in quantity of amino groups in the composition of peptides will lead to the fact that their electrostatic interac-tions with AEM become more significant and more stable when the pH of the treated solu-tion exceeds the pKa of amino groups (≈10). The adsorption of a protein monolayer by the CEM surface is most probable in acidic solutions [63,68] (pH 2). In an alkaline solution (pH 10), the protein acquires a negative charge and is repelled from the similarly charged CEM surface (Figure 24).

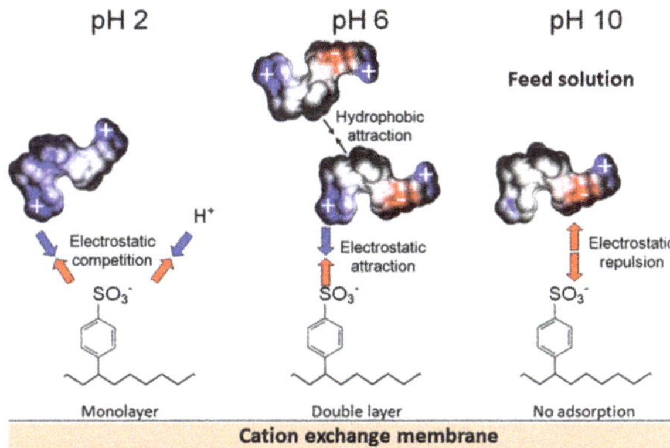

Figure 24. Schematic fouling mechanisms of CEM by peptides depending on their electrostatic charge and the pH of feeding whey solution. Adapted from [62].

The use of overlimiting current modes leads to alkalinization of the solution near the CEM surface and its acidification near the AEM surface due to water splitting [68]. The result of a shift in pH compared to bulk solution is the loss of electric charge by proteins or even the acquisition of a charge opposite to the charge of membrane fixed groups. These changes in the charge contribute to a reduction in IEM fouling with proteins. Low molec-ular weight amino acids and polybasic organic acid anions found in whey and other dairy products, as well as anthocyanins found in wine and juices, can enter the IEM.

A schematic representation of the electrostatic interactions of polyphenols (anthocyanins) with aromatic CEMs and aliphatic AEMs in bathing (external) solutions with pH 6 is shown in Figure 25. It has already been discussed in Section 2.1.2. that the pH of a CEM (or a cation exchange resins) internal solution is shifted to acidic values, while the pH of an AEM internal solution is shifted to alkaline values as compared to the external solution. The reason for this shift is the Donnan exclusion from ion exchange materials of hydroxyl ions (CEMs) or protons (AEMs), which are the products of water molecules and ampholytes protonation reactions [170–173]. The higher the membrane exchange capaci-ty, the stronger the Donnan effect [93]. The result of such a shift can be a change in the foulant electric charge inside the ion-exchange material as compared to the external solu-tion. Pismenskaya et al. [92,174] demonstrated this possibility using FTIR and color indi-cation of anthocyanin structure (Figure 12). For example, in an external solution with a pH of 6, anthocyanins have a guinoidal anhydrobase structure and have no electrical charge. At pH 4, which is established within an aromatic cation exchange resin (having negatively charged sulfonate groups), anthocyanins are mainly converted to carbinol pseudobase and hardly participate in electrostatic interactions. Inside the anion exchange resins, anthocyanins become singly charged (aromatic AV-17-8, LLC "CHIMIMPEX", Moscow, Russia) or doubly charged (aliphatic EDE-10, PJSC "Uralchimplast", Nizhny Tagil, Russia) anions and enter the electrostatic interactions with positively charged resin fixed groups. The result of this interaction is a higher sorption of anthocyanins by anion exchange resins than in the case of aromatic cation exchange resin (KU-2-8, LLC "CHIMIMPEX", Moscow, Russia) at the pH 3 of the external solution (Figure 26). At the external solution pH = 3, on the contrary, anthocya-nins inside the anion exchange resins do not have an electric charge, but inside the cation exchange resin they are cations. As a result, the fouling of cation-exchange materials un-der the conditions of industrial ED processing of wines and juices (pH 3) is much higher than that observed for anion-exchange materials. T

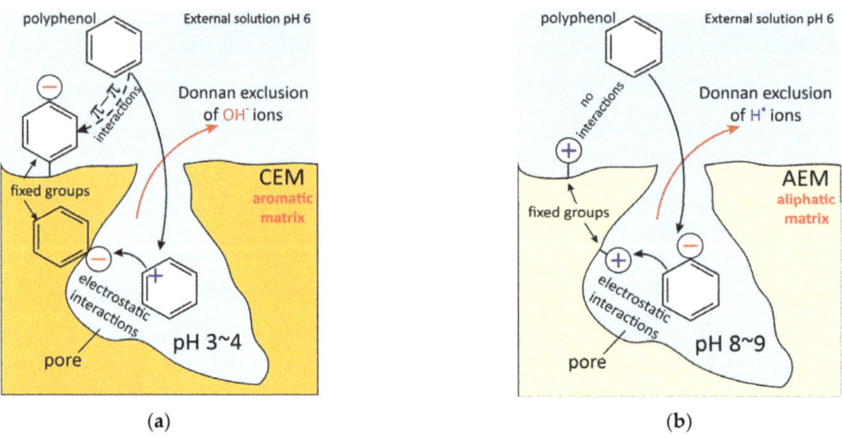

Figure 25. Schematic representation of the electrostatic interactions of polyphenols (anthocyanins) with aromatic CEMs (**a**) and aliphatic AEMs (**b**) in bathing solution with pH 6.

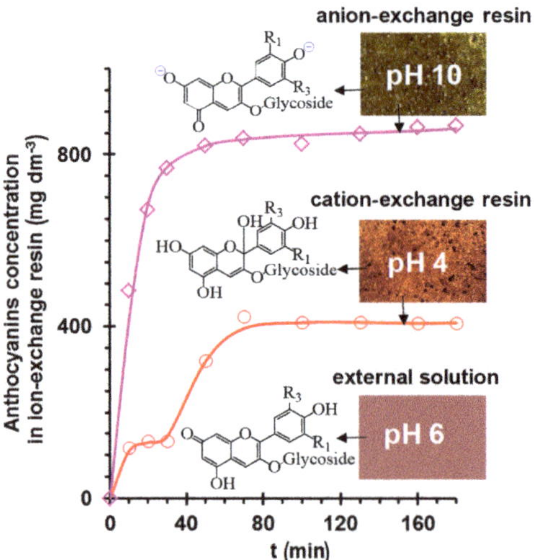

Figure 26. The anthocyanins adsorption by the ion-exchange resins vs time of it soaking in the aqueous solutions (pH 6) with anthocyanin concentration of 40 mg dm^{-3}. The insets show the color of the external solution and the color of resins equilibrated with this solution, as well as the structures of anthocyanin species that correspond to these colors. Adapted from [92].

The influence of electrostatic interactions on the sorption of anthocyanins by ion-exchange materials, depending on their exchange capacity and the external solution pH, is also considered in [175–180]. Similar results were obtained in the case of IEM [31,33,73].

It should also be mentioned that some of the multiply charged counterions contained in liquid media of the food industry enter the electrostatic interactions not with one, but simultaneously with two fixed groups, causing an effect equivalent to additional cross-linking of the ion-exchange matrix [93]. Such interactions are typical, for example, for tri-ply charged anions of phosphoric acid and weakly basic groups of AEMs, as well as for doubly charged calcium anions and sulfonate groups of CEMs. It is known [93,181–183] that the latter enter the specific (donor–acceptor) interactions, which, in particular, are expressed in the formation of weakly dissociating ion–ion associates "sulfo group - calcium ion".

Hydrophobic–hydrophobic π–π (stacking) interactions generally occur if both the foulant and the IEM material contain aromatic rings. These interactions are often a key contributor to the fouling of IEMs and ion exchange resins in wine and juice processing. Indeed, even under conditions that are unfavorable for electrostatic interactions (Figure 26), the sorption of anthocyanins by the KU-2-8 resin is only two times less than in the case of the AV-17-8 resin, which has a similar aromatic matrix and structure. This sorp-tion is provided by π–π (stacking) interactions between the aromatic rings of polyphenols and the KU-2-8 polymer matrix. Similar results are presented in works [173,176,184,185], where an increase in the sorption of anthocyanins and other polyphenols is noted during the transition from an aliphatic to an aromatic polymer matrix of ion-exchange resins and membranes.

Two plateaus on kinetic (Figure 26) and equilibrium sorption isotherms and the results of isotherm processing using the Freudlich model equation [186] allow us to con-clude that polymolecular adsorption of these substances [92] by ion exchange materials occurs due to the π–π (stacking) interaction of polyphenols with each other [75,187] or with other substances. Therefore, proanthocyanidins 2-mers, 3-mers, 4-mers and poly-mers are identified inside IEMs that were in contact with wine or cranberry juice [26,73]. These substances can probably form directly in membrane pores.

Table 2 summarizes ATR–FTIR data on CH-π and π–π interactions or hydrogen bonds formation between IEM materials and polyphenols that are contained in juice or wine.

Table 2. Assignment of characteristic bands of pristine and fouled CMX-Sb and AMX-Sb in the ATR–FTIR spectra and signs of foulant-matrix interactions [57].

	σ or Spectral Region (cm^{-1})	Characteristic Bands in Pristine IEMs	Indications of CH-π and π–π Interactions or Hydrogen Bonds in IEMs after ED Industrial Tartrate Stabilization of Wine
AMX-Sb	1200–1250 region Intense peak at 1240	N-C Stretching vibrations bands of functional sites [31,157,188]	Enlargement and intensification of bands by accumulation of phenolic acids [31]
	Doubled band at 1610 and 1625	N-H of functional sites [69,157]	Appearance of –COOH band at 1168 cm^{-1} and C=O band at 1710 cm^{-1} [69]
	1450, 1475, 1480, and 1510	Aromatic C=C stretching bands in aromatic ring of polystyrene (PS) [31]	Enlargement and intensification by accumulation of aromatic rings of phenolic compounds [41] Appearance of bands of the aromatic ring breathing modes and bands resulting from stretching and contracting of the C=C bonds in the range 1450–1615 cm^{-1} [157,189]
	Attached peaks at 2852, 2915 and 2967	Stretching vibrations bands of aliphatic –CH and –CH_2– bonds in functionalized PS [16,18]	Enlargement and intensification of bands by accumulation of phenolic compounds in polymer matrix Blue-shift of bands to higher σ by 10 to 20 cm^{-1} under CH-π interactions [187]
	3010	Stretching vibrations of aromatic –CH– [190]	
	3400	Stretching vibrations band of –OH bonds [16]	Enlargement of the band by the appearance of a hydrogen-bonded region 3050–3350 cm^{-1} added to the non-hydrogen bonded band [16] Red-shift of the band to lower σ by 30 cm^{-1} under hydrogen bonds between linked water in polymer matrix and O of phenolic compounds [32,191]

Table 2. *Cont.*

	σ or Spectral Region (cm^{-1})	Characteristic Bands in Pristine IEMs	Indications of CH-π and π–π Interactions or Hydrogen Bonds in IEMs after ED Industrial Tartrate Stabilization of Wine
CMX-Sb	1041	Stretching vibrations bands of S-O in $-SO_3^-$ sites [185]	Inhibition of $-SO_3^-$ functional sites by accumulation of colloidal particles of polyphenols or physical detachment of the sites from the polymer matrix by disruption of the sulfur-carbon bonds [32]
	1166	Stretching vibrations bands of SO_3-H [192]	
	1186	Stretching vibrations bands of SO_3-Na groups [159]	
	1485 and attached peaks at 1590 and 1650	Stretching bands of aromatic C=C bonds in aromatics rings of PS [16]	Intensification of bands by accumulation of phenolic compounds rich in aromatic rings and Peaks located at 1485 and 1590 cm^{-1} in new CMX are blue-shifted to higher wavenumbers by ~20 cm^{-1} under π-π and CH-π interactions [32,187]
	3400	Stretching vibrations band of –OH bonds [16,17]	Enlargement of the band by the appearance of a hydrogen-bonded region 3100–3300 cm^{-1} added to the non-hydrogen bonded band Red-shift of the band to lower σ by ~50 cm^{-1} under hydrogen bonds between linked water in polymer matrix and O of phenolic compounds [32]

Ion–dipole (hydrogen bonds) and dipole–dipole (Van der Waals) interactions are characteristic of foulants and IEMs which have oxygen-containing polar (hydroxyl, carbonic, phosphonic, sulphonic, etc.) groups and hydrogen in fixed groups (primary and secondary amines) or aliphatic chains (materials such as polyamide, polytetrafluoroethylene, polyvinyl chloride, etc.) [172,173,182]. According to [62], the hydrogen bonds and dipole–dipole interactions are the main reason for the formation of multilayered fouling on the CEM surface in protein-containing solutions at pH = 6 (Figure 24), that is, when there are no electrostatic interactions between proteins and fixed membrane groups. The aliphatic anion exchange resin EDE-10P sorbs 30% more polyphenols than the aromatic cation exchange resin KU-2-8 at an external solution pH = 3 [92], although anthocyanins inside EDE-10P are mostly in molecular form and are not able to enter the electrostatic in-teractions, while inside KU-2-8 anthocyanins are cations that interact with negatively charged fixed groups. The reason is the primary, secondary, and tertiary amines (which are fixed groups of the aliphatic resin) are able to form hydrogen bonds with hydroxyl groups of anthocyanins.

The ability of proteins and polyphenols (proanthocyanins and anthocyanins) to participate in almost all of the listed types of interactions with each other and with the IEM matrix predetermines the formation of colloidal particles in membrane pores and on their surface. It is known [71,193–195], for example, that polyphenols form high-molecular colloidal systems with carbonic acids, amino acids and saccharides, as well as with inorganic species, such as Ca^{2+} or Fe^{3+} [196] in wine and juices. Moreover, CEMs, which can contain Ca^{2+} or Fe^{3+} ions as counterions, have more favorable conditions for the formation and growth of colloidal particles inside pores (in situ) compared to AEM.

Perreault et al. [73] investigated the fouling of cation exchange membranes MK-40 (Shchekinoazot, Russia), CSE-fg (Astom, Shunan, Japan), CEM Type-II (Fujifilm, Tilburg, The Netherland) and CJMC-5 (Chemjoy Polymer Material Co. Ltd., Hefei, China) by polyphenols from cranberry juice (pH 2.45). They showed (according to the mechanisms described above) that CEMs with an aromatic matrix are more prone to fouling than CEMs with aliphatic matrix. Membranes with a lower exchange capacity lose it faster due to shielding of fixed groups by foulants. The thinner the membrane, and the more meso- and macropores it contains, the faster the fouling of its volume is completed. High molecular weight polyphenols mainly penetrate deep into the membrane through the macropores between the beads of the ion exchange resin and the inert binder, as well as through the extended macropores between the ion exchange material and the reinforcing cloth (if they take place) that reached the surface of the membrane.

3.2. Stretching of the Polymer ion Exchange Matrix

Typically, CEMs and AEMs after manufacturing are converted to Na^+ or H^+ and Cl^- or OH^- ionic forms, respectively. Before assembling of ED apparatuses, they swell in saline (NaCl) solutions. At the same time, most of the liquid media subjected to ED processing in the food industry contain highly hydrated anions of inorganic (carbonic, sulfuric, orthophosphoric, etc.) and organic (lactic, tartaric, malic, citric, etc.) acids, and also strongly hydrated cations, for example Ca^{2+}, Mg^{2+} [197]. These ions displace less hydrated Cl^- and Na^+ ions [194] in the pores of AEMs and CEMs. As a result, the fraction of bound water in the pores increases. Accordingly, the fraction of free water decreases in the internal IEM solution, which leads to an increase in osmotic pressure. This increase leads to the stretching of the elastic polymer matrix and, respectively, to increase in the membrane effective pore radius as compared to a state achieved upon contact of the IEM with less hydrated ions (Figure 27). The described phenomenon is well known [98] and has long been used to explain the change in the swelling of ion-exchange materials upon contact with an electrolyte and a solvent of various nature [198].

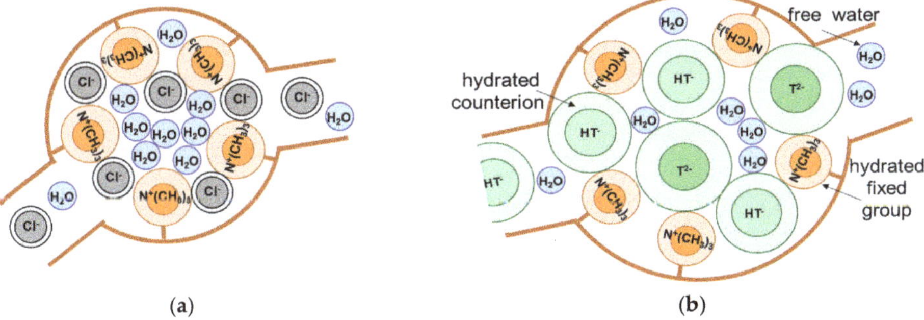

Figure 27. Scheme of free and bound water distribution and its effect on anion exchange membrane pore size in the case of weakly hydrated (**a**) or strongly hydrated (**b**) counterions: chlorides (Cl^-) or hydrotartrates (HT^-) and tartrates (T^{2-}). Adapted from [171].

The influence of the components of the treated solutions on the elastic matrix state is of particular importance for explaining and predicting the consequences of fouling de-pending on the foulant nature. Therefore, in recent years, additional studies have been carried out [171] using the standard contact porosimetry method. It was found that the amount of bound water in the pores of AEMs actually increases if replacing NaCl solution with KHT, $NaHCO_3$, NaH_2PO_4 solutions. In both homogeneous (AMX-Sb) and heterogeneous (MA-41, Shchekinoazot, Russia; FTAM-EDE, FUMATECH BWT GmbH, Germany) membranes, the greatest increase in water content (and increase in size) is observed for micro- and mesopores with a radius from 1 to 13 μm. Swelling of ion-exchange material in heterogeneous membranes, apparently, leads to a decrease in adhesion between this material, the inert binder and the reinforcing material. This leads to an increase in the free water content and in the size of macropores in the places of contact between the particles of the ion-exchange resin and the inert binder, as well as between the ion-exchange com-posite and the threads of the reinforcing cloth.

The described phenomenon is the key for explaining the change in the IEM thickness, d, and their destruction during long-term (thousands of hours) operation in ED processing of the food industry liquid media. An increase in AEM thickness is observed, for example, in liquid media containing anions of polybasic carboxylic acids and/or phosphoric acid [171]. Moreover, in the case of a heterogeneous membrane MA-41, which contains a poly-styrene matrix regularly crosslinked with divinylbenzene, a new stable state of the matrix (characterized by the cessation of increase in d) is reached within ≈50 h after replacing the NaCl solution with solutions of polybasic acid anions. In the case of homogeneous membranes containing a randomly crosslinked copolymer of divinylbenzene and poly-styrene, an increase in d is more significant and continues abruptly throughout the observation period (more than 200 h). Apparently, the high osmotic pressure arising in the internal solution of such membranes leads not only to the stretching of the polymer matrix, but also to the rupture of the "bridges" that cross-link the polymer. The result is not only an increase in thickness, but also a gradual destruction of IEM, which is accelerated with an increase in the membrane operation time in ED processing of dairy products [157], as well as wine or juices [16,31,32]. These changes entail a loss of membrane mechanical strength [16,18,31,157]. For example, Garcia-Vasquez et al. [157] showed that the Young's modulus (E), which can be related to the rigidity of a material, decreased by 20% for AEM (AMX-Sb, Astom, Shunan, Japan) at the end of their lifetime in a stack of ED used for whey demineralization. The breaking strength, which represents the membrane plasticity (break or/rupture point) decreased by 45%, and the area under the stress–strain curve decreased by almost 80%, which strongly indicated a loss of the material toughness.

In the case of industrial ED processing of wines and juices, which is carried out at pH < 3.5, CEMs absorb more polyphenols (PP) than AEMs (see Section 3.1). The continuous growth of colloidal particles, which are compacted during the periodic cleaning of membrane stacks with chemical reagents, leads to a more significant stretching of the CEM ion exchange matrix [31,32]. As a result, with the same duration of operation in an ED apparatus, the cation-exchange membrane CMX-Sb (Astom, Shunan, Japan) is more destructed than the anion-exchange membrane AMX-Sb, which has a similar polymer matrix and reinforcing cloth (Figure 28).

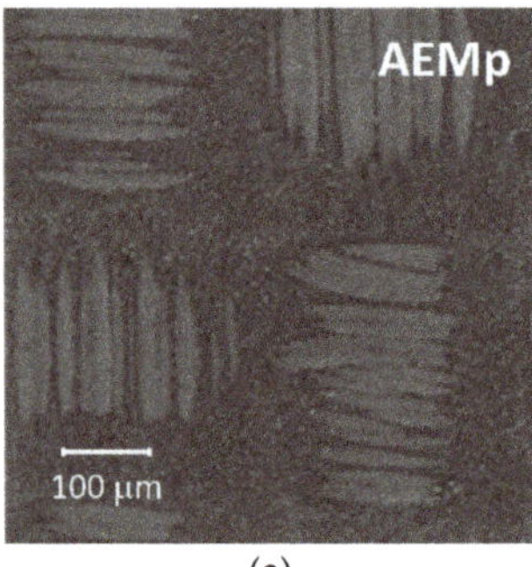

(a)

(b)

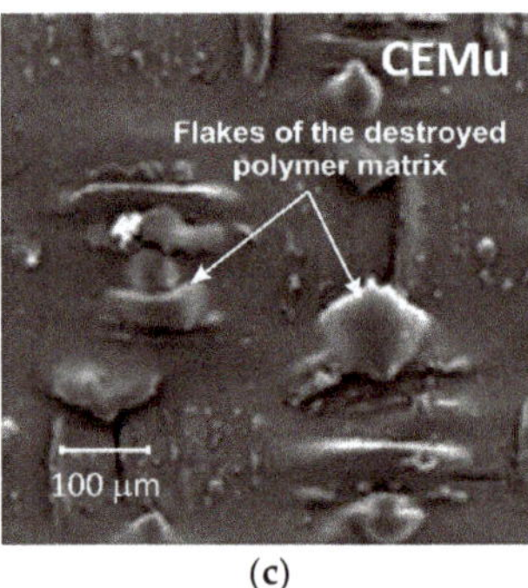

(c)

Figure 28. SEM images of the surface of AEMp pristine anion exchange membrane (**a**) as well as anion (AEMu) and cation (CEMu) exchange membranes used in industrial wine tartrate stabilization ED process (**b**,**c**) during 2500 and 2738 h, correspondently. Reconstructed from [31,32].

Conversely, prolonged contact of the IEM with a foulant, which contains a small amount of bound water, can cause the ion exchange matrix to become denser. For exam-ple, Vasil'eva et al. observed a reduction in pore size, a decrease in thickness and flattening of the surface of profiled CEMs used for dialysis and electrodialysis processing of solu-tions containing the amino acid (phenylalanine) [199,200].

Thus, the rate and the degree of the ion-exchange matrix stretching (and destruction) are determined not only by the foulant nature, but also by the nature of the ion-exchange matrix (aliphatic, aromatic), as well as by the degree of its crosslinking [16]. In addition, as will be shown in Section 4, the cleaning conditions and the chemical nature of the rea-gents play an important role in this process [16,201,202]. The combination of these factors is necessary to predict the geometric parameters of CEMs and AEMs, as well as their me-chanical strength and susceptibility to degradation in the food industry ED processes.

Before concluding, we group the main techniques presented here and used to study the different aspects related to the fouling of ion exchange membranes in Table 3.

Table 3. Summary table of the different techniques used to study the various aspects related to the fouling of ion-exchange membranes.

Technique	Application	Device Complexity	Interpretation Complexity	Use Frequency
	I: Identification V: Visualisation Q: Quantification		H: High M: Middle L: Low	
2D fluorescence/Fourier transform infrared correlation spectroscopy	V, I	H	H	L
31P nuclear magnetic resonance spectroscopy	I	H	M	L
Atomic force microscopy (AFM)		H	L	M
Classical optical microscopy	V	L	M	H
Combined with energy dispersive X-ray spectrometry (EDS)	I	H	H	L
Confocal laser scanning microscopy (CLSM)	V	H	M	L
Contact angle	Q	M	L	H
Fluorescence excitation-emission matrix (EEM)	Q, I	H	H	L
Fluorescence spectroscopy	I	H	M	L
Fourier transform-ion cyclotron resonance-mass spectrometry (FT-ICR-MS)	I	H	H	L
High-liquid performance chromatography (HPLC)	Q	M	L	M
High-resolution optical microscopy	V	M	M	M
Inductively coupled plasma optical emission spectrometry	I	M	H	L
Mass spectrometry (MS) coupled	I	H	M	L
Molybdate colorimetry inductively coupled plasma optical emission	I	M	M	L
Optical coherence tomography (OCT)	I	H	M	L
Optical microscopy combined with a color scale for pH indication	V	L	L	L
Raman spectroscopy	I	H	M	M
Reflectance–Fourier-transform infrared (ATR–FTIR)	I	M	M	H
Rutherford backscattering spectroscopy (RBS)	I	H	H	L
Scanning electrochemical microscopy (SECM)	V	M	L	M
Scanning electron microscopy (SEM)	V	H	L	H
Scanning ion conductance microscopy (SICM)	V	H	L	L
Size-exclusion (SEC)	I	M	L	M
Smear-prints	V	M	M	L
Sodium dodecyl sulphate–polyacrylamide gel electrophoresis (SDS-PAGE)	I	M	M	L
Standard contact porosimetry method	Q	L	M	L
Surface plasmon resonance (SPR)	I	H	H	L
Surface-enhanced Raman spectroscopy (SERS)	V, I	H	H	L
Synchrotron Fourier transform infrared mapping	V, I	H	H	L
Tip-enhanced Raman spectroscopy (TERS)	V, I	H	H	L

Table 3. *Cont.*

Technique	Application	Device Complexity	Interpretation Complexity	Use Frequency
	I: Identification V: Visualisation Q: Quantification		H: High M: Middle L: Low	
Total nitrogen content analysis Dumas method	Q	M	M	L
Total nitrogen content analysis LECO nitrogen quantification	Q	M	M	L
Ultra-high-liquid performance chromatography (UPLC)	Q	M	L	L
X-ray absorption fine structure (EXAFS)	I	H	M	L
X-ray diffraction (XRD)	I	H	L	L
X-ray photoelectron spectroscopy (XPS)	V, I	H	M	M
Zeta (the electrokinetic) potential	Q	H	M	L

4. Conclusions

Electrodialysis is a very attractive method to use in the food industry. The attractiveness of this method is primarily determined by the possibility of controlling the electrical charge of many substances (proteins, amino acids, particles of polybasic organic and inorganic acids, food dyes, etc.) by reagent-free pH control in the intermembrane space and within the IEM. This property provide unlimited possibilities in the separation and con-centration of valuable food and medicinal components, as well as in their purification from mineral impurities. The widespread use of ED in the food industry is constrained by the active interaction of the treated substances with ion-exchange membranes, which leads to a decrease in current efficiency, an increase in energy consumption and a de-crease in the life cycle of IEMs, which are the most expensive component of ED modules.

Recently, many new methods for studying fouling, as well as transport, mass trans-fer, and electrochemical characteristics of fouled membranes, have appeared. Their active use has led to a deeper understanding of interaction mechanisms of substances (constit-uent fluids of the food industry) with each other and with the IEM. It has been established that, in the case of ED application in the dairy industry and similar industries, the main reason for IEM fouling is electrostatic interactions of their fixed groups with proteins, as well as scaling of salts and hydroxides of alkaline earth metals (Ca^{2+}, Mg^{2+}, etc.).

In ED processing of liquid media of wineries, the tea industry and juice production, phenolic compounds (primarily anthocyanins) play a key role in fouling. Just like proteins, they can change their electrical charge depending on the pH of the environment. However, unlike the more massive proteins, anthocyanins can penetrate into the IEM and acquire an electrical charge that is different from their charge in solution. In addition, they actively enter the dipole–dipole (π–π, Van der Waals) interactions with each other and with the aromatic IEM matrix, which contributes to the formation of colloidal particles even in relatively small membrane pores. All IEM foulants in the food industry form hy-drogen bonds not only with fixed groups, but also with most of the polymers that make up membranes.

New knowledge has allowed broadening of our understanding of the reasons for the deterioration of IEM characteristics upon contact with liquid media of the food industry and more deliberate choosing of the appropriate ion exchange membranes and current modes. Thus, in the case of protein-containing solutions, this should be an IEM with a more hydrophobic surface in order to weaken the hydrogen bonding process. In the case of solutions containing phenolic compounds, it is better to use IEMs made from aliphatic materials, etc.

The main challenges for the near future are apparently the use of the accumulated knowledge (few π-bonds, low polar and more hydrophobic surfaces, a surface charge close to zero and minimal pH influence) to produce IEMs better adapted to the food industry (with the weakest protein-matrix and/or polyphenols–matrix interactions), or the selection of such membranes among the large number of IEMs that have recently appeared (Astom, Fujifilm, Mega (Auckland, New Zealand), Chimimpex (Wood Dale, IL, USA), HCPM (Wilmington, DE, USA), Du Pont (Wilmington, DE, USA), Hefei Chemjoy Polymer Materials Co. Ltd.). The production of these new IEMs can be achieved by searching for new material formulations (composites, co-polymers.) or by modifying the surfaces of existing IEMs. This last point is in itself a very vast domain.

Author Contributions: L.D.: Research design and investigation; Manuscript writing, revision and validation; Project administration. J.F., M.B. and V.S.: Methodology; Data analysis and curation; Draft preparation, writing and editing. C.L.: Methodology; Data analysis and curation; Validation. E.R. and L.B.: Research conceptualization and investigation; Manuscript writing, revision and validation. A.K.: Research conceptualization and investigation; Manuscript writing, revision and validation; Data curation; Visualization. N.P.: Research conceptualization and investigation; Manuscript writing, revision and validation; funding acquisition. All authors have read and agreed to the published version of the manuscript.

Funding: This work was financially supported by the Russian Foundation for Basic Research, project No 19-48-230024r_a.

Institutional Review Board Statement: Not applicable.

Informed Consent Statement: Not applicable.

Conflicts of Interest: The authors declare no conflict of interest.

Abbreviations

All other abbreviations have their usual meaning, or are sufficiently explained in the text.

AEM	Anion-exchange membrane
AFM	Atomic force microscopy
ARISA	Automated ribosomal intergenic spacer analysis
ATR–FTIR	Reflectance–Fourier-transform infrared spectroscopy
BSA	Bovine serum albumin
CEM	Cation-exchange membrane
CLSM	Confocal laser scanning microscopy
ED	Electrodialysis
EDS	Energy dispersive X-ray spectrometry
EEM	Excitation-emission matrix
EIS	Electrochemical impedance spectroscopy (spectrum)
EXAFS	X-ray absorption fine structure
FISH	Fluorescence in situ hybridization
FT-ICR-MS	Fourier transform-ion cyclotron resonance-mass spectrometry
HPLC	High-liquid performance chromatography
HPLC-MS	Mass spectrometry coupled to the HPLC
IEC	Ion exchange capacity
IEM	Ion-exchange membrane
MS	Mass spectrometry
MW	Moleculer weight
OCT	Optical coherence tomography
PARAFAC	Parallel factor analysis
PCR-DGGE	Polymerase chain reaction denaturing gradient gel electrophoresis
PDA	Polydopamine
PEF	Pulsed electric field

PP	Polyphenol
PS	Polystyrene
PVDF	Polyvinylidene fluoride
RBS	Rutherford backscattering spectroscopy
RO	Reverse osmosis
SDS	Sodium dodecyl sulphate
SDS-PAGE	Sodium dodecyl sulphate–polyacrylamide gel electrophoresis
SEC	Size-exclusion chromatography
SECM	Scanning electrochemical microscopy
SEM	Scanning electron microscopy
SERS	Surface-enhanced Raman spectroscopy
SICM	Scanning ion conductance microscopy
SPHS	Soy protein hydrolysate solution
SPR	Surface plasmon resonance
TERS	Tip-enhanced Raman spectroscopy
T-RFLP	Terminal restriction fragment length polymorphism
UF	Ultrafiltration
UPLC	Ultra-high-liquid performance chromatography
UPLC-MS	Mass spectrometry coupled to the UPLC
XPS	X-ray photoelectron spectroscopy
XRD	X-ray diffraction

References

1. Galama, A.H.; Saakes, M.; Bruning, H.; Rijnaarts, H.H.M.; Post, J.W. Seawater predesalination with electrodialysis. *Desalination* **2014**, *342*, 61–69. [CrossRef]
2. Regula, C.; Carretier, E.; Wyart, Y.; Gésan-Guiziou, G.; Vincent, A.; Boudot, D.; Moulin, P. Chemical cleaning/disinfection and ageing of organic UF membranes: A review. *Water Res.* **2014**, *56*, 325–365. [CrossRef]
3. Valero, F.; Arbós, R. Desalination of brackish river water using Electrodialysis Reversal (EDR): Control of the THMs formation in the Barcelona (NE Spain) area. *Desalination* **2010**, *253*, 170–174. [CrossRef]
4. Al-Amshawee, S.; Yunus, M.Y.B.M.; Azoddein, A.A.M.; Hassell, D.G.; Dakhil, I.H.; Abu Hasan, H. Electrodialysis desalination for water and wastewater: A review. *Chem. Eng. J.* **2020**, *380*, 122231. [CrossRef]
5. Liu, Y.; Yang, S.; Chen, Y.; Liao, J.; Pan, J.; Sotto, A.; Shen, J. Preparation of water-based anion-exchange membrane from PVA for anti-fouling in the electrodialysis process. *J. Membr. Sci.* **2019**, *570–571*, 130–138. [CrossRef]
6. Delyannis, E.-E. Status of solar assisted desalination: A review. *Desalination* **1987**, *67*, 3–19. [CrossRef]
7. Bauer, B.; Gerner, F.J.; Strathmann, H. Development of bipolar membranes. *Desalination* **1988**, *68*, 279–292. [CrossRef]
8. Fu, R.; Xu, T.; Yang, W.; Pan, Z. Preparation of a mono-sheet bipolar membrane by simultaneous irradiation grafting polymerization of acrylic acid and chloromethylstyrene. *J. Appl. Polym. Sci.* **2003**, *90*, 572–576. [CrossRef]
9. Qian, Z.; Miedema, H.; Sahin, S.; de Smet, L.C.P.M.; Sudhölter, E.J.R. Separation of alkali metal cations by a supported liquid membrane (SLM) operating under electro dialysis (ED) conditions. *Desalination* **2020**, *495*, 114631. [CrossRef]
10. Irfan, M.; Wang, Y.; Xu, T. Novel electrodialysis membranes with hydrophobic alkyl spacers and zwitterion structure enable high monovalent/divalent cation selectivity. *Chem. Eng. J.* **2020**, *383*, 123171. [CrossRef]
11. Pelletier, S.; Serre, É.; Mikhaylin, S.; Bazinet, L. Optimization of cranberry juice deacidification by electrodialysis with bipolar membrane: Impact of pulsed electric field conditions. *Sep. Purif. Technol.* **2017**, *186*, 106–116. [CrossRef]
12. Serre, E.; Rozoy, E.; Pedneault, K.; Lacour, S.; Bazinet, L. Deacidification of cranberry juice by electrodialysis: Impact of membrane types and configurations on acid migration and juice physicochemical characteristics. *Sep. Purif. Technol.* **2016**, *163*, 228–237. [CrossRef]
13. Bazinet, L.; Montpetit, D.; Ippersiel, D.; Amiot, J.; Lamarche, F. Identification of Skim Milk Electroacidification Fouling: A Microscopic Approach. *J. Colloid Interface Sci.* **2001**, *237*, 62–69. [CrossRef] [PubMed]
14. Mikhaylin, S.; Patouillard, L.; Margni, M.; Bazinet, L. Milk protein production by a more environmentally sustainable process: Bipolar membrane electrodialysis coupled with ultrafiltration. *Green Chem.* **2018**, *20*, 449–456. [CrossRef]
15. Gonçalves, F.; Fernandes, C.; Cameira dos Santos, P.; de Pinho, M.N. Wine tartaric stabilization by electrodialysis and its assessment by the saturation temperature. *J. Food Eng.* **2003**, *59*, 229–235. [CrossRef]
16. Ghalloussi, R.; Garcia-Vasquez, W.; Chaabane, L.; Dammak, L.; Larchet, C.; Deabate, S.V.; Nevakshenova, E.; Nikonenko, V.; Grande, D. Ageing of ion-exchange membranes in electrodialysis: A structural and physicochemical investigation. *J. Membr. Sci.* **2013**, *436*, 68–78. [CrossRef]

17. Ghalloussi, R.; Garcia-Vasquez, W.; Bellakhal, N.; Larchet, C.; Dammak, L.; Huguet, P.; Grande, D. Ageing of ion-exchange membranes used in electrodialysis: Investigation of static parameters, electrolyte permeability and tensile strength. *Sep. Purif. Technol.* **2011**, *80*, 270–275. [CrossRef]
18. Garcia-Vasquez, W.; Ghalloussi, R.; Dammak, L.; Larchet, C.; Nikonenko, V.; Grande, D. Structure and properties of heterogeneous and homogeneous ion-exchange membranes subjected to ageing in sodium hypochlorite. *J. Membr. Sci.* **2014**, *452*, 104–116. [CrossRef]
19. Al-Amoudi, A.; Lovitt, R.W. Fouling strategies and the cleaning system of NF membranes and factors affecting cleaning efficiency. *J. Membr. Sci.* **2007**, *303*, 4–28. [CrossRef]
20. Gautam, A.; Menkhaus, T.J. Performance evaluation and fouling analysis for reverse osmosis and nanofiltration membranes during processing of lignocellulosic biomass hydrolysate. *J. Membr. Sci.* **2014**, *451*, 252–265. [CrossRef]
21. Grossman, G.; Sonin, A.A. Membrane fouling in electrodialysis: A model and experiments. *Desalination* **1973**, *12*, 107–125. [CrossRef]
22. Mikhaylin, S.; Bazinet, L. Fouling on ion-exchange membranes: Classification, characterization and strategies of prevention and control. *Adv. Colloid Interface Sci.* **2016**, *229*, 34–56. [CrossRef] [PubMed]
23. Persico, M.; Bazinet, L. Fouling prevention of peptides from a tryptic whey hydrolysate during electromembrane processes by use of monovalent ion permselective membranes. *J. Membr. Sci.* **2018**, *549*, 486–494. [CrossRef]
24. Suwal, S.; Doyen, A.; Bazinet, L. Characterization of protein, peptide and amino acid fouling on ion-exchange and filtration membranes: Review of current and recently developed methods. *J. Membr. Sci.* **2015**, *496*, 267–283. [CrossRef]
25. Bdiri, M.; Larchet, C.; Dammak, L. A review on ion-exchange membranes fouling and antifouling during electrodialysis used in food industry: Cleanings and strategies of prevention. *Chem. Afr.* **2020**, *3*, 609–633. [CrossRef]
26. Bdiri, M.; Perreault, V.; Mikhaylin, S.; Larchet, C.; Hellal, F.; Bazinet, L.; Dammak, L. Identification of phenolic compounds and their fouling mechanisms in ion-exchange membranes used at an industrial scale for wine tartaric stabilization by electrodialysis. *Sep. Purif. Technol.* **2019**, *233*, 115995. [CrossRef]
27. Bazinet, L.; Degrandpre, Y.; Porter, A. Enhanced tobacco polyphenol electromigration and impact on membrane integrity. *J. Membr. Sci.* **2005**, *254*, 111–118. [CrossRef]
28. Casademont, C.; Sistat, P.; Ruiz, B.; Pourcelly, G.; Bazinet, L. Electrodialysis of model salt solution containing whey proteins: Enhancement by pulsed electric field and modified cell configuration. *J. Membr. Sci.* **2009**, *328*, 238–245. [CrossRef]
29. Ruiz, B.; Sistat, P.; Huguet, P.; Pourcelly, G.; Araya-Farias, M.; Bazinet, L. Application of relaxation periods during electrodialysis of a casein solution: Impact on anion-exchange membrane fouling. *J. Membr. Sci.* **2007**, *287*, 41–50. [CrossRef]
30. Lee, H.-J.; Hong, M.-K.; Han, S.-D.; Shim, J.; Moon, S.-H. Analysis of fouling potential in the electrodialysis process in the presence of an anionic surfactant foulant. *J. Membr. Sci.* **2008**, *325*, 719–726. [CrossRef]
31. Bdiri, M.; Dammak, L.; Larchet, C.; Hellal, F.; Porozhnyy, M.; Nevakshenova, E.; Pismenskaya, N.; Nikonenko, V. Characterization and cleaning of anion-exchange membranes used in electrodialysis of polyphenol-containing food industry solutions; comparison with cation-exchange membranes. *Sep. Purif. Technol.* **2019**, *210*, 636–650. [CrossRef]
32. Bdiri, M.; Dammak, L.; Chaabane, L.; Larchet, C.; Hellal, F.; Nikonenko, V.; Pismenskaya, N.D. Cleaning of cation-exchange membranes used in electrodialysis for food industry by chemical solutions. *Sep. Purif. Technol.* **2018**, *199*, 114–123. [CrossRef]
33. Sarapulova, V.; Nevakshenova, E.; Nebavskaya, X.; Kozmai, A.; Aleshkina, D.; Pourcelly, G.; Nikonenko, V.; Pismenskaya, N. Characterization of bulk and surface properties of anion-exchange membranes in initial stages of fouling by red wine. *J. Membr. Sci.* **2018**, *559*, 170–182. [CrossRef]
34. Bukhovets, A.; Eliseeva, T.; Oren, Y. Fouling of anion-exchange membranes in electrodialysis of aromatic amino acid solution. *J. Membr. Sci.* **2010**, *364*, 339–343. [CrossRef]
35. Mustafa, G.; Wyns, K.; Buekenhoudt, A.; Meynen, V. New insights into the fouling mechanism of dissolved organic matter applying nanofiltration membranes with a variety of surface chemistries. *Water Res.* **2016**, *93*, 195–204. [CrossRef]
36. Grossman, G.; Sonin, A.A. Experimental study of the effects of hydrodynamics and membrane fouling in electrodialysis. *Desalination* **1972**, *10*, 157–180. [CrossRef]
37. Akhondi, E.; Zamani, F.; Law, A.W.K.; Krantz, W.B.; Fane, A.G.; Chew, J.W. Influence of backwashing on the pore size of hollow fiber ultrafiltration membranes. *J. Membr. Sci.* **2017**, *521*, 33–42. [CrossRef]
38. Krahnstöver, T.; Hochstrat, R.; Wintgens, T. Comparison of methods to assess the integrity and separation efficiency of ultrafiltration membranes in wastewater reclamation processes. *J. Water Process Eng.* **2019**, *30*, 100646. [CrossRef]
39. Taghavijeloudar, M.; Park, J.; Han, M.; Taghavi, A. A new approach for modeling flux variation in membrane filtration and experimental verification. *Water Res.* **2019**, *166*, 115027. [CrossRef] [PubMed]
40. Judd, S. *The MBR Book: Principles and Applications of Membrane Bioreactors for Water and Wastewater Treatment*; Elsevier: Amsterdam, The Netherlands, 2011; p. 536. ISBN 9780080966823.
41. Spettmann, D.; Eppmann, S.; Flemming, H.C.; Wingender, J. Simultaneous visualisation of biofouling, organic and inorganic particle fouling on separation membranes. *Water Sci. Technol.* **2007**, *55*, 207–210. [CrossRef]
42. El Rayess, Y.; Mietton-Peuchot, M. Membrane Technologies in Wine Industry: An Overview. *Crit. Rev. Food Sci. Nutr.* **2016**, *56*, 2005–2020. [CrossRef] [PubMed]
43. Bleha, M.; Tishchenko, G.; Sumberova, V.; Kudela, V. Characteristic of the critical state of membranes in ED-desalination of milk whey. *Desalination* **1992**, *86*, 173–186. [CrossRef]

44. Korngold, E.; De Körösy, F.; Rahav, R.; Taboch, M.F. Fouling of anionselective membranes in electrodialysis. *Desalination* **1970**, *8*, 195–220. [CrossRef]
45. Lindstrand, V.; Sundström, G.; Jönsson, A.S. Fouling of electrodialysis membranes by organic substances. *Desalination* **2000**, *128*, 91–102. [CrossRef]
46. Park, J.S.; Lee, H.J.; Moon, S.H. Determination of an optimum frequency of square wave power for fouling mitigation in desalting electrodialysis in the presence of humate. *Sep. Purif. Technol.* **2003**, *30*, 101–112. [CrossRef]
47. Mikhaylin, S.; Nikonenko, V.; Pourcelly, G.; Bazinet, L. Intensification of demineralization process and decrease in scaling by application of pulsed electric field with short pulse/pause conditions. *J. Membr. Sci.* **2014**, *468*, 389–399. [CrossRef]
48. Asfar-Snir, M.; Gilron, J.; Oren, Y. Gypsum scaling on anion exchange membrane during Donnan exchange. *J. Membr. Sci.* **2014**, *455*, 384–391. [CrossRef]
49. Higa, M.; Tanaka, N.; Nagase, M.; Yutani, K.; Kameyama, T.; Takamura, K.; Kakihana, Y. Electrodialytic properties of aromatic and aliphatic type hydrocarbonbased anion-exchange membranes with various anion-exchange groups. *Polymer* **2014**, *55*, 3951–3960. [CrossRef]
50. Mikhaylin, S.; Nikonenko, V.; Pourcelly, G.; Bazinet, L. Hybrid bipolar membrane electrodialysis/ultrafiltration technology assisted by a pulsed electric field for casein production. *Green Chem.* **2016**, *18*, 307–314. [CrossRef]
51. Dai, R.; Li, Z.; Wang, T.; Ma, J.; Wang, Z. Techniques for understanding mechanisms underlying membrane fouling. In *Current Developments in Biotechnology and Bioengineering*; Elsevier: Amsterdam, The Netherlands, 2020; pp. 81–102. [CrossRef]
52. Oshchepkov, M.; Golovesov, V.; Ryabova, A.; Tkachenko, S.; Redchuk, A.; Rönkkömäki, H.; Rudakova, G.; Pervov, A.; Popov, K. Visualization of a novel fluorescent-tagged bisphosphonate behavior during reverse osmosis desalination of water with high sulfate content. *Sep. Purif. Technol.* **2021**, *255*, 117382. [CrossRef]
53. Oshchepkov, M.; Golovesov, V.; Ryabova, A.; Redchuk, A.; Tkachenko, S.; Pervov, A.; Popov, K. Gypsum crystallization during reverse osmosis desalination of water with high sulfate content in presence of a novel fluorescent-tagged polyacrylate. *Crystals* **2020**, *10*, 309. [CrossRef]
54. Hansma, P.; Drake, B.; Marti, O.; Gould, S.A.; Prater, C.B. The scanning ion-conductance microscope. *Science* **1989**, *243*, 641–643. [CrossRef]
55. Shi, X.; Qing, W.; Marhaba, T.; Zhang, W. Atomic force microscopy-scanning electrochemical microscopy (AFM-SECM) for nanoscale topographical and electrochemical characterization: Principles, applications and perspectives. *Electrochim. Acta* **2020**, *332*, 135472. [CrossRef]
56. Mareev, S.A.; Butylskii, D.Y.; Pismenskaya, N.D.; Larchet, C.; Dammak, L.; Nikonenko, V.V. Geometrical heterogeneity of homogeneous ion-exchange Neosepta membranes. *J. Membr. Sci.* **2018**, *563*, 768–776. [CrossRef]
57. Butylskii, D.Y.; Mareev, S.A.; Nikonenko, V.V.; Pismenskaya, N.D.; Larchet, C.; Dammak, L.; Grande, D.; Apel, P.Y. In situ investigation of electrical inhomogeneity of ion exchange membrane surface using scanning electrochemical microscopy. *Petrol. Chem.* **2016**, *56*, 1006–1013. [CrossRef]
58. Casademont, C.; Araya Farias, M.; Pourcelly, G.; Bazinet, L. Impact of electrodialytic parameters on cation migration kinetics and fouling nature of ion-exchange membranes during treatment of solutions with different magnesium/calcium ratios. *J. Membr. Sci.* **2008**, *325*, 570–579. [CrossRef]
59. Butylskii, D.Y. Study of Surface Morphology of Ion-Exchange Membranes and Its Influence on Their Electrochemical Characteristics. Ph.D. Thesis, Kuban State University, Krasnodar, Russia, 2019.
60. Park, J.-S.; Chilcott, T.C.; Coster, H.G.L.; Moon, S.-H. Characterization of BSA-fouling of ion-exchange membrane systems using a subtraction technique for lumped data. *J. Membr. Sci.* **2005**, *246*, 137–144. [CrossRef]
61. Kattan Readi, O.M. Membranes in the Biobased Economy: Electrodialysis of Amino Acids for the Production of Biochemical. Ph.D. Thesis, Universiteit Twente, Enschede, The Netherlands, 2013; p. 176.
62. Persico, M.; Mikhaylin, S.; Doyen, A.; Firdaous, L.; Hammami, R.; Chevalier, M.; Flahaut, C.; Dhulster, P.; Bazinet, L. Formation of peptide layers and adsorption mechanisms on a negatively charged cation-exchange membrane. *J. Colloid Interface* **2017**, *508*, 488–499. [CrossRef] [PubMed]
63. Langevin, M.-E.; Bazinet, L. Ion-exchange membrane fouling by peptides: A phenomenon governed by electrostatic interactions. *J. Membr. Sci.* **2011**, *369*, 359–366. [CrossRef]
64. Ayala-Bribiesca, E.; Araya-Farias, M.; Pourcelly, G.; Bazinet, L. Effect of concentrate solution pH and mineral composition of a whey protein diluate solution on membrane fouling formation during conventional electrodialysis. *J. Membr. Sci.* **2006**, *280*, 790–801. [CrossRef]
65. Persico, M.; Mikhaylin, S.; Doyen, A.; Firdaous, L.; Hammami, R.; Bazinet, L. How peptide physicochemical and structural characteristics affect anion-exchange membranes fouling by a tryptic whey protein hydrolysate. *J. Membr. Sci.* **2016**, *520*, 914–923. [CrossRef]
66. Harrigan, W.F.; McCance, M.E. *Laboratory Methods in Microbiology*; Academic Press: Cambridge, MA, USA, 2014; p. 374. ISBN 9781483274348.
67. Kozmai, A.; Sarapulova, V.; Sharafan, M.; Melkonian, K.; Rusinova, T.; Kozmai, Y.; Pismenskaya, N.; Dammak, L.; Nikonenko, V. Electrochemical impedance spectroscopy of anion-exchange membrane amx-sb fouled by red wine components. *Membranes* **2021**, *11*, 2. [CrossRef]

68. Persico, M.; Mikhaylin, S.; Doyen, A.; Firdaous, L.; Nikonenko, V.; Pismenskaya, N.; Bazinet, L. Prevention of peptide fouling on ion-exchange membranes during electrodialysis in overlimiting conditions. *J. Membr. Sci.* **2017**, *543*, 212–221. [CrossRef]
69. Lee, H.-J.; Hong, M.-K.; Han, S.-D.; Cho, S.-H.; Moon, S.-H. Fouling of an anion exchange membrane in the electrodialysis desalination process in the presence of organic foulants. *Desalination* **2009**, *238*, 60–69. [CrossRef]
70. Labbé, D.; Bazinet, L. Effect of membrane type on cation migration during green tea electromigration and equivalent mass transported calculation. *J. Membr. Sci.* **2006**, *275*, 220–228. [CrossRef]
71. Labbé, D.; Araya-Farias, M.; Tremblay, A.; Bazinet, L. Electromigration feasibility of green tea catechins. *J. Membr. Sci.* **2005**, *254*, 101–109. [CrossRef]
72. Faucher, M.; Serre, É.; Langevin, M.-È.; Mikhaylin, S.; Lutin, F.; Bazinet, L. Drastic energy consumption reduction and ecoefficiency improvement of cranberry juice deacidification by electrodialysis with bipolar membranes at semi-industrial scale: Reuse of the recovery solution. *J. Membr. Sci.* **2018**, *555*, 105–114. [CrossRef]
73. Perreault, V.; Sarapulova, V.; Tsygurina, K.; Pismenskaya, N.; Bazinet, L. Understanding of adsorption and desorption mechanisms of anthocyanins and proanthocyanidins on heterogeneous and homogeneous cation-exchange membranes. *Membranes* **2021**, *11*, 136. [CrossRef]
74. Evans, P.J.; Bird, M.R.; Rogers, D.; Wright, C.J. Measurement of polyphenol-membrane interaction forces during the ultrafiltration of black tea liquor. *Colloids Surf. A Physicochem. Eng. Asp.* **2009**, *335*, 148–153. [CrossRef]
75. Sholokhova, A.Y.; Eliseeva, T.V.; Voronyuk, I.V. Sorption of vanillin by highly basic anion exchangers under dynamic conditions. *Russ. J. Phys. Chem. A* **2018**, *92*, 2048–2052. [CrossRef]
76. Sosa-Fernandez, P.A.; Miedema, S.J.; Bruning, H.; Leermakers, F.A.M.; Post, J.W.; Rijnaarts, H.H.M. Effects of feed composition on the fouling on cation-exchange membranes desalinating polymer-flooding produced water. *J. Colloid Interface Sci.* **2021**, *584*, 634–646. [CrossRef] [PubMed]
77. Wang, T.; Yu, S.; Hou, L.A. Impacts of HPAM molecular weights on desalination performance of ion exchange membranes and fouling mechanism. *Desalination* **2017**, *404*, 50–58. [CrossRef]
78. Xia, Q.; Guo, H.; Ye, Y.; Yu, S.; Li, L.; Li, Q.; Zhang, R. Study on the fouling mechanism and cleaning method in the treatment of polymer flooding produced water with ion exchange membranes. *RSC Adv.* **2018**, *8*, 29947–29957. [CrossRef]
79. Jackson, R.S. *Wine Science: Principles and Applications*; Academic Press: Cambridge, MA, USA, 2014; p. 751. ISBN 9780123736468.
80. Suwal, S.; Roblet, C.; Amiot, J.; Bazinet, L. Presence of free amino acids in protein hydrolysate during electroseparation of peptides: Impact on system efficiency and membrane physicochemical properties. *Sep. Purif. Technol.* **2015**, *147*, 227–236. [CrossRef]
81. Ge, S.; Zhang, Z.; Yan, H.; Irfan, M.; Xu, Y.; Li, W.; Wang, Y. Electrodialytic Desalination of Tobacco Sheet Extract: Membrane Fouling Mechanism and Mitigation Strategies. *Membranes* **2020**, *10*, 245. [CrossRef] [PubMed]
82. Zhao, Z.; Shi, S.; Cao, H.; Li, Y.; Van der Bruggen, B. Comparative studies on fouling of homogeneous anion exchange membranes by different structured organics in electrodialysis. *J. Environ. Sci.* **2019**, *77*, 218–228. [CrossRef]
83. Gorzalski, A.S.; Donley, C.; Coronell, O. Elemental composition of membrane foulant layers using EDS, XPS, and RBS. *J. Membr. Sci.* **2017**, *522*, 31–44. [CrossRef]
84. Cheesman, A.W.; Turner, B.L.; Reddy, K.R. Interaction of phosphorus compounds with anion-exchange membranes: Implications for soil analysis. *Soil Sci. Soc. Am. J.* **2010**, *74*, 1607–1612. [CrossRef]
85. Chan, R.; Chen, V. Characterization of protein fouling on membranes: Opportunities and challenges. *J. Membr. Sci.* **2004**, *242*, 169–188. [CrossRef]
86. Labbe, J.P.; Quemerais, A.; Michel, F.; Daufin, G. Fouling of inorganic membranes during whey ultrafiltration: Analytical methodology. *J. Membr. Sci.* **1990**, *51*, 293–307. [CrossRef]
87. Merkel, A.; Fárová, H.; Voropaeva, D.; Yaroslavtsev, A.; Ahrné, L.; Yazdi, S.R. The impact of high effective electrodialytic desalination on acid whey stream at high temperature. *Int. Dairy J.* **2021**, *114*, 104921. [CrossRef]
88. Lee, H.; Im, S.J.; Lee, H.; Kim, C.M.; Jang, A. Comparative analysis of salt cleaning and osmotic backwash on calcium-bridged organic fouling in nanofiltration process. *Desalination* **2021**, *507*, 115022. [CrossRef]
89. Xie, M.; Luo, W.; Gray, S.R. Synchrotron Fourier transform infrared mapping: A novel approach for membrane fouling characterization. *Water Res.* **2017**, *111*, 375–381. [CrossRef] [PubMed]
90. Adusei-Gyamfi, J.; Ouddane, B.; Rietveld, L.; Cornard, J.-P.; Criquet, J. Natural organic matter-cations complexation and its impact on water treatment: A critical review. *Water Res.* **2019**, *160*, 130–147. [CrossRef]
91. Ribéreau-Gayon, P.; Glories, Y.; Maujean, A.; Dubourdieu, D. *Handbook of Enology: The Chemistry of Wine, Stabilization and Treatments*; John Wiley & Sons Ltd.: Chichester, UK, 2006; p. 451. ISBN 100470010371.
92. Pismenskaya, N.; Sarapulova, V.; Klevtsova, A.; Mikhaylin, S.; Bazinet, L. Adsorption of anthocyanins by cation and anion exchange resins with aromatic and aliphatic polymer matrices. *Int. J. Mol. Sci.* **2020**, *21*, 7874. [CrossRef] [PubMed]
93. Helfferich, F. *Ion Exchange*; McGraw-Hill: New York, NY, USA, 1962; p. 624. ISBN 0486687848.
94. Eker, F.; Cao, X.; Nafie, L.; Schweitzer-Stenner, R. Tripeptides adopt stable structures in water. A combined polarized visible Raman, FTIR, and VCD spectroscopy study. *J. Am. Chem. Soc.* **2002**, *124*, 14330–14341. [CrossRef] [PubMed]
95. Virtanen, T.; Parkkila, P.; Koivuniemi, A.; Lahti, J.; Viitala, T.; Kallioinen, M.; Mänttäri, M.; Bunker, A. Characterization of membrane-foulant interactions with novel combination of Raman spectroscopy, surface plasmon resonance and molecular dynamics simulation. *Sep. Purif. Technol.* **2018**, *205*, 263–272. [CrossRef]

96. Chen, W.; Qian, C.; Hong, W.-L.; Cheng, J.-X.; Yu, H.-Q. Evolution of membrane fouling revealed by label-free vibrational spectroscopic imaging. *Environ. Sci. Technol.* **2017**, *51*, 9580–9587. [CrossRef]
97. Chen, W.; Qian, C.; Zhou, K.-G.; Yu, H.-Q. Molecular spectroscopic characterization of membrane fouling: A critical review. *Chem* **2018**, *4*, 1492–1509. [CrossRef]
98. Sun, W.; Yue, D.; Song, J.; Nie, Y. Adsorption removal of refractory organic matter in bio-treated municipal solid waste landfill leachate by anion exchange resins. *Waste Manag.* **2018**, *81*, 61–70. [CrossRef]
99. Xu, H.; Yan, Z.; Cai, H.; Yu, G.; Yang, L.; Jiang, H. Heterogeneity in metal binding by individual fluorescent components in a eutrophic algae-rich lake. *Ecotoxicol. Environ. Saf.* **2013**, *98*, 266–272. [CrossRef]
100. Shi, L.; Xie, S.; Hu, Z.; Wu, G.; Morrison, L.; Croot, P.; Zhan, X. Nutrient recovery from pig manure digestate using electrodialysis reversal: Membrane fouling and feasibility of long-term operation. *J. Membr. Sci.* **2019**, *573*, 560–569. [CrossRef]
101. Chon, K.; Jeong, N.; Rho, H.; Nam, J.Y.; Jwa, E.; Cho, J. Fouling characteristics of dissolved organic matter in fresh water and seawater compartments of reverse electrodialysis under natural water conditions. *Desalination* **2020**, *496*, 114478. [CrossRef]
102. Guan, Y.F.; Qian, C.; Chen, W.; Huang, B.C.; Wang, Y.J.; Yu, H.Q. Interaction between humic acid and protein in membrane fouling process: A spectroscopic insight. *Water Res.* **2018**, *145*, 146–152. [CrossRef]
103. Tang, J.; Zhuang, L.; Yu, Z.; Liu, X.; Wang, Y.; Wen, P.; Zhou, S. Insight into complexation of Cu(II) to hyperthermophilic compost-derived humic acids by EEM-PARAFAC combined with heterospectral two dimensional correlation analyses. *Sci. Total Environ.* **2019**, *656*, 29–38. [CrossRef]
104. Peiris, R.H.; Ignagni, N.; Budman, H.; Moresoli, C.; Legge, R.L. Characterizing natural colloidal/particulate-protein interactions using fluorescence-based techniques and principal component analysis. *Talanta* **2012**, *99*, 457–463. [CrossRef] [PubMed]
105. Shi, L.; Hu, Z.; Simplicio, W.S.; Qiu, S.; Xiao, L.; Harhen, B.; Zhan, X. Antibiotics in nutrient recovery from pig manure via electrodialysis reversal: Sorption and migration associated with membrane fouling. *J. Membr. Sci.* **2020**, *597*, 117633. [CrossRef]
106. Ray, S.K.; Truong, H.B.; Arshad, Z.; Shin, H.S.; Hur, J. Recent advances in the characterization and the treatment methods of effluent organic matter. *Membr. Water Treat.* **2020**, *11*, 257–274. [CrossRef]
107. Park, J.-S.; Lee, H.-J.; Choi, S.-J.; Geckeler, K.E.; Cho, J.; Moon, S.-H. Fouling mitigation of anion exchange membrane by zeta potential control. *J. Colloid Interface Sci.* **2003**, *259*, 293–300. [CrossRef]
108. Kim, D.H.; Moon, S.-H.; Cho, J. Investigation of the adsorption and transport of natural organic matter (NOM) in ion-exchange membranes. *Desalination* **2003**, *151*, 11–20. [CrossRef]
109. Lee, H.-J.; Moon, S.-H. Enhancement of electrodialysis performances using pulsing electric fields during extended period operation. *J. Colloid Interface Sci.* **2005**, *287*, 597–603. [CrossRef]
110. Persico, M.; Daigle, G.; Kadel, S.; Perreault, V.; Pellerin, G.; Thibodeau, J.; Bazinet, L. Predictive models for determination of peptide fouling based on the physicochemical characteristics of filtration membranes. *Sep. Purif. Technol.* **2020**, *240*, 116602. [CrossRef]
111. Yamato, N.; Kimura, K.; Miyoshi, T.; Watanabe, Y. Difference in membrane fouling in membrane bioreactors (MBRs) caused by membrane polymer materials. *J. Membr. Sci.* **2006**, *280*, 911–919. [CrossRef]
112. Madrigal Carballo, S.; Rodriguez, G.; Vega Baudrit, J.; Krueger, C.G. MALDI-TOF mass spectrometry of oligomeric food polyphenols. *Int. Food Res. J.* **2013**, *20*, 2023–2034. Available online: http://hdl.handle.net/20.500.12337/3329 (accessed on 13 October 2021).
113. Alecu, A.; Albu, C.; Litescu, S.C.; Eremia, S.A.; Radu, G.L. Phenolic and anthocyanin profile of Valea Calugareasca red wines by HPLC-PDA-MS and MALDI-TOF analysis. *Food Anal. Methods* **2016**, *9*, 300–310. [CrossRef]
114. Chan, R.; Chen, V.; Bucknall, M.P. Quantitative analysis of membrane fouling by protein mixtures using MALDI-MS. *Biotechnol. Bioeng.* **2004**, *85*, 190–201. [CrossRef] [PubMed]
115. Flemming, H.C.; Wingender, J. The biofilm matrix. *Nat. Rev. Microbiol.* **2010**, *8*, 623–633. [CrossRef] [PubMed]
116. Allen, A.; Semião, A.C.; Habimana, O.; Heffernan, R.; Safari, A.; Casey, E. Nanofiltration and reverse osmosis surface topographical heterogeneities: Do they matter for initial bacterial adhesion? *J. Membr. Sci.* **2015**, *486*, 10–20. [CrossRef]
117. Vaselbehagh, M.; Karkhanechi, H.; Takagi, R.; Matsuyama, H. Biofouling phenomena on anion exchange membranes under the reverse electrodialysis process. *J. Membr. Sci.* **2017**, *530*, 232–239. [CrossRef]
118. Surman, S.B.; Walker, J.T.; Goddard, D.T.; Morton, L.H.G.; Keevil, C.W.; Weaver, W.; Skinner, A.; Hanson, K.; Caldwell, D.; Kurtz, J. Comparison of microscope techniques for the examination of biofilms. *J. Microbiol. Methods* **1996**, *25*, 57–70. [CrossRef]
119. Luo, H.; Xu, P.; Jenkins, P.E.; Ren, Z. Ionic composition and transport mechanisms in microbial desalination cells. *J. Membr. Sci.* **2012**, *409–410*, 16–23. [CrossRef]
120. Luo, H.; Xu, P.; Ren, Z. Long-term performance and characterization of microbial desalination cells in treating domestic wastewater. *Bioresour. Technol.* **2012**, *120*, 187–193. [CrossRef]
121. Křivčík, J.; Neděla, D.; Válek, R. Ion-exchange membrane reinforcing. *Desalin. Water Treat.* **2015**, *56*, 3214–3219. [CrossRef]
122. Lencki, R.W.; Riedl, K. Effect of fractal flocculation behavior on fouling layer resistance during apple juice microfiltration. *Food Res. Int.* **1999**, *32*, 279–288. [CrossRef]
123. Lu, X.; Peng, Y.; Ge, L.; Lin, R.; Zhu, Z.; Liu, S. Amphiphobic PVDF composite membranes for anti-fouling direct contact membrane distillation. *J. Membr. Sci.* **2016**, *505*, 61–69. [CrossRef]
124. Nagaraj, V.; Skillman, L.; Li, D.; Xie, Z.; Ho, G. Culturable bacteria from a full-scale desalination plant: Identification methods, bacterial diversity and selection of models based on membrane-biofilm community. *Desalination* **2019**, *457*, 103–114. [CrossRef]

125. Zhi, W.; Ge, Z.; He, Z.; Zhang, H. Methods for understanding microbial community structures and functions in microbial fuel cells: A review. *Bioresour. Technol.* **2014**, *171*, 461–468. [CrossRef] [PubMed]
126. Vanysacker, L.; Declerck, P.; Bilad, M.R.; Vankelecom, I.F.J. Biofouling on microfiltration membranes in MBRs: Role of membrane type and microbial community. *J. Membr. Sci.* **2014**, *453*, 394–401. [CrossRef]
127. Gao, D.; Fu, Y.; Ren, N. Tracing biofouling to the structure of the microbial community and its metabolic products: A study of the three-stage MBR process. *Water Res.* **2013**, *47*, 6680–6690. [CrossRef]
128. Wu, B.; Yi, S.; Fane, A.G. Microbial behaviors involved in cake fouling in membrane bioreactors under different solids retention times. *Bioresour. Technol.* **2011**, *102*, 2511–2516. [CrossRef] [PubMed]
129. Kocherginskaya, S.A.; Cann, I.K.; Mackie, R.I. *Denaturing Gradient Gel Electrophoresis*; Springer: Dordrecht, The Netherlands, 2005; p. 119. ISBN 9781402037900. [CrossRef]
130. Ghosh, A.; Bhadury, P. *Methods of Assessment of Microbial Diversity in Natural Environments*; Elsevier Inc.: Amsterdam, The Netherlands, 2018; p. 14. [CrossRef]
131. Chen, C.H.; Fu, Y.; Gao, D.W. Membrane biofouling process correlated to the microbial community succession in an A/O MBR. *Bioresour. Technol.* **2015**, *197*, 185–192. [CrossRef] [PubMed]
132. Kniggendorf, A.K.; Nogueira, R.; Kelb, C.; Schadzek, P.; Meinhardt-Wollweber, M.; Ngezahayo, A.; Roth, B. Confocal Raman microscopy and fluorescent in situ hybridization—A complementary approach for biofilm analysis. *Chemosphere* **2016**, *161*, 112–118. [CrossRef] [PubMed]
133. Karygianni, L.; Follo, M.; Hellwig, E.; Burghardt, D.; Wolkewitz, M.; Anderson, A.; Al-Ahmad, A. Microscope-based imaging platform for large-scale analysis of oral biofilms. *Appl. Environ. Microbiol.* **2012**, *78*, 8703–8711. [CrossRef] [PubMed]
134. Lawrence, J.R.; Swerhone, G.D.W.; Leppard, G.G.; Araki, T.; Zhang, X.; West, M.M.; Hitchcock, A.P. Scanning transmission X-ray, laser scanning, and transmission electron microscopy mapping of the exopolymeric matrix of microbial biofilms. *Appl. Environ. Microbiol.* **2003**, *69*, 5543–5554. [CrossRef] [PubMed]
135. Dubois, M.; Gilles, K.A.; Hamilton, J.K.; Rebers, P.A.; Smith, F. Colorimetric method for determination of sugars and related substances. *Anal. Chem.* **1956**, *28*, 350–356. [CrossRef]
136. Sweity, A.; Ying, W.; Ali-Shtayeh, M.S.; Yang, F.; Bick, A.; Oron, G.; Herzberg, M. Relation between EPS adherence, viscoelastic properties, and MBR operation: Biofouling study with QCM-D. *Water Res.* **2011**, *45*, 6430–6440. [CrossRef]
137. Yan, Z.; Yang, H.; Qu, F.; Zhang, H.; Rong, H.; Yu, H.; Van der Bruggen, B. Application of membrane distillation to anaerobic digestion effluent treatment: Identifying culprits of membrane fouling and scaling. *Sci. Total Environ.* **2019**, *688*, 880–889. [CrossRef]
138. Johnson, D.J.; Galliano, F.; Deowan, S.A.; Hoinkis, J.; Figoli, A.; Hilal, N. Adhesion forces between humic acid functionalised colloidal probes and polymer membranes to assess fouling potential. *J. Membr. Sci.* **2015**, *484*, 35–46. [CrossRef]
139. Johnson, D.; Hilal, N. Characterisation and quantification of membrane surface properties using atomic force microscopy: A comprehensive review. *Desalination* **2015**, *356*, 149–164. [CrossRef]
140. Krisilova, E.V.; Eliseeva, T.V.; Oros, G.Y. Estimation of effect of amino acid's sorption on surface state of ion-exchange membranes using atomic-force microscopy data. *Prot. Met. Phys. Chem. Surf.* **2011**, *47*, 39–42. [CrossRef]
141. Donald, A.M. The use of environmental scanning electron microscopy for imaging wet and insulating materials. *Nat. Mater.* **2003**, *2*, 511–516. [CrossRef]
142. Reichert, U.; Linden, T.; Belfort, G.; Kula, M.-R.; Thömmes, J. Visualising protein adsorption to ion-exchange membranes by confocal microscopy. *J. Membr. Sci.* **2002**, *199*, 161–166. [CrossRef]
143. Vaselbehagh, M.; Karkhanechi, H.; Takagi, R.; Matsuyama, H. Effect of polydopamine coating and direct electric current application on anti-biofouling properties of anion exchange membranes in electrodialysis. *J. Membr. Sci.* **2016**, *515*, 98–108. [CrossRef]
144. Herzberg, M.; Pandit, S.; Mauter, M.S.; Oren, Y. Bacterial biofilm formation on ion exchange membranes. *J. Membr. Sci.* **2020**, *596*, 117564. [CrossRef]
145. Bauer, A.; Wagner, M.; Saravia, F.; Bartl, S.; Hilgenfeldt, V.; Horn, H. In-situ monitoring and quantification of fouling development in membrane distillation by means of optical coherence tomography. *J. Membr. Sci.* **2019**, *577*, 145–152. [CrossRef]
146. Liu, X.; Chen, G.; Tu, G.; Li, Z.; Deng, B.; Li, W. Membrane fouling by clay suspensions during NF-like forward osmosis: Characterization via optical coherence tomography. *J. Membr. Sci.* **2020**, *602*, 117965. [CrossRef]
147. Li, W.; Liu, X.; Wang, Y.N.; Chong, T.H.; Tang, C.Y.; Fane, A.G. Analyzing the evolution of membrane fouling via a novel method based on 3D optical coherence tomography imaging. *Environ. Sci. Technol.* **2016**, *50*, 6930–6939. [CrossRef] [PubMed]
148. Fortunato, L.; Jeong, S.; Leiknes, T. Time-resolved monitoring of biofouling development on a flat sheet membrane using optical coherence tomography. *Sci. Rep.* **2017**, *7*, 15. [CrossRef] [PubMed]
149. Trinh, T.A.; Li, W.; Han, Q.; Liu, X.; Fane, A.G.; Chew, J.W. Analyzing external and internal membrane fouling by oil emulsions via 3D optical coherence tomography. *J. Membr. Sci.* **2018**, *548*, 632–640. [CrossRef]
150. Merino-Garcia, I.; Kotoka, F.; Portugal, C.A.M.; Crespo, J.G.; Velizarov, S. Characterization of poly(Acrylic) acid-modified heterogenous anion exchange membranes with improved monovalent permselectivity for RED. *Membranes* **2020**, *10*, 134. [CrossRef] [PubMed]

151. Li, Y.; Shi, S.; Cao, H.; Xu, B.; Zhao, Z.; Cao, R.; Chang, J.; Duan, F.; Wen, H. Anion exchange nanocomposite membranes modified with graphene oxide and polydopamine: Interfacial structure and antifouling applications. *ACS Appl. Nano Mater.* **2020**, *3*, 588–596. [CrossRef]
152. Pintossi, D.; Saakes, M.; Borneman, Z.; Nijmeijer, K. Electrochemical impedance spectroscopy of a reverse electrodialysis stack: A new approach to monitoring fouling and cleaning. *J. Power Sources* **2019**, *444*, 227302. [CrossRef]
153. Zhang, L.; Jia, H.; Wang, J.; Wen, H.; Li, J. Characterization of fouling and concentration polarization in ion exchange membrane by in-situ electrochemical impedance spectroscopy. *J. Membr. Sci.* **2020**, *594*, 117443. [CrossRef]
154. Femmer, R.; Martí-Calatayud, M.C.; Wessling, M. Mechanistic modeling of the dielectric impedance of layered membrane architectures. *J. Membr. Sci.* **2016**, *520*, 29–36. [CrossRef]
155. Guo, H.; Xiao, L.; Yu, S.; Yang, H.; Hu, J.; Liu, G.; Tang, Y. Analysis of anion exchange membrane fouling mechanism caused by anion polyacrylamide in electrodialysis. *Desalination* **2014**, *346*, 46–53. [CrossRef]
156. Cai, M.; Xie, C.; Zhong, H.; Tian, B.; Yang, K. Identification of anthocyanins and their fouling mechanisms during non-thermal nanofiltration of blueberry aqueous extracts. *Membranes* **2021**, *11*, 200. [CrossRef]
157. Huhtamäki, T.; Tian, X.; Korhonen, J.T.; Ras, R.H.A. Surface-wetting characterization using contact-angle measurements. *Nat. Protoc.* **2018**, *13*, 1521–1538. [CrossRef]
158. Drelich, J.W.; Boinovich, L.; Chibowski, E.; Volpe, C.D.; Hołysz, L.; Marmur, A.; Siboni, S. Contact angles: History of over 200 years of open questions. *Surf. Innov.* **2019**, *8*, 3–27. [CrossRef]
159. Mikhaylin, S.; Nikonenko, V.; Pismenskaya, N.; Pourcelly, G.; Choi, S.; Kwon, H.J.; Han, J.; Bazinet, L. How physico-chemical and surface properties of cation-exchange membrane affect membrane scaling and electroconvective vortices: Influence on performance of electrodialysis with pulsed electric field. *Desalination* **2016**, *393*, 102–114. [CrossRef]
160. Garcia-Vasquez, W.; Dammak, L.; Larchet, C.; Nikonenko, V.; Pismenskaya, N.; Grande, D. Evolution of anion-exchange membrane properties in a full scale electrodialysis stack. *J. Membr. Sci.* **2013**, *446*, 255–265. [CrossRef]
161. Zhang, W.; Wahlgren, M.; Sivik, B. Membrane characterization by the contact angle technique: II. Characterization of UF-membranes and comparison between the captive bubble and sessile drop as methods to obtain water contact angles. *Desalination* **1989**, *72*, 263–273. [CrossRef]
162. Fouco, A.; Zwijnenberg, H.; Galier, S.; Balmann, H.R.; De Luca, G. Structural properties of cation exchange membranes: Characterization, electrolyte effects and solute transfer. *J. Membr. Sci.* **2016**, *520*, 45–53. [CrossRef]
163. Xie, H.; Saito, T.; Hickner, M.A. Zeta potential of ion-conductive membranes by streaming current measurements. *Langmuir* **2011**, *27*, 4721–4727. [CrossRef]
164. Zhang, Y.; Xu, T. An experimental investigation of streaming potentials through homogeneous ion-exchange membranes. *Desalination* **2006**, *190*, 256–266. [CrossRef]
165. Li, D. Microfluidic methods for measuring zeta potential. *Interface Sci. Technol.* **2004**, *2*, 617–640. [CrossRef]
166. Lee, H.-J.; Choi, J.-H.; Cho, J.; Moon, S.-H. Characterization of anion exchange membranes fouled with humate during electrodialysis. *J. Membr. Sci.* **2002**, *203*, 115–126. [CrossRef]
167. Sabbatovskii, K.G.; Vilenskii, A.I.; Sobolev, V.D. Electrosurface properties of poly(ethylene terephthalate) films irradiated by heavy ions and track membranes based on these films. *Colloid J.* **2016**, *78*, 573–575. [CrossRef]
168. Sedkaoui, Y.; Szymczyk, A.; Lounici, H.; Arous, O. A new lateral method for characterizing the electrical conductivity of ion-exchange membranes. *J. Membr. Sci.* **2016**, *507*, 34–42. [CrossRef]
169. Nebavskaya, K.A.; Sarapulova, V.V.; Sabbatovskiy, K.G.; Sobolev, V.D.; Pismenskaya, N.D.; Sistat, P.; Cretin, M.; Nikonenko, V.V. Impact of ion exchange membrane surface charge and hydrophobicity on electroconvection at underlimiting and overlimiting currents. *J. Membr. Sci.* **2017**, *523*, 36–44. [CrossRef]
170. Franck-Lacaze, L.; Sistat, P.; Huguet, P. Determination of the pKa of poly (4-vinylpyridine)-based weak anion exchange membranes for the investigation of the side proton leakage. *J. Membr. Sci.* **2009**, *326*, 650–658. [CrossRef]
171. Ramírez, P.; Alcaraz, A.; Mafé, S.; Pellicer, J. Donnan equilibrium of ionic drugs in pH-dependent fixed charge membranes: Theoretical modeling. *J. Colloid Interface Sci.* **2002**, *253*, 171–179. [CrossRef] [PubMed]
172. Sarapulova, V.; Nevakshenova, E.; Pismenskaya, N.; Dammak, L.; Nikonenko, V. Unusual concentration dependence of ion-exchange membrane conductivity in ampholyte-containing solutions: Effect of ampholyte nature. *J. Membr. Sci.* **2015**, *479*, 28–38. [CrossRef]
173. Belashova, E.D.; Pismenskaya, N.D.; Nikonenko, V.V.; Sistat, P.; Pourcelly, G. Current-voltage characteristic of anion-exchange membrane in monosodium phosphate solution. Modelling and experiment. *J. Membr. Sci.* **2017**, *542*, 177–185. [CrossRef]
174. Pismenskaya, N.; Sarapulova, V.; Nevakshenova, E.; Kononenko, N.; Fomenko, M.; Nikonenko, V. Concentration dependences of diffusion permeability of anion-exchange membranes in sodium hydrogen carbonate, monosodium phosphate and potassium hydrogen tartrate solutions. *Membranes* **2019**, *9*, 170. [CrossRef] [PubMed]
175. Liu, S.; Wang, J.; Huang, W.; Tan, X.; Dong, H.; Goodman, B.A.; Du, H.; Lei, F.; Diao, K. Adsorption of phenolic compounds from water by a novel ethylenediamine rosin-based resin: Interaction models and adsorption mechanisms. *Chemosphere* **2019**, *214*, 821–829. [CrossRef]
176. Kammerer, J.; Boschet, J.; Kammerer, D.R.; Carle, R. Enrichment and fractionation of major apple flavonoids, phenolic acids and dihydrochalcones using anion exchange resins. *LWT—Food Sci. Technol.* **2011**, *44*, 1079–1087. [CrossRef]

177. Hashim, H.; Wan Ahmad, W.Y.; Zubairi, S.I.; Maskat, M.Y. Effect of pH on adsorption of organic acids and phenolic compounds by amberlite ira 67 resin. *J. Teknol.* **2018**, *81*, 69–81. [CrossRef]
178. Lasanta, C.; Caro, I.; Pérez, L. The influence of cation exchange treatment on the final characteristics of red wines. *Food Chem.* **2013**, *138*, 1072–1078. [CrossRef]
179. Caetano, M.; Valderrama, C.; Farran, A.; Cortina, J.L. Phenol removal from aqueous solution by adsorption and ion exchange mechanisms onto polymeric resins. *J. Colloid Interf. Sci.* **2009**, *338*, 402–409. [CrossRef]
180. Ibeas, V.; Correia, A.C.; Jordão, A.M. Wine tartrate stabilization by different levels of cation exchange resin treatments: Impact on chemical composition, phenolic profile and organoleptic properties of red wines. *Food Res. Int.* **2015**, *69*, 364–372. [CrossRef]
181. Geise, G.M.; Freeman, B.D.; Paul, D.R. *Comparison of the Permeation of $MgCl_2$ versus NaCl in Highly Charged Sulfonated Polymer Membranes*; American Chemical Society: Washington, DC, USA, 2011; pp. 239–245. ISBN 9780841226180.
182. Geise, G.M.; Paul, D.R.; Freeman, B.D. Fundamental water and salt transport properties of polymeric materials. *Prog. Polym. Sci.* **2014**, *39*, 1–42. [CrossRef]
183. Cassady, H.J.; Cimino, E.C.; Kumar, M.; Hickner, M.A. Specific ion effects on the permselectivity of sulfonated poly(ether sulfone) cation exchange membranes. *J. Membr. Sci.* **2016**, *508*, 146. [CrossRef]
184. Kammerer, J.; Schweizer, C.; Carle, R.; Kammerer, D.R. Recovery and fractionation of major apple and grape polyphenols from model solutions and crude plant extracts using ion exchange and adsorbent resins: Recovery and fractionation of polyphenols. *Int. J. Food Sci. Technol.* **2011**, *46*, 1755–1767. [CrossRef]
185. Shuang, C.; Wang, J.; Li, H.; Li, A.; Zhou, Q. Effect of the chemical structure of anion exchange resin on the adsorption of humic acid: Behavior and mechanism. *J. Colloid Interface Sci.* **2015**, *437*, 163–169. [CrossRef]
186. Ebadi, A.; Soltan, J.S.; Khudiev, M.A. What is the correct form of BET isotherm for modeling liquid phase adsorption? *Adsorption* **2009**, *15*, 65–73. [CrossRef]
187. Kammerer, D.R.; Kammerer, J.; Carle, R. Resin adsorption and ion exchange to recover and fractionate polyphenols. In *Polyphenols in Plants*; Elsevier: Amsterdam, The Netherlands, 2014; pp. 219–230. [CrossRef]
188. Persico, M.; Dhulster, P.; Bazinet, L. Redundancy analysis for determination of the main physicochemical characteristics of filtration membranes explaining their fouling by peptides. *J. Membr. Sci.* **2018**, *563*, 708–717. [CrossRef]
189. Smith, B.C. *Infrared Spectral Interpretation: A Systematic Approach*; CRC Press: Boca Raton, FL, USA, 1998; p. 304. ISBN 9780849324635.
190. Su, Z.; Cocinero, E.J.; Stanca-Kaposta, E.C.; Davis, B.G.; Simons, J.P. Carbohydrate-aromatic interactions: A computational and IR spectroscopic investigation of the complex, methyl α-l-fucopyranoside-toluene, isolated in the gas phase. *Chem. Phys. Lett.* **2009**, *471*, 17–21. [CrossRef]
191. Jialin, L.; Yazhen, W.; Changying, Y.; Guangdou, L.; Hong, S. Membrane catalytic deprotonation effects. *J. Membr. Sci.* **1998**, *147*, 247–256. [CrossRef]
192. Das, S.; Kumar, P.; Dutta, K.; Kundu, P.P. Partial sulfonation of PVdF-co-HFP: A preliminary study and characterization for application in direct methanol fuel cell. *Appl. Energy* **2014**, *113*, 169–177. [CrossRef]
193. Audinos, R. Fouling of ion-selective membranes during electrodialysis of grape must. *J. Membr. Sci.* **1989**, *41*, 115–126. [CrossRef]
194. Araya-Farias, M.; Bazinet, L. Effect of calcium and carbonate concentrations on anionic membrane fouling during electrodialysis. *J. Colloid Interface Sci.* **2006**, *296*, 242–247. [CrossRef]
195. Bazinet, L.; Araya-Farias, M. Effect of calcium and carbonate concentrations on cationic membrane fouling during electrodialysis. *J. Colloid Interface Sci.* **2005**, *281*, 188–196. [CrossRef]
196. Andersen, O.M.; Markham, K.R. *Flavonoids: Chemistry, Biochemistry and Applications*; CRC Press: Boca Raton, FL, USA, 2005; p. 1256. ISBN 9780849320217.
197. Luo, T.; Abdu, S.; Wessling, M. Selectivity of ion exchange membranes: A review. *J. Membr. Sci.* **2018**, *555*, 429–454. [CrossRef]
198. Barragán, V.M.; Villaluenga, J.P.G.; Godino, M.P.; Izquierdo-Gil, M.A.; Ruiz-Bauzá, C.; Seoane, B. Swelling and electro-osmotic properties of cation-exchange membranes with different structures in methanol–water media. *J. Power Sources* **2008**, *185*, 822–827. [CrossRef]
199. Vasil'eva, V.I.; Goleva, E.A.; Selemenev, V.F. Features of the sorption of phenylalanine by profiled ion-exchange membranes. *Russ. J. Phys. Chem. A* **2016**, *90*, 2035–2043. [CrossRef]
200. Vasil'eva, V.; Goleva, E.; Pismenskaya, N.; Kozmai, A.; Nikonenko, V. Effect of surface profiling of a cation-exchange membrane on the phenylalanine and NaCl separation performances in diffusion dialysis. *Sep. Purif. Technol.* **2019**, *210*, 48–59. [CrossRef]
201. Garcia-Vasquez, W.; Dammak, L.; Larchet, C.; Nikonenko, V.; Grande, D. Effects of acid–base cleaning procedure on structure and properties of anion-exchange membranes used in electrodialysis. *J. Membr. Sci.* **2016**, *507*, 12–23. [CrossRef]
202. Chaabane, L.; Bulvestre, G.; Larchet, C.; Nikonenko, V.; Deslouis, C.; Takenouti, H. The influence of absorbed methanol on the swelling and conductivity properties of cation-exchange membranes: Evaluation of nanostructure parameters. *J. Membr. Sci.* **2008**, *323*, 167–175. [CrossRef]

Review

A Review on Ion-Exchange Membranes Fouling during Electrodialysis Process in Food Industry, Part 2: Influence on Transport Properties and Electrochemical Characteristics, Cleaning and Its Consequences

Natalia Pismenskaya [1], Myriam Bdiri [2], Veronika Sarapulova [1], Anton Kozmai [1], Julie Fouilloux [2], Lassaad Baklouti [3], Christian Larchet [2], Estelle Renard [2] and Lasâad Dammak [2,*]

[1] Department of Physical Chemistry, Kuban State University, 149 Stavropolskaya Str., 350040 Krasnodar, Russia; n_pismen@mail.ru (N.P.); vsarapulova@gmail.com (V.S.); kozmay@yandex.ru (A.K.)
[2] Institut de Chimie et des Matériaux Paris-Est (ICMPE), Université Paris-Est Créteil, CNRS, ICMPE, UMR 7182, 2 Rue Henri Dunant, 94320 Thiais, France; myriem.bdiri@u-pec.fr (M.B.); julie.fouilloux@u-pec.fr (J.F.); larchet@u-pec.fr (C.L.); e.renard@u-pec.fr (E.R.)
[3] Department of Chemistry, College of Sciences and Arts at Al Rass, Qassim University, Ar Rass 51921, Saudi Arabia; bakloutilassaad@yahoo.fr
* Correspondence: dammak@u-pec.fr; Tel.: +33-145171786

Abstract: Ion-exchange membranes (IEMs) are increasingly used in dialysis and electrodialysis processes for the extraction, fractionation and concentration of valuable components, as well as reagent-free control of liquid media pH in the food industry. Fouling of IEMs is specific compared to that observed in the case of reverse or direct osmosis, ultrafiltration, microfiltration, and other membrane processes. This specificity is determined by the high concentration of fixed groups in IEMs, as well as by the phenomena inherent only in electromembrane processes, i.e., induced by an electric field. This review analyzes modern scientific publications on the effect of foulants (mainly typical for the dairy, wine and fruit juice industries) on the structural, transport, mass transfer, and electrochemical characteristics of cation-exchange and anion-exchange membranes. The relationship between the nature of the foulant and the structure, physicochemical, transport properties and behavior of ion-exchange membranes in an electric field is analyzed using experimental data (ion exchange capacity, water content, conductivity, diffusion permeability, limiting current density, water splitting, electroconvection, etc.) and modern mathematical models. The implications of traditional chemical cleaning are taken into account in this analysis and modern non-destructive membrane cleaning methods are discussed. Finally, challenges for the near future were identified.

Keywords: ion-exchange membrane; food industry; fouling; transport; mechanical and electrochemical properties; modelling and experiment; cleaning

Citation: Pismenskaya, N.; Bdiri, M.; Sarapulova, V.; Kozmai, A.; Fouilloux, J.; Baklouti, L.; Larchet, C.; Renard, E.; Dammak, L. A Review on Ion-Exchange Membranes Fouling during Electrodialysis Process in Food Industry, Part 2: Influence on Transport Properties and Electrochemical Characteristics, Cleaning and Its Consequences. *Membranes* **2021**, *11*, 811. https://doi.org/10.3390/membranes11110811

Academic Editor: Marek Gryta

Received: 6 October 2021
Accepted: 19 October 2021
Published: 25 October 2021

1. Introduction

Ion exchange membranes (IEMs), which are the heart of dialysis and electrodialysis units, are increasingly used in the food industry for the extraction, fractionation and concentration of valuable components, as well as reagent-free control of liquid media pH (whey, wine, fruit juices, etc.) [1–6]. Electrodialysis (ED) compares favorably with reverse (RO), direct (FO) osmosis, ultra- (UF) and micro (MF)-filtration by the ability to separate substances not only by particle size, but also by their electrical charge. Many components of whey or blood serum of animals, as well as wines, juices and waste products from these industries change their electrical charge depending on the environment pH. These are proteins, amino acids, anions of polybasic inorganic and organic acids, dyes (anthocyanins) and other polyphenols. Electrodialysis, as well as electrophoresis with IEMs, are some of the processes where this charge (hence, and the efficiency of valuable component separation) can be controlled by reagent less pH change at the surfaces of cation-exchange (CEMs) and anion-exchange (AEMs) membranes [7,8], or by use of bipolar membranes [9]. Such control is carried out by enhancing or suppressing water splitting, the intensity of which depends on the electric field strength and the catalytic activity of membrane fixed groups in relation to the water dissociation reaction [10,11].

Fouling of IEMs is in many respects similar to that described in reviews for RO, FO, UF, and MF processes [12,13]. However, there are also some peculiarities. One of these is steric hindrance during the large organic particles transport in IEMs [14]. The pore sizes of IEM materials, as a rule, do not exceed 40 nm [15,16]. This problem is being actively solved through the use of porous ion-exchange membranes, the introduction of which has been actively carried out recently [17]. Another problem is intense electrostatic interactions between the polar groups of the foulants and the fixed groups of IEMs. Moreover, the degree of this interaction cannot always be predicted based on the pH of the feeding solutions. The reason is acidification (CEMs) or alkalization (AEMs) of the inner membrane solution compared to bathing solution due to Donnan exclusion of hydroxyl ions (CEMs) or protons (AEMs) as co-ions from membranes [18]. These ions are products of the protonation–deprotonation reactions of the above substances when they enter the membrane. In the case of whey and blood serum of animals, which contain only a small amount of aromatic substances, fouling is mainly determined by these electrostatic interactions and the formation of hydrogen bonds between IEMs and foulants [19–21]. In the case of wine, juices, and tea, π-π (stacking) interactions between the aromatic rings of polyphenols (PP, contained in these liquid media) and the aromatic matrix of IEMs are added to the above mentioned interactions [22–25]. Moreover, even a small amount of PP can trigger the formation of colloidal aggregates in the pores and on the membrane surface.

This review focuses on the analysis of modern scientific publications, which investigated the effect of foulants and chemical cleaning agents on the structural, transport, mass transfer and electrochemical characteristics of cation and anion exchange membranes used in dialysis and electrodialysis processing of liquid media in the food industry. Much attention is paid to modern theoretical concepts and mathematical models that can be used to interpret experimental data and predict the behavior of fouled IEMs. The implications of traditional chemical cleaning are taken into account in this analysis and modern non-destructive membrane cleaning methods are discussed.

2. Impact of Traditional Cleaning Methods on the Chemical Structure of IEMs

As mentioned above, the vast majority of liquid media in the food industry are a breeding ground for microorganisms. This fact, together with the need to counteract the growth of colloidal structures inside IEM and on its surface, necessitate regular cleaning of membrane stacks of ED apparatuses [26]. Most often, HCl, NaOH, hypochlorides and other oxidants (peracetic acid, and P3 Active Oxonia® (Ecolab, Saint Paul, MN, USA) solutions) are used for these purposes [27–34]. Moreover, cleaning is carried out at a high temperature (60–90 °C) [35–37]. A number of articles have been devoted to the influence of these procedures on the structure and chemical properties of membranes. The results of these works can be briefly summarized as follows (Figure 1). Thermochemical desulfonation (detachment of sulfonate groups) in aqueous solutions is possible in the case of CEMs [38]. Fixed amino groups of AEMs are much more susceptible to destruction under cleaning conditions. Methylated ammonium-type fixed groups are the least stable in an alkaline environment [39] or in the presence of hypochlorite and hydrogen peroxide [27]. As a result of the reaction called Stevens rearrangement [40], as well as other reactions described in detail in [41], some of the quaternary amino groups are converted into secondary and tertiary amines. In addition, there is the detachment of fixed groups, rupture of the carbon chains of the polymer matrix, and partial destruction of the copolymer of polystyrene and divinylbenzene, which make up the ion-exchange matrix of most aromatic membranes [42]. The result of such reactions is the appearance on the surface and inside the IEMs of cavities, the walls of which do not have an electric charge. Hydroxyl ions influence not only on the ion-exchange material, but also on polyvinyl chloride, PVC (an inert filler and reinforcing cloth of most membranes made by the paste method [43]), according to the 2E elimination reaction [44] mechanism. Hydroxyl ions pull out β-hydrogen from PVC; the chloride ions are eliminated and double bonds are formed, forming black polyenes [45–47]. Such degradation is more significant for AEMs [48], because hydroxyl ions are counter ions, and their concentration in the internal solution of AEMs is high. In CEM, for which hydroxyl ions are co-ions and are excluded from the membrane due to the Donnan effect, PVC degradation is observed only at very high external solution pH [46,48]. In addition, even at neutral pH of the solution adjacent to the membrane, under the action of a high electric field strength, free radicals are formed on the PVC surface, which initiate the breaking of macrochains with the formation of oxidation products [49]. Grains of PVC, not fixed by the ion-exchange material (which is the cause of the IEM destruction), are washed out by the solution that flows over the membrane. Operation of AEM in intense current regimes of ED tartrate- and phosphate-containing solutions processing intensifies the process of AEM degradation in comparison with NaCl solutions [50].

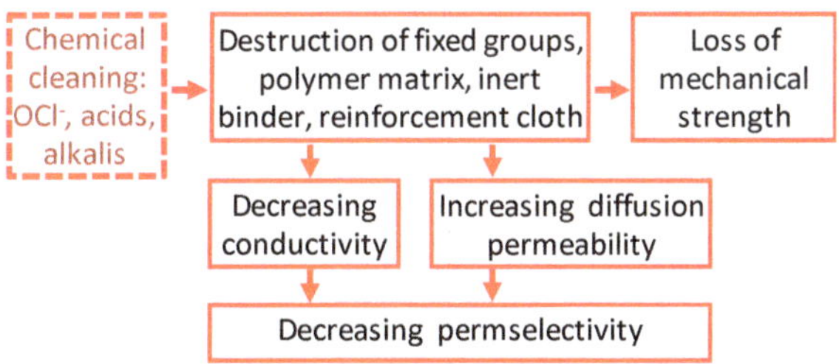

Figure 1. Influence of traditional chemical cleaning on the structure, mechanical strength and transport characteristics of ion-exchange membranes.

Vasil'eva et al. [38] found that the contact of heterogeneous membranes with acid, alkali or even water at high temperatures leads to an increase in the gaps (macropores) between the beads of the ion exchange resin and low-pressure polyethylene, which is an inert binder in heterogeneous CEMs and AEMs. The polyethylene degradation in water occurs due to the washing out of the antioxidants [51] and hydrolytic oxidation of the polymer with the formation of carboxyl groups [52]. Dissolution of reinforcing cloth made of nylon and similar materials is possible by treating membranes with acids [38].

Thus, the cleaning by traditional chemical reagents leads not only to the partial loss of the AEM and CEM fixed groups, but also to the formation of cavities and/or microcavities on the surface and in the volume of the IEM. These cavities are often not electrically charged and can be filled with colloidal particles of foulants. As will be shown in the following chapters, the application of traditional cleaning with oxidizing agents and other harsh chemicals sometimes does as much detriment to IEM performance as fouling. Moreover, the consequences of such hard cleaning contribute to the intensification of fouling. If membranes are used for a long time in industrial electrodialysis, it is not always possible to separate the effect of fouling and cleaning on equilibrium (water content and exchange capacity) and transport (conductivity, diffusion permeability, selectivity) characteristics of IEMs, as well as on their behavior in an electric field. Therefore, we will analyze the impact of both fouling and hard cleaning in the following chapters.

3. Water Content and Exchange Capacity of Fouled IEMs

The ion exchange capacity, IEC, is defined as the number of functional sites (in mmole) per mass (in gram) or volume (in cm^3) of dried or swollen membrane [53]. In all publications related to the study of IEMs fouling, organic matter always has a tendency to decrease the IEC by total or partial inhibition of the functional sites and accumulation of organic particles that form aggregates of more or less dense colloidal particles [21,54,55].

In addition, cleaning of membrane stacks at high temperatures and using oxidizing agents result in the elimination of fixed groups or, in the case of AEM, the conversion of strongly basic quaternary amino groups to weakly basic secondary or tertiary amino groups.

As discussed in Section 2, the degree of reduction in IEC of fouled membranes as compared to a pristine one is determined by the cleaning method (chemical nature of the substances used, temperature); the chemical nature of the fixed groups, polymer matrix and IEM inert filler; pH of the working solution; and current modes (under-limiting, overlimiting) of the ED operation.

In the processes of milk (acid or sweet whey, whey protein isolate or hydrolysate, etc.) or animal serum electrodialysis, the decrease in IEC is mainly determined by electrostatic interactions of membranes with highly hydrated colloidal structures that form proteins, carboxylic acids, amino acids, and mineral salts (sodium, potassium, calcium, phosphorus and magnesium). To a greater extent, the IEC of CEMs decreases in acidic solutions, while the IEC of AEMs increases in alkaline and neutral solutions [26,56–58]. These colloidal (network) structures fill both the pores of the ion-exchange material (the walls of which contain polar fixed groups) and the pores with non-charged walls (the reasons for the appearance of these pores were discussed in Section 2), contributing to an increase in water uptake. The structure of such fouled membranes is shown in Figure 2. The changes in IEC, water uptake and membrane thickness are not dramatic. For example, Garcia-Vasquez et al. [58] observed a 20% decrease in IEC, 30% increase in water uptake and <2% increase in thickness of the AMX-Sb anion exchange membrane (Astom, Osaka, Japan) after several thousand hours of the sweet milk whey ED demineralization.

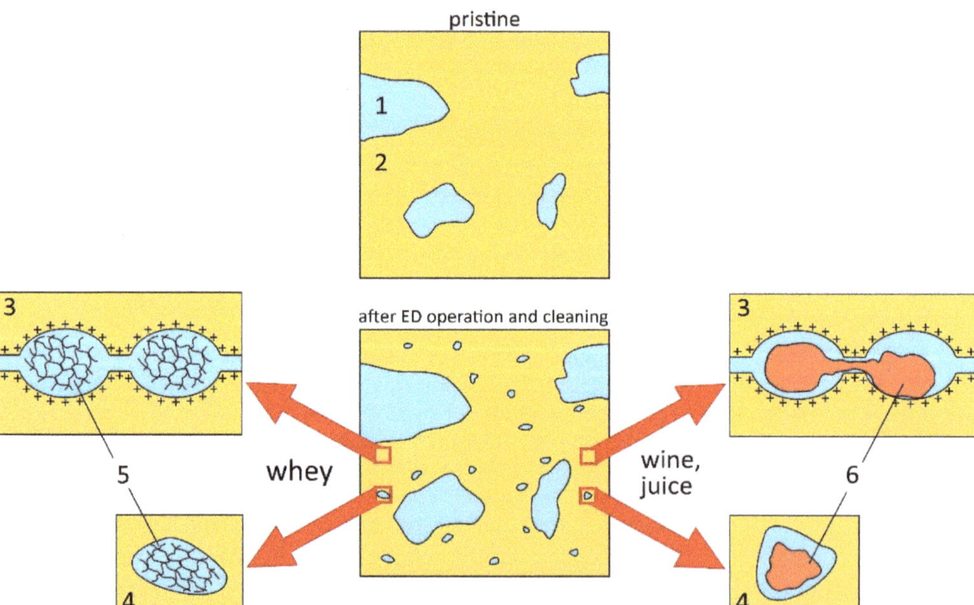

Figure 2. The proposed structure of AEMs that contain PVC as an inert filler, before and after their use in the ED process of wine tartrate stabilization: (1) macro-defects, filled with an external solution; (2) nanoporous medium; (3) nanopore with fixed groups on the walls, filled with an external solution; (4) nanopore with non-electrically charged walls filled with an external solution; (5) reticulated hydraulically permeable colloidal structures; (6) hydraulically impermeable colloidal aggregates. Adapted from [58,59].

The scenario and consequences of IEM fouling change fundamentally in the case of juices, tea and wine that contain polyphenols (PP). As mentioned in [18,22–24], the main role in fouling is played by the π-π (stacking) interactions of polyphenols with each other and with the IEM materials. Colloidal particles formed inside the pores are compacted during long-term use of IEM (also, as a result of cleaning) and turn into hydraulically impermeable agglomerates. As a result, the effective IEM pore diameter decreases [22,23,59–63] (Figure 2).

For example, Ghalloussi et al. [59] have shown that the decrease in IEC of Neosepta CEMs (Astom, Osaka, Japan) after several thousand hours of operation in ED is from 3% to 84%. In this and in later works [22,23,59–61], it was found that the most significant decrease in IEC is observed for membranes with an aromatic ion-exchange matrix and is accompanied by an insignificant (7–16%) decrease in water uptake. On the contrary, membranes with an aliphatic ion-exchange matrix demonstrate a less noticeable decrease in IEC, but their water uptake can increase by 30–40% (taking into account the formation of large cracks and cavities filled with water). The thickness of some AEMs and CEMs increased by a factor of 1.6–2.1 [22,59]; the linear dimensions of other membranes do not undergo significant changes [18,22,60–64]. For example, after 2600 ± 100 h of operation in industrial ED processing of polyphenol (PP)-containing liquids, the thickness of aromatic membranes made by the paste method increases by 52% (CMX-Sb) and 11% (AMX-Sb). The reasons for the different behavior of these membranes are the ability of anthocyanins and other PPs to change their electrical charge depending on the environment pH. Industrial ED processing of wines and juices is carried out at pH < 3.5, where anthocyanins and other polyphenols are typically cations (PP^+). Low pH values within the cation exchange membrane contribute to maintain the PP^+ positive electrical charge as it enters CEMs [18,23]. The PP^+ cations enter the electrostatic interactions with the fixed groups of CEMs, which are negatively charged, and screen these groups. In contrast, the pH of the internal solution of AEMs is neutral or slightly alkaline [18,60]. Upon entry into AEMs, PP^+ lose their electrical charge or become anions that do not enter the electrostatic interactions with positively charged fixed groups of AEMs or are excluded from them as co-ions.

4. Transport Characteristics of Fouled IEMs

Electrical resistance (R_m) or conductivity (κ_m), as well as diffusion permeability of IEMs, determine the energy consumption for the ED process and the resulting concentration of desalinated and concentrated solutions. Knowing these parameters, it is possible to calculate the counterion and co-ion transport numbers in IEM (t_{m1} and t_{mA}, respectively) [65], that determine membrane selectivity [66]:

$$t_{m1} = \frac{1}{2} + \sqrt{\frac{1}{4} - \frac{(z_1|z_A|)P^*F^2C}{(z_1 + |z_A|)RT\kappa_m}}\ ,\ t_{mA} = 1 - t_{m1} \quad (1)$$

where z_1 and z_A are the charge numbers of the counterion and co-ion, respectively; C is an electrolyte concentration; F is the Faraday constant; R is the gas constant; T is the temperature. The relation between the differential (or local) diffusion permeability coefficient, P^*, and the integral diffusion permeability coefficient of membranes (P_m) is given as $P_m = \frac{1}{C}\int_0^C P^*dC$ [67,68]. It leads to the following equation [69]:

$$P^* = P_m + C\frac{dP}{dC} \quad (2)$$

In practice, for calculation of P^*, it is more convenient to use the relationship between P^* and P in the form [69]:

$$P^* = P_m(\beta + 1) \quad (3)$$

with $\beta = dlogP_m/dlogC$, which is the slope of the $logP_m$ vs. $logC$ dependence. The integral diffusion permeability coefficient of membranes (is usually defined as [68]:

$$P = j\frac{d}{C} \quad (4)$$

where j is the electrolyte flux density. This electrolyte diffuses through IEM from the compartment with electrolyte solution with concentration C into the compartment with initially distilled water; d is the membrane thickness. Experimental methods for determining the resistance (electrical conductivity), diffusion permeability and selectivity of IEM are summarized in reviews [66,68,70]. For example, the differential method with a clip cell [71,72] seems to be the simplest and most reliable to determine the electrical conductivity after removing the IEM from the ED apparatus. The electrolyte solutions used in most publications are mainly NaCl or KCl solutions at 0.1–1.0 M [67,68,73–75], or even at lower concentration [70,72] for both homogeneous or heterogeneous IEMs. Membranes, after their operation in the food industry, are also tested in these solutions [23,60–62,64,76]. This opens the possibility of using known mathematical models to determine the structural parameters of IEMs or the effect of fouling on the development of water splitting and electroconvection. This is due to the fact that most of the already developed models do not take into account the possible interactions of foulants with the matrix and fixed groups of membranes, as well as the possible change in the electric charge of amino acids, polybasic acid anions, anthocyanins, etc. in IEMs as compared to the external solution [18,23,77,78].

Monitoring the resistance of the membrane stack or individual membranes in situ (in operando) is carried out by applying Ohm's law when a constant electric field is applied [79] or by periodically measuring the impedance of the membrane system in an alternating current mode in the high frequency range (>1 kHz) [80,81]. To determine the resistance of the membrane system, the initial sections in current–voltage characteristics and ohmic sections in chronopotentiograms [82,83] are used. Another option is to find the length of the segment on the abscissa axis, cut off by the high-frequency arc of electrochemical impedance spectrum represented in the complex plane [79,80].

Voltage and current readings from the indicator on the power supply [79] are commonly used in ED industrial application. However, the values obtained in this way include not only the actual IEM resistance, but also the resistance of the diffusion layers adjacent to the membrane, which is controlled by the degree of concentration polarization development (see Section 5). Therefore, such assessments should be treated with caution.

4.1. Modelling

Most of the "classical" [84–87] and modern [88–92] models are intended to describe transport phenomena in ideally homogeneous IEMs. The microheterogeneous model developed by Gnusin, Nikonenko, Zabolotskii et al. [68,74,93] seems to be the most suitable for real membrane systems. It is widely used in the literature [14,48,70,75,94–99] to interpret the experimental concentration dependences of membrane conductivity, diffusion permeability, and to determine the relationship "structure–transport properties" of IEMs. The model is based on a simplified representation of the IEM structure, according to which the membrane can be considered as a multiphase (in the simplest case, two-phase) system. The properties of such a system are determined by the properties of individual phases, which is consistent with the fundamentals of the effective medium theory [100]. The gel phase (denoted by the index g) is a microporous swollen medium. It includes a polymer matrix carrying fixed groups, as well as a charged solution of mobile counterions (and, to a lesser extent, co-ions), which balance the charge of fixed groups. There are two regions within the gel phase. A "pure" gel consists of an electrically charged double electrical layer and side polymer chains with hydrated fixed groups. The hydrophobic regions are formed by agglomerates of polymer chains that do not contain fixed groups, as well as an inert binder and reinforcing cloth. The second phase, the inter-gel space (denoted by index "int") is an electrically neutral solution (identical to the external solution). It is located in the central part of the meso- and macropores, and also fills in the membrane structural defects. A membrane fragment that includes both phases is shown in Figure 3.

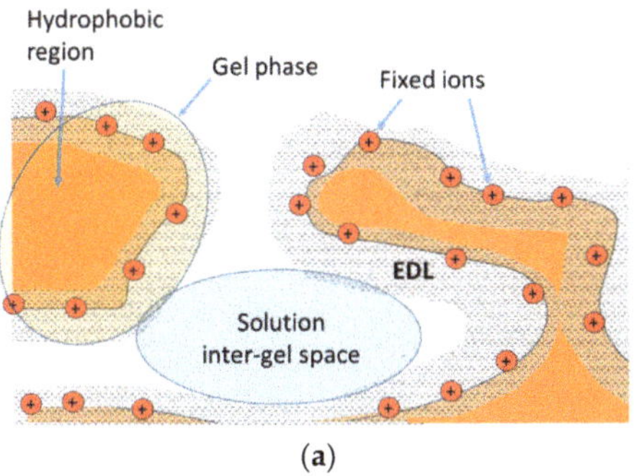

(a)

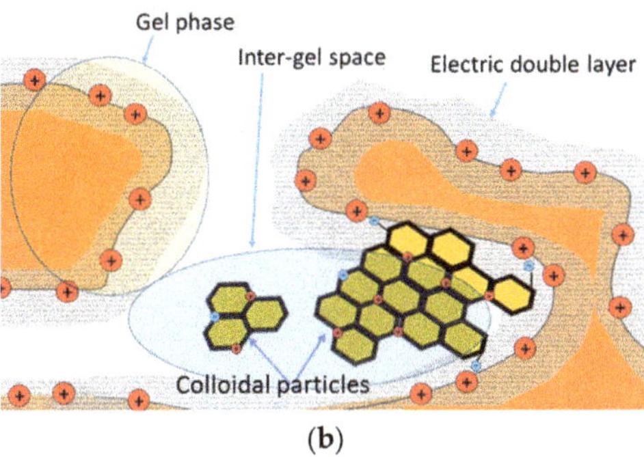

(b)

Figure 3. Schematic IEM structure represented in terms of microheterogeneous model [68] (**a**), and the same model, which takes into account colloidal particles (**b**) formed by polyphenols. EDL is the electric double layer. Adapted from [60].

In accordance with the approach that combines the effective medium theory and nonequilibrium thermodynamics [101], the flux of ions of sort i (j_i) in a two-phase IEM is proportional to the gradient of the electrochemical potential $d\mu_i/dx$ [68,102]:

$$J_i = -L_i^* d\mu_i/dx \tag{5}$$

where L_i^* is the effective conductivity coefficient characterizing a multiphase medium, which is equal to:

$$L_i^* = \left[f_1\left(L_i^g\right)^\alpha + f_2\left(L_i^{int}\right)^\alpha\right]^{1/\alpha} \tag{6}$$

The conductivity coefficients L_i^g and L_i^{int} refer to the gel phase and the inter-gel solution phase, respectively. The structural parameter α takes values from 1 to -1, respectively, with a parallel and sequential arrangement of phases; the sum of the volume fraction of the gel phase (f_1) and the inter-gel space (f_2) is equal to one ($f_1 + f_2 = 1$). The values L_i^g and L_i^{int} are determined by the ion diffusion permeability coefficients D_i^g and D_i^{int}, and ion concentration, C, in the corresponding phase:

$$L_i^g = D_i^g C_i^g/RT \tag{7}$$

$$L_i^{int} = D_i^s C_i^s/RT \tag{8}$$

The exchange capacity of the membrane gel phase (the concentration of fixed groups in the gel phase), Q_g, is related to the membrane exchange capacity, Q, by the equation: $Q_g = Q/f_1$. From Equations (5)–(8), simple expressions are obtained to determine the membrane conductivity (κ_m), diffusion permeability (P_m) and counterion transport numbers (t_{m-}) (for example, in AEM):

$$\kappa_m = (z_+ L_{m+} + z_- L_{m-})F^2 \tag{9}$$

$$t_{m-} = \frac{L_{m-}}{L_{m-} + L_{m+}} \tag{10}$$

$$P_m = 2t_{m-} L_{m+} \frac{RT}{C} \tag{11}$$

where index "+" relates to a cation (co-ion in the AEM).

The simple relationship between the membrane conductivity (κ_m) and the conductivities of the gel phase (κ_g) and electroneutral solution (κ_s) filling the inter-gel spaces is of the greatest interest:

$$\kappa_m = \left(f_1\kappa_g^\alpha + f_2\kappa_{int}^\alpha\right)^{1/\alpha} \tag{12}$$

If $|\alpha|$ is not too far from zero (<0.2), and the external solution concentration is in the range 0.1 C_{iso} < C < 10C_{iso} (C_{iso} is the electrolyte concentration at the isoconductance point: $\kappa_m = \kappa_g = \kappa_s$), Equation (12) may be approximated as:

$$log(\kappa_m) \approx f_2 \log(C) + const \tag{13}$$

where const $\approx (1 - f_2)\log(\kappa_g)$.

It is believed that in all the cases under consideration, fouling leads to a decrease in the membrane IEC. In the case of formation of hydraulically permeable reticular colloidal structures (milk whey) [76] in the pores:

$$L_i^{int} = (\gamma D_i^s)C_i^s/RT \tag{14}$$

where γ is a coefficient reflecting the ratio of ion mobility in the intergel space of fouled membrane and in an external solution.

In the case of contact of IEMs with polyphenol-containing solutions (wine, juices, tea extracts), it is believed that dense hydraulically impermeable conglomerates of colloidal particles (denoted by index "*cp*") are formed in the meso- and macropores, but do not penetrate into the nanopores. These changes in the structure can lead to a possible change in the ion transport path, which is manifested in a change in the structural parameter from α to β. In this case, from Equation (12), it is easy to obtain an equation for κ_m [60]:

$$\kappa_m = \left[f_1\kappa_g^\alpha + f_2\left(f_{2s}^{int}\right)^{\alpha/\beta}\kappa_s^\alpha\right]^{1/\alpha} \tag{15}$$

Equation (15) can be also approximated in the same way as Equation (12):

$$log(\kappa_m) \approx f_2\left(f_{2s}^{int}\right)^{\alpha/\beta} log(C) + const \tag{16}$$

A similar approach was used to describe the transport characteristics of IEMs with conglomerates of various nanoparticles immobilized in their meso- and macropores [103].

The use of the basic [68] and modified microheterogeneous model [60,76,103] develops a theoretical basis for explaining the effect of foulants on the membrane transport characteristics and makes it possible to predict the tendency of IEM to fouling depending on membrane structure and component composition of processed solutions in the food industry.

4.2. Interpretation of Experimental Data Using a Microheterogeneous Model

IEMs conductivity, as a rule, decreases after prolonged exposure to liquid media typical for the food industry [22,23,59–61,104,105]. In the case where a membrane is in contact with amino acids, carboxylic acids or anions of polybasic inorganic acids, the decrease in conductivity is not dramatic [58,104] and can be reversible if a dense layer of proteins or mineral precipitate does not form on the IEM surface [28,29,106]. The decrease in κ_m is most often caused by electrostatic interactions and hydrogen bonds of foulants with fixed EIM groups.

The f_2 values of IEMs increase as a result of the polymer matrix stretching when strongly hydrated ions enter it [16], or due to its destruction as a result of cleaning (see Section 2), as well as due to the operation of membranes in intensive current regimes. For example, Garcia-Vasquez et al. showed that after several thousand hours of the Neosepta AEM (Astom, Osaka, Japan) used in ED demineralization of whey, κ_m decreases by a factor of 1.3 and f_2 increases by 25% as compared to pristine AEM.

In the case of polyphenol containing liquid media, the membrane conductivity (and other transport characteristics) undergo more significant changes. For example, after exposure to green tea, κ_m of AEMs dropped by at least 50% (AMX-Sb, AFN, Astom, Osaka, Japan, and PC-400 D) [107,108]. After contact with wine and cranberry juice, κ_m

of AEMs decreased by a factor of 1.6–1.9 [59], 4 [22] (AEMs Neosepta), 3 [105] (MA-41P, Shchekinoazot, Pervomaisky, Russia) and 4 (AMX, AMX-Sb, Astom, Osaka, Japan) times. The conductivity of CEMs decreased by 2 [22,61], 1.1 (MK-40, Shchekinoazot, Pervomaisky, Russia), 1.2 (Fuji CEM Type II, Fujifilm, the Netherland; CSE-fg, Astom, Osaka, Japan) and 1.7 (CJMC-5, ChemJoi, Hefei, China) [23] times. The decrease in conductivity is primarily caused by the loss of IEM in exchange capacity due to not only electrostatic, but also π-π (stacking) interactions of foulants with ion-exchange materials.

It is important to note that the experimental data treatment on conductivity using a microheterogeneous model demonstrates a clear decrease in the volume fraction of the gel phase f_2 in the case of CEMs. This result provided additional evidence for the formation of agglomerates of hydraulically impermeable colloidal particles in the pores of cation exchange membranes.

As for AEMs, in some studies [59,105], an increase in f_2 is observed, while in [22] the volume fraction of the intergel space decreases. To clarify the reasons for this seeming inconsistency, Bdiri et al. studied the fouling kinetics of aromatic AEM soaked in synthetic solution which contained tartaric, acetic, lactic acids, KCl, $CaCl_2$ and a high concentration of polyphenols (5 g.L^{-1}), sufficiently greater than that in juices and wine [109,110]. Processing these data (Figure 4) using a modified microheterogeneous model showed that both the conductivity and the volume fraction of the gel phase decrease with an increase in the time of membrane contact with the phenol-containing solution (Table 1).

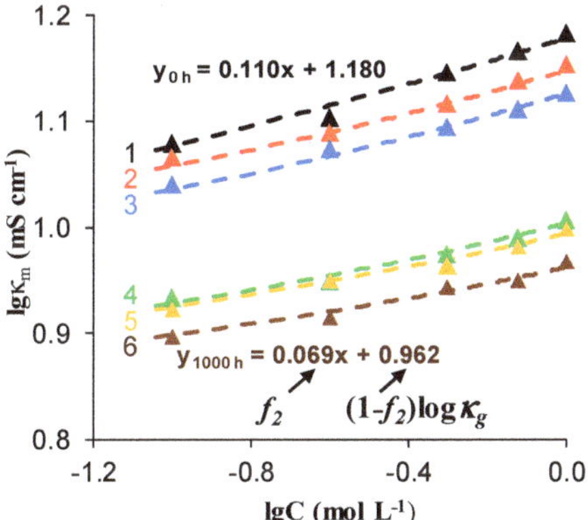

Figure 4. The conductivity of the AEMs soaked in the synthetic polyphenol-containing solution vs. NaCl concentration. Markers show the experimental data measured for sample soaking duration 0 (1), 24 (2), 100 (3), 500 (4), 750 (5) and 1000 (6) hours; dashed lines are the results of calculations according to the modified microheterogeneous model. Adapted from [60].

Table 1. Parameters of the modified microheterogeneous model [60] calculated for the AEM samples soaked with the synthetic polyphenol containing solution.

AEM Soaking Duration, h	f_{2app}	f_2	f_{2s}	f_{cp}	D^g_-/D^g_- $(h = 0)$	d/d (h = 0)	IEC_{sw} (mmol cm^{-3})
0	0.11	0.09	0.09	0	1.00	1.00	2.30
24	0.087	0.09	0.07	0.02	1.26	1.02	1.95
100	0.084	0.12	0.07	0.05	1.44	1.05	1.84
500	0.073	0.16	0.06	0.1	1.47	1.08	1.77
750	0.071	0.18	0.05	0.13	1.74	1.08	1.72
1000	0.069	0.20	0.045	0.155	1.88	1.08	1.70

Pristine membrane, which have not been soaked in the model solution (h = 0) does not contain colloidal particles. However, $f_{2s} = f_2 = 0.09$ values are slightly lower than the apparent volume fraction of the inter-gel spaces, $f_{2app} = 0.11$ (the slope of the log(κ_m) vs. log(C) dependence, Figure 4). Colloidal particles are formed in the inter-gel spaces with an increase in soaking time. They displace a part of the electroneutral solution that results in decreasing f_{2s}, and, as a consequence, f_{2app}. A part of AEM functional groups becomes blocked by the colloidal particles that lead to IEC decreases with soaking duration. The increase in counterion diffusion coefficient in gel phase D^g_- (Table 1) should be due to an increase in membrane swelling degree. This parameter can be estimated by the ratio of the fouled and pristine AEMs thickness, $d/d_h = 0$. Higher swelling indicates an increase in the size of micropores that promotes counterion mobility. Apparently, higher swelling of IEMs in the considered liquid media is caused by relatively small and highly hydrated [66] organic acid species, such as tartaric acid.

Faster destruction of anion-exchange membranes during cleaning in ED industrial processes and in the case of application of intensive current modes is due to the more alkaline media inside AEM compared to CEM (see Section 2). If an increase in f_2 caused by this factor is not compensated by a decrease in the inter-gel space due to the formation of agglomerates of colloidal particles, we observe an increase in the volume fraction of the gel phase and an increase in the AEM diffusion permeability. Note also, the diffusion permeability of AEMs in juices and wine may increase slightly if colloidal particles do not form agglomerates (no cleaning) and are completely or partially destroyed by concentrated saline solutions, which are usually used to study diffusion permeability [76,103]. The main types of interaction of polyphenols with ion-exchange membranes and the effect of this fouling on the transport characteristics of IEMs are summarized in the Figure 5.

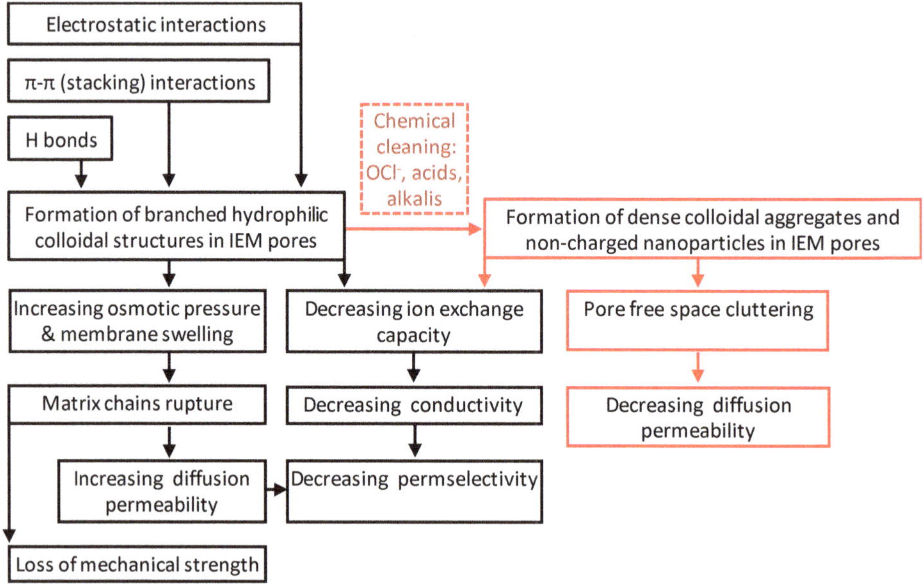

Figure 5. The main types of interaction of polyphenols with ion-exchange membranes and the effect of this fouling on the transport characteristics of IEMs.

5. Effect of Fouling on IEMs Behavior in an Electric Field

5.1. Theoretical Background

Many of the researchers note that fouling leads to a decrease in the current efficiency (η) in ED of liquid media (including in the food industry) [108,111–115]. It is known [116,117] that the current efficiency of the target component (for definiteness, the salt anion) is determined by the difference between the effective transport numbers of this anion through

AEM (as a counterion) and through CEM (as a co-ion), which form the ED desalination channel. For simplicity, we neglect the transfer of the target component through CEM and consider a solution that contains only one salt, for example, NaCl. In this case, the current efficiency is close to the value of the counterion effective transport number through AEM (T_1), which is equal to the ratio of the counterion partial current density, i_1, to the total current density in the membrane system, i:

$$\eta \approx \frac{i_1}{i} = T_1 = 1 - T_A - T_w \quad (17)$$

According to Equation (17), the more the effective counterion transport number (T_1) differs from unity, the higher the effective transport numbers of co-ions—salt cations (T_A) and water splitting (T_w) products, for example, hydroxyl ions (T_{OH}) (generated at the AEM/depleted solution interface). If the difference in the concentration of the target electrolyte at the releasing and receiving membrane surfaces is not too large, then $T_A \approx t_{mA}$ [118]. According to Equation (1), the t_{mA} value is determined by the ratio of the conductivity to the differential coefficient of the membrane diffusion permeability. Thus, a decrease in conductivity and an increase in membrane diffusion permeability due to fouling (see Section 4) can actually lead to a decrease in current efficiency.

The effect of fouling on electric current-induced phenomena, i.e., water splitting (WS) and electroconvection (EC), has been intensively studied recently. Reviews of works devoted to these phenomena can be found in [11,119–121]. Interest in these phenomena stems from the fact that they affect (a) the value of T_w (T_{OH} for AEM and T_H for CEM); and (b) precipitation and fouling. WS causes a local pH shift at the IEM surfaces [122], which leads to intensification of precipitation or, on the contrary, dissolution of already formed precipitates [123]. EC intensively mixes the solution at the IEM / depleted diffusion layer interface, counteracting the fixation of precipitates on the membrane surface [124] and shifting WS toward higher potential drops [125].

The term "water splitting" implies the generation of H^+ and OH^- ions at the IEM/ depleted solution interface or at the bipolar CEM/AEM interface with the participation of fixed membrane groups. The catalytic activity of fixed groups in relation to water dissociation is determined by the values of the protonation–deprotonation reactions of these groups, the electric field strength at the membrane/solution interface and the rate of reaction product (H^+ and OH^- ions) removal from the reaction zone [10,126–128] with a thickness of several nanometers.

It is known [10,126] that carboxyl, phosphoric acid groups, as well as secondary and primary amines, have high catalytic activity, while fixed sulfonate groups and quaternary ammonium groups have low catalytic activity with respect to WS. A high electric field strength promotes a special orientation of water molecules favorable for the proton transfer and "stretching" of the H–OH [129] bonds.

In the case of monopolar membranes, the electric field strength required for WS is achieved at current densities (i) close to the limiting value (i_{lim}) or higher [119,126,128]. An increase in the current density, as well as the presence of cation-exchange and anion-exchange layers that form a bipolar junction, contribute to an increase in the electric field strength in the reaction zone and accelerate the removal of H^+ and OH^- ions from the reaction zone [119,126,127].

It is important to note that many of the liquid media components in the food industry (ammonium, amino acids, polybasic carboxylic and inorganic acids, etc.) are an additional source of protons or hydroxyl ions at the IEM/solution interface [10,77,78,130–132]. The point is that they are involved in protonation–deprotonation reactions in solution and in the membrane. If the products of these reactions (H^+ or OH^- ions) are co-ions, they are excluded from the IEM due to the Donnan effect.

This phenomenon intensifies with an increase in the current density due to a decrease in the concentration of the depleted solution at the IEM surface [78,133]. It is known [10,126,134] that some of the precipitated substances, for example, magnesium hydroxide (which is a traditional component of milk whey) participate in protonation–deprotonation reactions and intensify the generation of H^+ and OH^- ions.

The term "electroconvection" denotes a quite diverse phenomena, which manifest themselves in the appearance of vortex structures in a depleted solution at the surface of monopolar IEMs. An indispensable condition for the EC development is the presence of a tangential component of the electric field, as well as an electric double layer (under-limiting current regimes) and/or an extended space charge region (over-limiting current regimes) [135–137].

Therefore, the degree of EC development depends on the surface charge [138], as well as its electrical (alternation of conductive and non-conductive areas) [136,139–141] and geometric (waviness, roughness) [135,137,140,142] inhomogeneity. Besides, the development of EC is enhanced by an increase in surface hydrophobicity, which facilitates fluid slip along the IEM/solution interface [124,138].

5.2. Voltammetry

Current–voltage characteristic (CVC) analysis is the most common method for assessing the effect of fouling on WS, EC, electrical resistance, T_1 and IEM conductive surface area [64,111,143–149]. As a rule, these curves (Figure 6) are obtained in flow pass or non-flowing ED laboratory cells, in which the electric current increases slowly (0.01 mA/s) and the potential drop between the Luggin capillaries is recorded. Luggin capillaries are in solution on both sides of the membrane under study. The group of researchers led by Seung-Hyeon Moon was apparently one of the first to use independent experiments to interpret the CVC, and proposed an algorithm for assessing the effect of fouling on IEM behavior in an electric field [111,146,150,151], which is now used by many researchers. For example, in [146], studies of cation and anion exchange membranes CMX and AMX (Astom, Osaka, Japan) were carried out in a solution of a strong electrolyte (0.01 M KCl) and in the same solution containing a foulant (0.01 M KCl and 1% BSA). The authors [146] concluded that change in the slope of the initial section of the CVC (Region I in Figure 6b) obtained in a solution with foulant vs. a solution that does not contain it, show the foulant effect on the membrane resistance (R) or conductivity (1/R). Note, however, that the value of R [111,146], which is found from the slope of the initial section of the CVC (Figure 6), includes not only the IEM resistance and the resistance of the foulant layer on its surface, but also the resistance of the solution between the Luggin capillaries. To exclude the effect of this solution on resistance, it is necessary to obtain a CVC under similar conditions, but without a membrane, or use other methods for measuring the electrical resistance of membranes, which are reviewed in [70,72].

The length of the inclined plateau in CVC (Region II in Figure 6b), or rather, the potential drop between points *a* and *b*, determined by the intersection of tangents to Region I–Region II and Region II–Region III in CVC, can correspond to the onset of nonequilibrium electroconvection, which develops according to the Rubinstein–Zaltzman mechanism [152,153]. The ratio of the conductivities (1/R) determined from the initial (Region I in Figure 6b) and over-limiting (Region III in Figure 6b) sections of the CVC is an indicator of the degree of EC and/or WS development, in particular, due to the appearance of a foulant layer on the membrane surface [111,146,154,155]; the conductivity of the membrane system increases in an over-limiting state ($R_3/R_1 < 1$) due to the development of WS and EC. Comparing the "plateau length" and R_3/R_1 values of pristine and fouled membranes, it is easy to assess the effect of foulants on WS and EC.

Point *a* (Figure 6) corresponds to the experimental value of the limiting current density, $i_{lim}{}^{exp}$. According to the Peirce equation [117], modified by the Seung-Hyeon Moon group [156], this current is determined by the ratio of the limiting current, I_{limc}^{exp}, to the conducting (not screened by foulant) IEM surface, S_c:

$$i_{lim}^{exp} = \frac{I_{\text{limc}}^{exp}}{S_c} = \frac{FD_1C_1}{\delta}\left(1 - \frac{z_1}{z_A}\right) = \frac{FDC}{\delta(T_1 - t_1)} \tag{18}$$

where $S_c = \varepsilon S$; S is the IEM surface polarized by electric current, ε is the surface fraction not occupied by the foulant (or other nonconducting substance); D_1, C_1, t_1 are the diffusion coefficient, concentration and transport number of the counterion in the solution; z_1 and z_A are the electric charges of the counterion and co-ion, respectively; D and C are the diffusion coefficient and the concentration of the electrolyte in the solution; T_1 is the counterion transport number (determined by Equation (1); and δ is the depleted diffusion layer thickness.

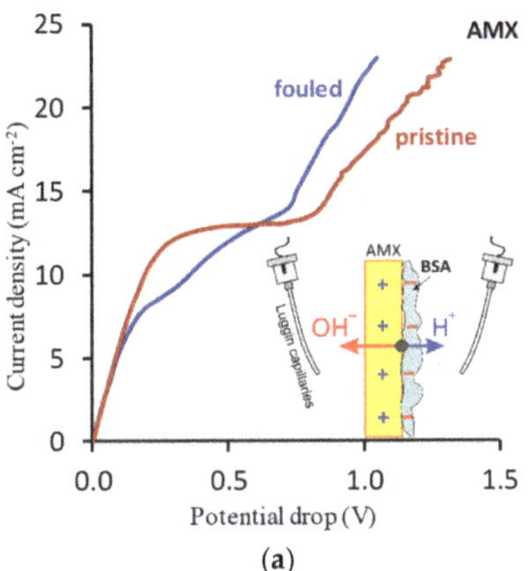

(**a**)

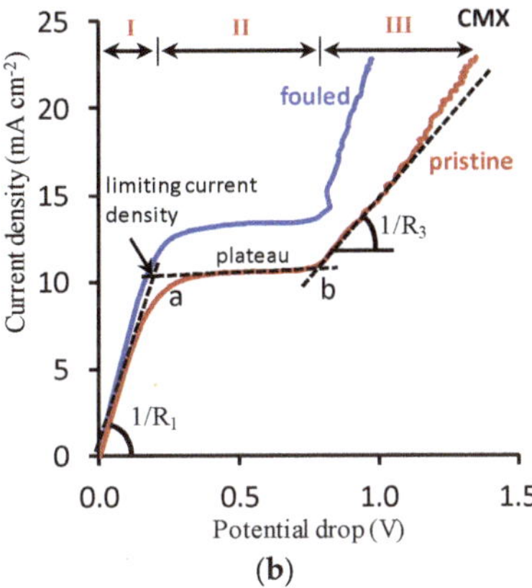

(**b**)

Figure 6. CVC of pristine and fouled by BSA anion exchange AMX (**a**) and cation exchange CMX (**b**) membranes. Insert in (**a**) shows the scheme of the mechanism of water splitting enhancement due to the formation of a bipolar junction between positively charged fixed AMX groups and negatively charged BSA. Explanations can be found in the text. Constructed using data from [146].

Thus, from the value of $i_{lim}{}^{exp}$ found by graphical treatment of CVC (Figure 6b), it is easy to estimate the effect of foulant on the depleted diffusion layer thickness if the counterion transport numbers in the pristine and fouled membranes are known, or to solve the inverse problem [157]. In particular, in articles [111,146], T_1 was found as the difference $1 - T_W$. The values of T_H (for CEM) and T_{OH} (for AEM) were determined by the Hittorff method, i.e., using the change in the concentration of water splitting products in the desalination compartment and adjoining concentration compartments of electrodialyzer. Equation (18) also makes it possible to find the real limiting current density on the conducting areas of the IEM surface, if the value of ε is known. Using the proposed algorithm, Park et al. [146] concluded that bovine serum albumin, BSA (whose molecules in the working solution were charged negatively), does not enter the electrostatic interactions with the negatively charged CMX surface, but, nevertheless, partially screens it. Since BSA contains aromatic amino acids [158,159], it can be assumed that the reason for this screening was π-π (stacking) interactions and the formation of hydrogen bonds between the membrane surface and the foulant [56,160]. Electrostatic interactions between the positively charged AMX surface and the negatively charged BSA species cause more significant screening of the AEM surface than the CEM surface [146]. The formation of a bipolar boundary (see insert on the Figure 6a) is facilitated by WS, whose products (H^+ and OH^- ions) have high diffusion coefficients as compared to other components of the processed solution. These H^+ and OH^- ions reduce the conductivity of fouled membrane and ajoining solution in comparison with that observed in the case of the pristine AMX membrane. In addition, Park et al. [151] found a 30% increase in the diluted diffusion layer thickness at the surface of BSA-fouled AMX membrane as compared to the pristine one, as well as fouled and pristine CMX membranes. Besides, $i_{lim}{}^{exp}$ increased after membrane fouling by BSA in the case of the CEM, but decreased in the case of the AEM (Figure 6). The authors of [151] did not explain these changes, because the understanding that foulants can significantly affect the surface properties responsible for the development of EC and WS, and that the equilibrium EC can affect the i_{lim}^{exp} value [161,162], began to take shape only in recent years.

To reveal this effect, along with CVC measurements, it is required to know the chemical composition of foulants, water contact angle and zeta potential, which allow estimation the degree of hydrophobicity and the magnitude of the surface charge. The use of 2D and 3D SEM, AFM, SECM and various types of optical microscopy provides insight into the electrical and geometric inhomogeneity of the IEM surface, as well as (analysis of IEM cross section) into the foulant film thickness. For example, the use of these methods along with the CVC measurements (Figure 7) and the simultaneous recording of the pH difference at the inlet and outlet of the desalination channel (which contained the pristine or fouled in red wine AMX-Sb membrane), made it possible to establish that the degree of WS and EC development is determined by the time of the AEM contact with the foulant [64].

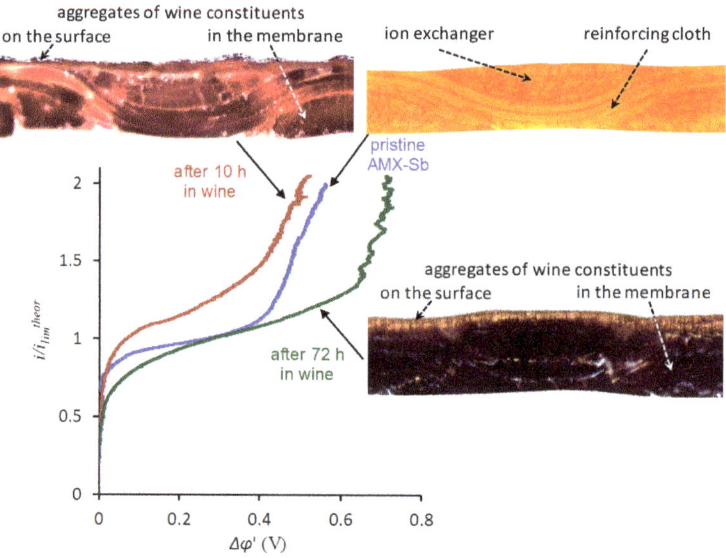

Figure 7. Corrected current–voltage characteristics after ohmic component subtraction, obtained in 0.02 NaCl solution for pristine and fouled in red wine AMX-Sb membrane samples. The insets show optical images of cross-sections of these samples. The value of i_{lim}^{theor} was calculated using the Lévêque equation [163]. Adapted from [64].

Thus, after 10 h of contact with wine, the ohmic resistance of AMX-Sb doubles. At the same time, WS decreases, EC intensifies, and mass transfer by chloride ions increases in comparison with the pristine membrane. The reason is the formation of the island-like structure of the more hydrophilic foulant layer on the more hydrophobic pristine membrane surface. Moreover, foulant aggregates (colloidal structures of anthocyanins, tannins, polysaccharides and other wine components) have a lower conductivity than the pristine membrane. The formation of such an electrical inhomogeneity, which is accompanied by an increase in the geometric heterogeneity of the surface, is the reason for the appearance of EC, which develops according to the mechanism of electroosmosis of the first kind [161]. This type of EC contributes to an increase in the limiting current density, i_{lim}^{exp}, found using the CVC graphical treatment. Note that theoretical estimates made for the case of natural convection of the feed solution [139] predict the maximum development of EC if 50% of the membrane surface is screened by a non-conductive substance. For forced convection of the feed solution, both theory [141] and experiment [140] give the maximum development of EC if 10–20% of the IEM surface does not conduct an electric current.

An increase in the time of contact with wine leads to an enrichment of the volume and surface of AMX-Sb with polyphenols. The result of this enrichment is a sharp drop in the electrical resistance of the fouled samples in comparison with the pristine membrane and the formation on its surface of an almost continuous layer of foulant, 2–3 μm thick. Moreover, the electric charge of the foulant surface is much lower than that of the pristine membrane. The result of such surface changes is a weakening of EC and enhancement of WS.

Belashova et al. [164] applied the same complex approach in studying the consequences of precipitation after the operation of the CMX-Sb cation exchange membrane (Astom, Osaka, Japan) in a solution (pH = 12) containing Na_2CO_3 (1000 mg/L), KCl (800 mg/L), $CaCl_2$ (800 mg/L) and $MgCl_2$ (452 mg/L), and in the same solution from which Na_2CO_3 was excluded to prevent the formation of carbonate precipitates. The precipitate formation was carried out in an over-limiting current mode. Simultaneously with CVCs obtained in 0.02 M NaCl solution, pH differences were measured in the ED desalination channel formed by the pristine or fouled membrane under study. Precipitates on the membrane surface were studied by SEM, EDS and X-ray diffraction analysis (XRD) of the surface. These researchers showed that the water splitting intensity is influenced by the localization of substances in the precipitate layer. In particular, in the case of deposition of magnesium hydroxide and calcium mixture on the CMX-Sb surface, WS is enhanced in comparison with the pristine membrane. The appearance of a dense film of calcium carbonate on top of this mixture prevents the contact of hydroxide ions with water molecules. As a result, water splitting is suppressed and EC increases.

5.3. Chronopotentiometry

Chronopotentiometry is another rapidly developing method. It consists of applying a direct electric current to the membrane system and recording a potential drop in the same electrochemical cells as in the case of CVCs. A review of the possibilities of this method, including its application to assessment the effect of inorganic and organic foulants on membrane behavior in an electric field, is described in detail in a recent review [83].

This method is most widely used [146,165–169] to determine the fraction of the conducting surface ε, which the Seung-Hyeon Moon group proposed to find [155,156] by the value of the transition time, τ, in the initial section of chronopotentiogramm (ChP), using the Sand equation adapted for IEM [170]:

$$\varepsilon = \frac{2i\tau^{1/2}(T_1 - t_1)}{Cz_1F\left(\pi D^{1/2}\right)} \tag{19}$$

However, in later works, Mareev et al. [171] showed that Sand equation (which was used to determine ε) is applicable to real IEMs only when the limiting current is exceeded by a factor of 1.5–2. This limitation is due to the fact that the Sand equation assumes the direction of streamlines strictly perpendicular to the membrane surface. This condition is met only if the dimensions of the electrical inhomogeneity of the surface are commensurate with or exceed the depleted diffusion layer thickness. This is why SEM, SECM and optical microscopy appear to be more reliable methods for determining ε.

Note that the possibilities of chronopotentiometry are not limited to the conducting surface fraction assessment. The ChP analysis makes it possible to follow up (including on-line) the effect of foulants on the counterion [83] and co-ions transport numbers in the membrane phase [172]. In addition, chronopotentiometry allows determination of the current densities at which intensive fouling or scaling of membranes begins, and estimation of the time interval required for the start of this process [123,173–175]. In the case of precipitation or foulant layer formation on the IEM surface, the ChP takes a special shape. In particular, an increase in the potential drop takes place some time after reaching the steady-state [123,174,175] (Figure 8). This increase is caused by an amplification in the concentration polarization of the membrane system due to a decrease in the IEM conductive surface fraction caused by precipitation. A decrease in the potential drop after switching off the current and an increase in the time required for the diffusion layer relaxation as compared to the pristine membrane (Figure 8) are other indicators of the precipitate presence on the surface [123,173–175]. Apparently, this change in the ChP shape is due to chemical reactions between substances that are part of the precipitates (for example, hydroxides of magnesium, iron and other metals) and WS products (for example, protons in the case of a CEM), which are released from the IEM at the moment the current is switched off.

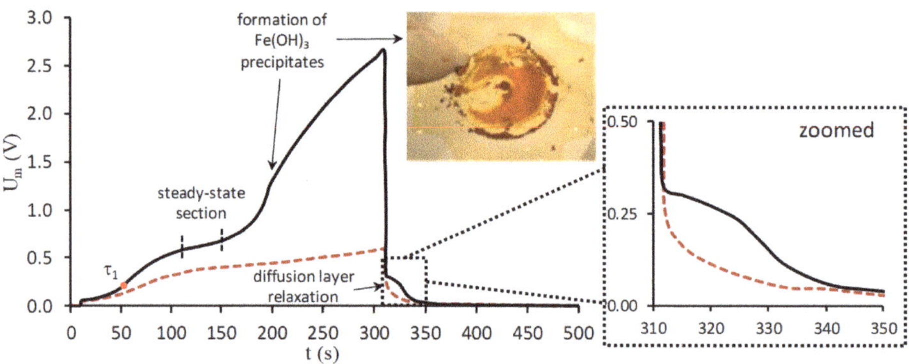

Figure 8. Chronopotentiometric curve of Nafion-117 membrane in 0.01 M $Fe_2(SO_4)_3$ solution in overlimiting current regime. Dashed line shows a typical shape of the curve in NaCl solution. The inset shows the precipitate found on the studied membrane surface after obtaining the chronopotentiometric curve (optical image). Adapted from [176].

In cases where the influence of the surface electrical heterogeneity on the transition time value can be neglected, the $i\tau^{0.5} - i$ (Sand coordinates) dependence is a straight line parallel to the abscissa axis if the counterion transport in the membrane system is carried out under diffusion control. A decrease in the experimental values of $i\tau^{0.5}$ in comparison with the values calculated by the Sand equation sometimes is an indicator of the conversion of the counterions of weak electrolytes in the near-membrane solution into a molecular form. For example, this phenomenon is observed in the presence of an amino acid (lysine) [177] or phosphoric acid anions [178] in a solution due to the participation of these substances in protonation–deprotonation reactions. The sloped line of the same dependence indicates the limitation of mass transfer by chemical reactions that occur in the near-membrane solution or on the IEM [179] surface. The appearance of an additional inclined section with another transition time on the ChP is characteristic of AEMs that operate in solutions containing phosphoric acid anions, polybasic carboxylic acids [132,154,178] or ammonium cations [77,180]. The presence of such a section is an indicator of a shift in the AEM internal solution pH to the alkaline values. Such a shift results in the formation of concentration profiles in the membrane of multiply charged ions of inorganic or organic acids [132,154,178], as well as NH_3 molecules [77,180]. This knowledge allows us to predict the prospects for IEM fouling by similar substances, which are often found in liquid media of the food industry.

5.4. *Electrochemical Impedance Spectroscopy*

Electrochemical impedance spectroscopy (EIS) is a very useful method for studying transport phenomena in membrane systems [132,181–183] and various aspects of IEM fouling by components of food production liquid media [21,144]. Measurements of electrochemical impedance spectra can be carried out in the same electrochemical cells as in the case of CVC and ChP. Recording of impedance spectra is carried out at a given direct current, as well as an alternating current (AC) having a significantly lower amplitude. As a rule, the overall AC frequency range is from 10^{-3} Hz to 10^5 Hz.

Park et al. [146], were apparently the first to use EIS to identify the WS intensity in the pristine and BSA-fouled IEM by analyzing the parameters of Gerischer impedance arc.

In work [62], a brief overview of modern studies devoted to the use of EIS is given and, using the example of AMX-Sb samples in contact with wine, the capabilities of this method are demonstrated. In particular, it was shown that the foulant island-type layer (formed after 10 h of AMX-Sb contact with wine; sample AMX-Sb_{w10}, Figure 9) contribution to an increase in the fouled membrane electrical resistance does not exceed 15%. Moreover, the presence of this membrane in a 0.02 M NaCl solution during the impedance spectra measurement decreases this contribution, apparently, due to the partial destruction of colloidal structures in the foulant layer.

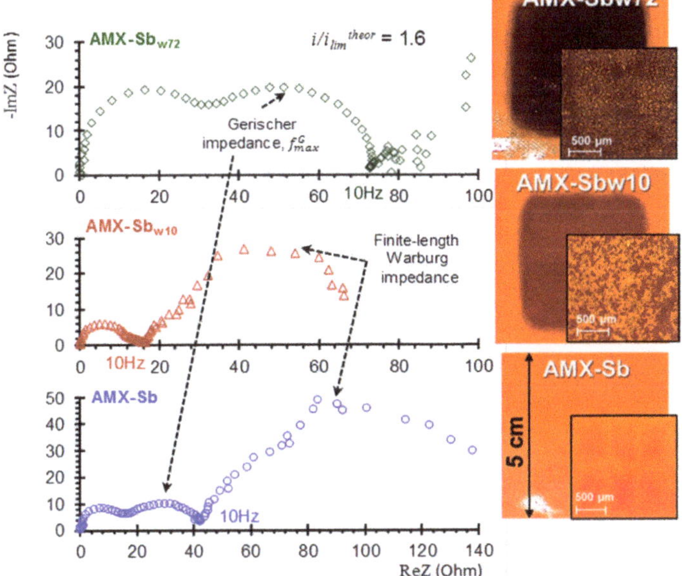

Figure 9. Electrochemical impedance spectra of pristine anion exchange membrane AMX-Sb and the same membrane after contact with wine for 10 (AMX-Sb_{w10}) and 72 (AMX-Sb_{w72}) hours. The data were obtained in 0.02 M NaCl solution; on the right are optical images of the studied samples. Adapted from [62].

The value of the foulant layer electric capacitance increases by two orders of magnitude as compared to the pristine membrane. These data were obtained by analyzing the high-frequency impedance spectra (or, rather, the superposition of the locus that characterizes the membrane and the arcs that characterize the foulant layer) using the electrical equivalent circuits method. Such an analysis is used, for example, for bilayer IEMs or membranes with modified surfaces [184–186], as well as in situ monitoring of membrane fouling in the course of natural water ED [80,81].

The treatment of low-frequency impedance spectra (Warburg-type impedance) using the previously developed concepts [187,188] made it possible to calculate the depleted diffusion layer thicknesses, and to establish that in overlimiting current modes this thickness decreases with an increasing current. Moreover, this decrease is most significant in the case of the AMX-Sb_{w10} as compared to the pristine AMX-Sb membrane. These results confirmed the development of more intensive equilibrium and non-equilibrium EC in the case of AMX-Sb_{w10}, of which the surface contains island-type distributed foulants.

In the analysis of the medium-frequency impedance spectra (Gerischer-type impedance) of the pristine and wine-contacted membranes, model concepts were applied, which were developed to determine the effective WS constant in systems with bipolar and monopolar membranes [189,190]:

$$\chi = \frac{2\pi f^{G}_{max}}{\sqrt{3}} \tag{20}$$

Equation (20) allows estimation of the effective constant of the WS reaction χ if the value of f^{G}_{max} (frequency corresponding to the maximum in the Gerischer impedance spectrum) is known. These estimates showed that the sizes of the Gerischer arc (Figure 9) and the χ values increase in the series: AMX-Sb$_{w10}$ < AMX-Sb << AMX-Sb$_{w72}$.

The combination of EIS with biochemical analysis of the studied samples showed [62] that the intensification of WS in the case of the AMX-Sb$_{w72}$ is facilitated by biofouling. Water splitting intensification is explained by the presence of native structures (membranes, RNA, etc.) in microorganisms, which contain, for example, phosphonate groups, characterized by very high catalytic activity with respect to WS reaction [10,126].

Note that in recent years, electrochemical methods traditionally used to study the fouling effect on the behavior of membrane systems in an electric field have been supplemented with modern tools for WS and EC visualization. For example, Slouka et al. [191,192], using optic microscopy with fluorescein and rhodamine dyes (Figure 10), proved the ability of RNA adsorbed on the IEM surface to intensify WS and suppress EC.

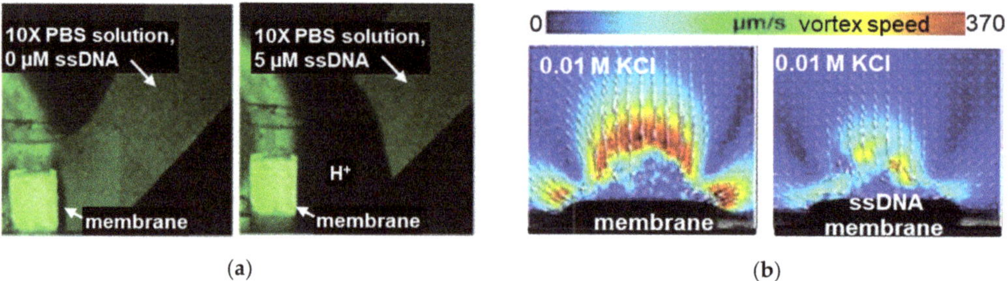

Figure 10. Results of visualization of the front of protons formed due to water splitting (**a**) and electroconvective vortices (**b**) at the surface of the anion-exchange membrane. On the left is the pristine membrane; on the right is a membrane on the surface of which single-stranded DNA (ssDNA, which contains dissociated phosphonate groups) is adsorbed. Adapted from [191,192].

Figure 11 summarizes the effect of volume and surface fouling of ion exchange membranes on their behavior in an electric field.

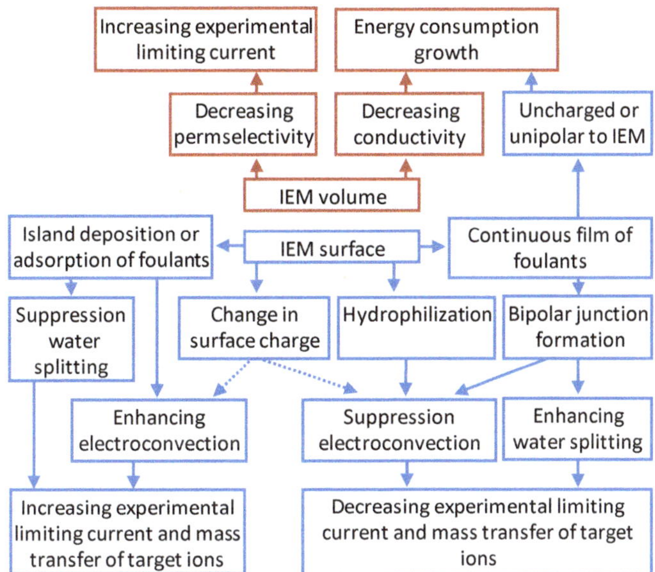

Figure 11. The scheme of the effect of fouling of the volume and surface of ion-exchange membranes on their behavior in an electric field.

6. Non-Destructive Methods of Membrane Fouling Control

The detrimental effect of traditional cleaning agents (acids, alkalis and oxidants) on the equilibrium and transport characteristics of membranes (see Sections 2–4) contributes to the search for alternative, non-destructive methods of fouling remediation. The application of these methods is based on knowledge gained from studying the mechanisms of IEM fouling in an electric field and in its absence. Note that the frequency of cleaning must also be carefully considered to avoid disrupting the separation process. There are so many different problems, each of which requires an individual answer depending on the materials of which the membranes are made, the type and composition of the solution being treated, as well as the hydrodynamic and electrical modes of electrodialysis. These problems, traditional and alternative cleaning procedures and the peculiarities of its implementation, depending on the chemical nature of the membranes, as well as the component composition of the processed liquid media, were discussed in detail in a recent review [193]. Therefore, below we will only give a few examples of this diversity.

The exposure to solutions of NaCl or other salts leads to partial destruction of colloidal particles on the surface and inside the IEM due to the salting-out effect. The higher the salt concentration in the regenerating solution, the closer the regenerated membrane conductivity to the conductivity of the pristine membrane [60,61,105]. The removal of colloidal particles destruction products from the IEM pores is apparently limited by their diffusion towards the external solution; the thinner the membrane and the larger its pores, the faster and more completely the regeneration process proceeds. Nevakshenova et al. showed [105] that cleaning of an AEM after its contact with wine using concentrated (58.5 g/L, pH 5.5) NaCl solution allowed almost completely restoration of AEM selectivity if the duration of its contact with wine does not exceed 70 h. In the case of a thin (140 ± 10 μm) homogeneous AMX-Sb membrane, the transport numbers of the Cl^- counterion in the samples of pristine and regenerated membranes differ by less than 1%. In the case of a thick (530 ± 10 μm) heterogeneous membrane MA-41P (OJSC Shchekinoazot, Pervomaisky, Russia), this difference increases to 3%. At the same time, the use of saline solutions for IEM regeneration, which have worked for a long time in industrial wine stabilization processes and have already been cleaned with acids and/or alkalis, gives less encouraging results [60,61]. For example, the conductivity of such CEMs and AEMs (Astom, Osaka, Japan) reaches, respectively, 60% and 45% of the pristine membranes conductivity, and the value of f_2 can be increased, respectively, by 12% and 23%. The fact is that NaCl solutions with neutral pH values have little effect on the electrostatic and π-π (stacking) interactions of polyphenols with ion-exchange membranes.

These solutions are not able to destroy agglomerates of polyphenol-containing nanoparticles that are formed in the pores of IEMs due to cleaning with acids and alkalis. The use of regenerating solutions that contain more hydrated Ca^{2+}, Mg^{2+}, and SO_4^{2-} ions could provide a more complete destruction of colloidal particles, as is observed, for example, in the case of regeneration of ultrafiltration membranes, used, for example, in BSA separation [194]. Meanwhile, after the use of such solutions, the conductivity of CEMs and AEMs does not increase, but decreases in comparison with conductivity achieved after prolonged ED stabilization of wine [60,61]. The reason for this decrease is the specific interactions of Ca^{2+} ions (and to a lesser extent Mg^{2+}) with fixed sulfonate groups of cation-exchange membranes, which lead to a loss in their exchange capacity and a decrease in the surface charge [172], as well as the formation of complexes with polyphenols by these ions [195].

Application of acidified (pH = 3.5) aqueous ethanol solution (12%) recovers IEC by 33% and nearly doubles the conductivity of CEMs and AEMs fouled with polyphenol. This result is apparently due to the destruction of hydrogen bonds in the presence of ethanol. Moreover, a more noticeable regeneration is achieved in the case of AEMs due to the acquisition of a positive electric charge by polyphenols in an acidic medium, as well as due to the electrostatic repulsion of these particles from the positively charged fixed groups of the anion-exchange membrane.

The use of a regenerating solution that contains equal volumes of acetonitril, methanol, isopropyl and distilled water allows an even more complete recovery of phenolic monomeric and polymeric substances (more than 30 items) from CEMs and AEMs that were in contact with wine [26] or cranberry juice [23]. Such recovery is due to the destruction of hydrogen bonds and π-π (stacking) interactions of polyphenols with the material of IEMs, as well as with proteins, amino acids, saccharides and other substances found in juices and wine. As expected, a more complete recovery of phenolic substances is achieved in the case of membranes having an aliphatic matrix (CEM Type-ll, CJMC-5) [23].

Regeneration of both the aromatic (CSE-fg, MK-40) and aliphatic (CEM Type-ll, CJMC-5) cation exchange membranes is facilitated by alkalization of an organic solvents mixture to pH = 10. The result of such alkalization is the initiation of electrostatic repulsion of negatively charged sulfonate fixed groups of IEMs and polyphenols, which acquire a negative charge. However, the use of this solution, as well as of saline solutions, does not provide complete regeneration of ion-exchange membranes. Therefore, the question about the optimal solution for the mild chemical regeneration of IEMs after their contact with juices, wines, tea and other phenolic-containing liquid media of the food industry remains open.

Another promising method for the regeneration of IEMs after their contact with liquid media of the food industry is the use of enzymatic agents specific to certain types of substances. They performed well in the case of cleaning of ultrafiltration membranes fouled by proteins [196] or in the case of reducing membrane fouling in bioreactors [197]. Bdiri et al. [198] first tested protease and β-glucanase, as well as polyphenol oxidase, which enter the specific interactions with proteins, polysaccharides, and phenolic compounds, respectively, as well as mixtures of these enzymes. They showed that the use of these cleaning agents can increase the conductivity and exchange capacity of the wine-fouled CMX-Sb cation-exchange membrane by at least 25% and 12%, respectively. The most encouraging results have been obtained with Tyrosinase, which is active against polyphenols.

All previous examples dealt with the remediation of fouling phenomena. However, another alternative approach exists, and consists of preventing fouling using surface modifications or special hydrodynamic and electrical modes of electrodialysis [3,199]. The most studied is the use of pulsed electric fields (PEF) to prevent fouling in the processing of whey and similar liquids [3,200]. In some works [201], there is evidence of a decrease in energy consumption and a decrease in fouling during ED separation of citric and malic acids from cranberry juice. Fouling by organic and inorganic substances when using PEF can be prevented due to a decrease in concentration polarization in the membrane system, a decrease in pH changes in near-membrane solutions, and ensuring favorable conditions for the development of electroconvection. The main advances in this field are comprehensively reviewed in a recent work by Bazinet and Geoffroy [3]. Therefore, we will not focus on this topic.

As to surface modification, it may be chemical or geometrical. Many kinds of chemical modification have been proposed in order to improve IEMs anti-organic fouling properties in recent studies (see for example [202–204]). Most of them focus on increasing the hydrophilicity of the AEMs surface [202] and imparting an electrical charge, which is the opposite to that of the AEMs fixed groups [203,204]. Extensive reviews on the modification of commercial membranes, as well as synthesis of selective and anti-organic fouling IEMs, can be found in [205–208].

The geometrical method consists of creating profiles of a few hundred micrometers on the surfaces of the IEMs in order to increase the exchange area, to enhance the hydrodynamic turbulence phenomena and to stimulate electroconvection [209]. These modification reduce concentration polarization and/or enhances mixing of the solution at the IEMs surface, preventing the formation of fouling layers. Details of membrane surface profiling can be found in the review [210].

7. Conclusions

The study of the consequences of ion-exchange membrane fouling and the development of methods of fouling control in dialysis and electrodialysis processes of food industry liquid media processing is increasingly attracting the attention of scientists.

In recent years, a certain algorithm for studying the characteristics of ion-exchange membranes has been developed, which includes comprehensive studies of exchange capacity, water content, conductivity, diffusion permeability before and after the operation of membranes in industrial or laboratory dialysis and electrodialysis processes. To interpret these data, the microheterogeneous model is increasingly being used. The recent modifications of this model take into account the possibility of formation in the pores of IEMs of hydraulically permeable (dairy industry) or impermeable (wine and juice industry) agglomerates of colloidal particles, the surface charge of which is identical or opposite to the electric charge of the pore walls. The use of this and a number of other known models makes it possible to estimate changes in the structural parameters of IEMs, the diffusion coefficients of counterions and coions, as well as the transport numbers of these ions (membrane selectivity) caused by interactions of foulants with the polymer matrix and fixed groups of IEMs.

The behavior of fouled IEM in an electric field (i.e., in electrodialysis processes) is largely determined by changes in electrical resistance, hydrophilic/hydrophobic balance, as well as the roughness of its surface and the foulants distribution morphology on this surface. These characteristics drive the development of electroconvection (which promotes stirring of the solution and counteracts fouling) and water splitting (which often enhances fouling due to local pH changes at the surface of the IEMs and within the membrane).

To study these phenomena, voltammetry, chronopotentiometry and impedance spectroscopy are most widely used. Moreover, the electrochemical impedance spectra obtained in the presence of a direct current in the membrane system provide the most complete information, including the resistance and thickness of foulant layers on the membrane surface, effective reaction rate constants of H^+ and OH^- ions generation, as well as the thickness of diffusion layers, which depends on the degree of electroconvection development. The use of modern models describing the behavior of membrane systems in overlimiting current modes has shed light on the reasons for the enhancement of water splitting during IEM fouling by proteins, polyphenols, or biofouling. It turned out that in some cases (island localization of polyphenols on the IEM surface), fouling can promote the increase in useful mass transfer due to the intensification of electroconvection.

The idea was finally formed that traditional IEM cleaning methods using acids, alkalis and/or oxidants are not only harmful to the environment, but also contribute to the destruction of membranes and an increase in fouling. Mild chemical reagents (aqueous salt solutions, ensims, mixtures of polar and non-polar organic solvents and water), as well as controlling the membrane regeneration process by varying the solution pH, seem to be more promising along with the use of a specially designed anti-organic fouling membranes.

In order for the results of an experimental study of the foulants effect on the structure, transport and mass transfer properties of IEMs to be more informative, it is necessary to improve the existing and develop new mathematical models that will more deeply take into account the nature of the foulants interactions with materials from which membranes are made, as well as consider the specifics of many components of food industry liquid media; the ability to change the structure and electrical charge depending on the environment pH.

Another challenge is to develop environmentally sound ways to counter biofouling and find solutions for the regeneration of fouled membranes (for example, using enzymes specific for key toxic substances), as well as to determine the optimal current regimes that will ensure suppression of fouling, low energy consumption and high current efficiency in ED processes in the food industry. The use of pulsed electric fields seems to be the most promising for achieving this goal.

Author Contributions: L.D.: Research design and investigation; Manuscript writing, revision and validation; Project administration. J.F., M.B. and V.S.: Methodology; Data analysis and curation; Draft preparation, writing and editing. C.L.: Methodology; Data analysis and curation; Validation. E.R. and L.B.: Research conceptualization and investigation; Manuscript writing, revision and validation. A.K.: Research conceptualization and investigation; Manuscript writing, revision and validation; Data curation; Visualization. N.P.: Research conceptualization and investigation; Manuscript writing, revision and validation; Funding acquisition. All authors have read and agreed to the published version of the manuscript.

Funding: This work was financially supported by the Kuban Science Foundation, project No MFI-20.1/130.

Institutional Review Board Statement: Not applicable.

Informed Consent Statement: Not applicable.

Data Availability Statement: Not applicable.

Conflicts of Interest: The authors declare no conflict of interest.

Abbreviations

All other abbreviations have their usual meaning, or are sufficiently explained in the text.

AEM	Anion-exchange membrane
BSA	Bovine serum albumin
CEM	Cation-exchange membrane
ChP	Chronopotentiogramm
CVC	Current–voltage characteristic
EC	Electroconvection
ED	Electrodialysis
EDS	Energy dispersive X-ray spectrometry
EIS	Electrochemical impedance spectroscopy (spectrum)
FO	Forward osmosis
IEC	Ion exchange capacity
NF	Nanofiltration
PEF	Pulsed electric field
PP	Polyphenol
PVC	Polyvinyl chloride

RNA	Ribonucleic acid
RO	Reverse osmosis
SECM	Scanning electrochemical microscopy
SEM	Scanning electron microscopy
UF	Ultrafiltration
WS	Water splitting
XRD	X-ray diffraction

References

1. Wang, Y.; Jiang, C.; Bazinet, L.; Xu, T. *Electrodialysis-Based Separation Technologies in the Food Industry*; Academic Press: Cambridge, MA, USA, 2019; pp. 349–381. [CrossRef]
2. Wang, Y.; Yu, J. Membrane separation processes for enrichment of bovine and caprine milk oligosaccharides from dairy byproducts. *Compr. Rev. Food Sci. Food Saf.* **2021**, *20*, 3667–3689. [CrossRef] [PubMed]
3. Bazinet, L.; Geoffroy, T.R. Electrodialytic processes: Market overview, membrane phenomena, recent developments and sustainable strategies. *Membranes* **2020**, *10*, 221. [CrossRef] [PubMed]
4. Gonçalves, F.; Fernandes, C.; Liu, G.; Wu, D.; Chen, G.; Halim, R.; Liu, J.; Deng, H. Comparative study on tartaric acid production by two-chamber and three-chamber electro-electrodialysis. *Sep. Purif. Technol.* **2021**, *263*, 118403. [CrossRef]
5. Cameira dos Santos, P.; de Pinho, M.N. Wine tartaric stabilization by electrodialysis and its assessment by the saturation temperature. *J. Food Eng.* **2003**, *59*, 229–235. [CrossRef]
6. El Rayess, Y.; Mietton-Peuchot, M. Membrane Technologies in Wine Industry: An Overview. *Crit. Rev. Food Sci. Nutr.* **2016**, *56*, 2005–2020. [CrossRef]
7. Kattan Readi, O.M. *Membranes in the Biobased Economy: Electrodialysis of Amino Acids for the Production of Biochemical*; Universiteit Twente: Enschede, The Netherlands, 2013; p. 176. ISBN 978-94-6108-414-9.
8. Shaposhnik, V.A.; Eliseeva, T.V. Barrier effect during the electrodialysis of ampholytes. *J. Membr. Sci.* **1999**, *161*, 223–227. [CrossRef]
9. Hülber-Beyer, É.; Bélafi-Bakó, K.; Nemestóthy, N. Low-waste fermentation-derived organic acid production by bipolar membrane electrodialysis—An overview. *Chem. Pap.* **2021**, *75*, 5223–5234. [CrossRef]
10. Simons, R. Electric field effects on proton transfer between ionizable groups and water in ion exchange membranes. *Electrochim. Acta* **1984**, *29*, 151–158. [CrossRef]
11. Pärnamäe, R.; Mareev, S.; Nikonenko, V.; Melnikov, S.; Sheldeshov, N.; Zabolotskii, V.; Hamelers, H.V.M.; Tedesco, M. Bipolar membranes: A review on principles, latest developments, and applications. *J. Membr. Sci.* **2021**, *617*, 118538. [CrossRef]
12. Nguyen, T.-T.; Adha, R.S.; Lee, C.; Kim, D.-H.; Kim, I.S. Quantifying the influence of divalent cations mass transport on critical flux and organic fouling mechanism of forward osmosis membrane. *Desalination* **2021**, *512*, 115146. [CrossRef]
13. Al-Amoudi, A.; Lovitt, R.W. Fouling strategies and the cleaning system of NF membranes and factors affecting cleaning efficiency. *J. Membr. Sci.* **2007**, *303*, 4–28. [CrossRef]
14. Chandra, A.; Bhuvanesh, E.; Chattopadhyay, S. Physicochemical interactions of organic acids influencing microstructure and permselectivity of anion exchange membrane. *Colloid Surface A* **2018**, *560*, 260–269. [CrossRef]
15. Kononenko, N.; Nikonenko, V.; Grande, D.; Larchet, C.; Dammak, L.; Fomenko, M.; Volfkovich, Y. Porous structure of ion exchange membranes investigated by various techniques. *Adv. Colloid Interfac.* **2017**, *246*, 196–216. [CrossRef] [PubMed]
16. Pismenskaya, N.; Sarapulova, V.; Nevakshenova, E.; Kononenko, N.; Fomenko, M.; Nikonenko, V. Concentration dependences of diffusion permeability of anion-exchange membranes in sodium hydrogen carbonate, monosodium phosphate and potassium hydrogen tartrate solutions. *Membranes* **2019**, *9*, 170. [CrossRef] [PubMed]
17. Sun, L.; Chen, Q.; Lu, H.; Wang, J.; Zhao, J.; Li, P. Electrodialysis with porous membrane for bioproduct separation: Technology, features, and progress. *Food Res. Int.* **2020**, *137*, 109343. [CrossRef] [PubMed]
18. Pismenskaya, N.; Sarapulova, V.; Klevtsova, A.; Mikhaylin, S.; Bazinet, L. Adsorption of anthocyanins by cation and anion exchange resins with aromatic and aliphatic polymer matrices. *Int. J. Mol. Sci.* **2020**, *21*, 7874. [CrossRef]
19. Langevin, M.-E.; Bazinet, L. Ion-exchange membrane fouling by peptides: A phenomenon governed by electrostatic interactions. *J. Membr. Sci.* **2011**, *369*, 359–366. [CrossRef]
20. Suwal, S.; Doyen, A.; Bazinet, L. Characterization of protein, peptide and amino acid fouling on ion-exchange and filtration membranes: Review of current and recently developed methods. *J. Membr. Sci.* **2015**, *496*, 267–283. [CrossRef]
21. Mikhaylin, S.; Bazinet, L. Fouling on ion-exchange membranes: Classification, characterization and strategies of prevention and control. *Adv. Colloid Interfac.* **2016**, *229*, 34–56. [CrossRef]
22. Bdiri, M.; Perreault, V.; Mikhaylin, S.; Larchet, C.; Hellal, F.; Bazinet, L.; Dammak, L. Identification of phenolic compounds and their fouling mechanisms in ion-exchange membranes used at an industrial scale for wine tartaric stabilization by electrodialysis. *Sep. Purif. Technol.* **2019**, *233*, 115995. [CrossRef]
23. Perreault, V.; Sarapulova, V.; Tsygurina, K.; Pismenskaya, N.; Bazinet, L. Understanding of adsorption and desorption mechanisms of anthocyanins and proanthocyanidins on heterogeneous and homogeneous cation-exchange membranes. *Membranes* **2021**, *11*, 136. [CrossRef]

24. Sholokhova, A.Y.; Eliseeva, T.V.; Voronyuk, I.V. Sorption of vanillin by highly basic anion exchangers under dynamic conditions. *Russ. J. Phys. Chem. A* **2018**, *92*, 2048–2052. [CrossRef]
25. Evans, P.J.; Bird, M.R.; Rogers, D.; Wright, C.J. Measurement of polyphenol-membrane interaction forces during the ultrafiltration of black tea liquor. *Colloids Surf. A Physicochem. Eng. Aspect.* **2009**, *335*, 148–153. [CrossRef]
26. Bdiri, M.; Larchet, C.; Dammak, L. A review on ion-exchange membranes fouling and antifouling during electrodialysis used in food industry: Cleanings and strategies of prevention. *Chem. Afr.* **2020**, *3*, 609–633. [CrossRef]
27. Garcia-Vasquez, W.; Ghalloussi, R.; Dammak, L.; Larchet, C.; Nikonenko, V.; Grande, D. Structure and properties of heterogeneous and homogeneous ion-exchange membranes subjected to ageing in sodium hypochlorite. *J. Membr. Sci.* **2014**, *452*, 104–116. [CrossRef]
28. Ge, S.; Zhang, Z.; Yan, H.; Irfan, M.; Xu, Y.; Li, W.; Wang, Y. Electrodialytic Desalination of Tobacco Sheet Extract: Membrane Fouling Mechanism and Mitigation Strategies. *Membranes* **2020**, *10*, 245. [CrossRef]
29. Shi, L.; Xie, S.; Hu, Z.; Wu, G.; Morrison, L.; Croot, P.; Zhan, X. Nutrient recovery from pig manure digestate using electrodialysis reversal: Membrane fouling and feasibility of long-term operation. *J. Membr. Sci.* **2019**, *573*, 560–569. [CrossRef]
30. Chon, K.; Jeong, N.; Rho, H.; Nam, J.Y.; Jwa, E.; Cho, J. Fouling characteristics of dissolved organic matter in fresh water and seawater compartments of reverse electrodialysis under natural water conditions. *Desalination* **2020**, *496*, 114478. [CrossRef]
31. Ghafari, M.; Mohona, T.M.; Su, L.; Lin, H.; Plata, D.L.; Xiong, B.; Dai, N. Effects of peracetic acid on aromatic polyamide nanofiltration membranes: A comparative study with chlorine. *Env. Sci. Water Res. Technol.* **2021**, *7*, 306–320. [CrossRef]
32. Beyer, F.; Laurinonyte, J.; Zwijnenburg, A.; Stams, A.J.; Plugge, C.M. Membrane fouling and chemical cleaning in three full-scale reverse osmosis plants producing demineralized water. *J. Eng.* **2017**, *2017*, 6356751. [CrossRef]
33. Šímová, H.; Kysela, V.; Černín, A. Demineralization of natural sweet whey by electrodialysis at pilot-plant scale. *Desalin. Water Treat.* **2010**, *14*, 170–173. [CrossRef]
34. Xia, Q.; Qiu, L.; Yu, S.; Yang, H.; Li, L.; Ye, Y.; Liu, G. Effects of alkaline cleaning on the conversion and transformation of functional groups on ion-exchange membranes in polymer-flooding wastewater treatment: Desalination performance, fouling behavior, and mechanism. *Environ. Sci. Technol.* **2019**, *53*, 14430–14440. [CrossRef] [PubMed]
35. Aumeier, B.M.; Yüce, S.; Wessling, M. Temperature enhanced backwash. *Water Res.* **2018**, *142*, 18–25. [CrossRef] [PubMed]
36. Peng, H.; Tremblay, A.Y. Membrane regeneration and filtration modeling in treating oily wastewaters. *J. Membr. Sci.* **2008**, *324*, 59–66. [CrossRef]
37. Lee, S.; Elimelech, M. Salt cleaning of organic-fouled reverse osmosis membranes. *Water Res.* **2007**, *41*, 1134–1142. [CrossRef]
38. Vasilyeva, V.I.; Pismenskaya, N.D.; Akberova, E.M.; Nebavskaya, K.A. Effect of thermomechanical treatment on the surface morphology and hydrophobicity of heterogeneous ion exchange membranes. *Rus. J. Phys. Chem. A* **2014**, *88*, 1293–1299. [CrossRef]
39. Maurya, S.; Shin, S.-H.; Kim, M.-K.; Yun, S.-H.; Moon, S.-H. Stability of composite anion exchange membranes with various functional groups and their performance for energy conversion. *J. Membr. Sci.* **2013**, *443*, 28–35. [CrossRef]
40. Pine, S.H. The base-promoted rearrangements of quaternary ammonium salts. *Org. React.* **2011**, 403–464. [CrossRef]
41. Merle, G.; Wessling, M.; Nijmeijer, K. Anion exchange membranes for alkaline fuel cells: A review. *J. Membr. Sci.* **2011**, *377*, 1–35. [CrossRef]
42. Dammak, L.; Larchet, C.; Grande, D. Ageing of ion-exchange membranes in oxidant solutions. *Sep. Purif. Technol.* **2009**, *69*, 43–47. [CrossRef]
43. Mizutani, Y.; Yamane, R.; Motomura, H. Studies of ion exchange membranes. XXII. Semicontinuous preparation of ion exchange membranes by the "Paste Method". *Bull. Chem. Soc. Jpn.* **1965**, *38*, 689–694. [CrossRef]
44. Higa, M.; Tanaka, N.; Nagase, M.; Yutani, K.; Kameyama, T.; Takamura, K.; Kakihana, Y. Electrodialytic properties of aromatic and aliphatic type hydrocarbonbased anion-exchange membranes with various anion-exchange groups. *Polymer* **2014**, *55*, 3951–3960. [CrossRef]
45. Sata, T. *Ion Exchange Membranes: Preparation, Characterization, Modification and Application*; Royal Society of Chemistry: London, UK, 2004; p. 314, ISBN-13: 978-0854045907.
46. Doi, S.; Taniguchi, I.; Yasukawa, M.; Kakihana, Y.; Higa, M. Effect of alkali treatment on the mechanical properties of anion-exchange membranes with a poly(vinyl chloride) backing and binder. *Membranes* **2020**, *10*, 344. [CrossRef]
47. Doi, S.; Takumi, N.; Kakihana, Y.; Higa, M. Alkali attack on cation-exchange membranes with polyvinyl chloride backing and binder: Comparison with anion-exchange membranes. *Membranes* **2020**, *10*, 228. [CrossRef]
48. Doi, S.; Yasukawa, H.; Kakihana, Y.; Higa, M. Alkali attack on anion exchange membranes with PVC backing and binder: Effect on performance and correlation between them. *J. Membr. Sci.* **2019**, *573*, 85–96. [CrossRef]
49. Minsker, K.S.; Zaikov, G.E. Achievements and research tasks for polyvinylchloride aging and stabilization. *J. Vinyl Add. Technol.* **2001**, *7*, 222–234. [CrossRef]
50. Rybalkina, O.A.; Tsygurina, K.A.; Sarapulova, V.V.; Mareev, S.A.; Nikonenko, V.V.; Pismenskaya, N.D. Evolution of current–voltage characteristics and surface morphology of homogeneous anion-exchange membranes during the electrodialysis desalination of alkali metal salt solutions. *Membr. Membr. Technol.* **2019**, *1*, 107–119. [CrossRef]
51. Amato, L.; Gilbert, M.; Caswell, A. Degradation studies of crosslinked polyethylene. II Aged in water. *Plast. Rubber Compos.* **2005**, *34*, 179–187. [CrossRef]
52. Henry, J.L.; Garton, A. Thermal oxidation of polyethylene in aqueous environments. *Polym. Prepr. ACS* **1989**, *30*, 183–184.

53. Helfferich, F. *Ion. Exchange*; McGraw-Hill: New York, NY, USA, 1962; p. 624. ISBN 0-486-68784-8.
54. Audinos, R. Fouling of ion-selective membranes during electrodialysis of grape must. *J. Membr. Sci.* **1989**, *41*, 115–126. [CrossRef]
55. Araya-Farias, M.; Bazinet, L. Effect of calcium and carbonate concentrations on anionic membrane fouling during electrodialysis. *J. Colloid Interface Sci.* **2006**, *296*, 242–247. [CrossRef] [PubMed]
56. Persico, M.; Mikhaylin, S.; Doyen, A.; Firdaous, L.; Hammami, R.; Chevalier, M.; Flahaut, C.; Dhulster, P.; Bazinet, L. Formation of peptide layers and adsorption mechanisms on a negatively charged cation-exchange membrane. *J. Colloid Interface* **2017**, *508*, 488–499. [CrossRef]
57. Persico, M.; Mikhaylin, S.; Doyen, A.; Firdaous, L.; Hammami, R.; Bazinet, L. How peptide physicochemical and structural characteristics affect anion-exchange membranes fouling by a tryptic whey protein hydrolysate. *J. Membr. Sci.* **2016**, *520*, 914–923. [CrossRef]
58. Garcia-Vasquez, W.; Dammak, L.; Larchet, C.; Nikonenko, V.; Pismenskaya, N.; Grande, D. Evolution of anion-exchange membrane properties in a full scale electrodialysis stack. *J. Membr. Sci.* **2013**, *446*, 255–265. [CrossRef]
59. Ghalloussi, R.; Garcia-Vasquez, W.; Chaabane, L.; Dammak, L.; Larchet, C.; Deabate, S.V.; Nevakshenova, E.; Nikonenko, V.; Grande, D. Ageing of ion-exchange membranes in electrodialysis: A structural and physicochemical investigation. *J. Membr. Sci.* **2013**, *436*, 68–78. [CrossRef]
60. Bdiri, M.; Dammak, L.; Larchet, C.; Hellal, F.; Porozhnyy, M.; Nevakshenova, E.; Pismenskaya, N.; Nikonenko, V. Characterization and cleaning of anion-exchange membranes used in electrodialysis of polyphenol-containing food industry solutions; comparison with cation-exchange membranes. *Sep. Purif. Technol.* **2019**, *210*, 636–650. [CrossRef]
61. Bdiri, M.; Dammak, L.; Chaabane, L.; Larchet, C.; Hellal, F.; Nikonenko, V.; Pismenskaya, N.D. Cleaning of cation-exchange membranes used in electrodialysis for food industry by chemical solutions. *Sep. Purif. Technol.* **2018**, *199*, 114–123. [CrossRef]
62. Kozmai, A.; Sarapulova, V.; Sharafan, M.; Melkonian, K.; Rusinova, T.; Kozmai, Y.; Pismenskaya, N.; Dammak, L.; Nikonenko, V. Electrochemical impedance spectroscopy of anion-exchange membrane amx-sb fouled by red wine components. *Membranes* **2021**, *11*, 2. [CrossRef]
63. Belashova, E.D.; Pismenskaya, N.D.; Nikonenko, V.V.; Sistat, P.; Pourcelly, G. Current-voltage characteristic of anion-exchange membrane in monosodium phosphate solution. Modelling and experiment. *J. Membr. Sci.* **2017**, *542*, 177–185. [CrossRef]
64. Sarapulova, V.; Nevakshenova, E.; Nebavskaya, X.; Kozmai, A.; Aleshkina, D.; Pourcelly, G.; Nikonenko, V.; Pismenskaya, N. Characterization of bulk and surface properties of anion-exchange membranes in initial stages of fouling by red wine. *J. Membr. Sci.* **2018**, *559*, 170–182. [CrossRef]
65. Larchet, C.; Dammak, L.; Auclair, B.; Parchikov, S.; Nikonenko, V. A simplified procedure for ion-exchange membrane characterization. *New J. Chem.* **2004**, *28*, 1260. [CrossRef]
66. Luo, T.; Abdu, S.; Wessling, M. Selectivity of ion exchange membranes: A review. *J. Membr. Sci.* **2018**, *555*, 429–454. [CrossRef]
67. Berezina, N.P.; Kononenko, N.A.; Dyomina, O.A.; Gnusin, N.P. Characterization of ion-exchange membrane materials: Properties vs structure. *Adv. Colloid Interfac.* **2008**, *139*, 3–28. [CrossRef] [PubMed]
68. Zabolotsky, V.I.; Nikonenko, V.V. Effect of structural membrane inhomogeneity on transport properties. *J. Membr. Sci.* **1993**, *79*, 181–198. [CrossRef]
69. Gnusin, N.P.; Berezina, N.P.; Shudrenko, A.A.; Ivina, A.P. Electrolyte diffusion across ion-exchange membranes. *Russ. J. Phys. Chem. A* **1994**, *68*, 506–510.
70. Sedkaoui, Y.; Szymczyk, A.; Lounici, H.; Arous, O. A new lateral method for characterizing the electrical conductivity of ion-exchange membranes. *J. Memb. Sci.* **2016**, *507*, 34–42. [CrossRef]
71. Lteif, R.; Dammak, L.; Larchet, C.; Auclair, B. Conductivité électrique membranaire: Étude de l'effet de la concentration, de la nature de l'électrolyte et de la structure membranaire. *Eur. Polym. J.* **1999**, *35*, 1187–1195. [CrossRef]
72. Karpenko, L.V.; Demina, O.A.; Dvorkina, G.A.; Parshikov, S.B.; Larchet, C.; Auclair, B.; Berezina, N.P. Comparative study of methods used for the determination of electroconductivity of ion-exchange membranes. *Russ. J. Electrochem.* **2001**, *37*, 287–293. [CrossRef]
73. Mareev, S.A.; Butylskii, D.Y.; Pismenskaya, N.D.; Larchet, C.; Dammak, L.; Nikonenko, V.V. Geometrical heterogeneity of homogeneous ion-exchange Neosepta membranes. *J. Membr. Sci.* **2018**, *563*, 768–776. [CrossRef]
74. Gnusin, N.P.; Berezina, N.P.; Kononenko, N.A.; Dyomina, O.A. Transport structural parameters to characterize ion exchange membranes. *J. Membr. Sci.* **2004**, *243*, 301. [CrossRef]
75. Tuan, L.X.; Mertens, D.; Buess-Herman, C. The two-phase model of structure microheterogeneity revisited by the study of the CMS cation exchange membrane. *Desalination* **2009**, *240*, 351–357. [CrossRef]
76. Porozhnyy, M.V.; Sarapulova, V.V.; Pismenskaya, N.D.; Huguet, P.; Deabate, S.; Nikonenko, V.V. Mathematical modeling of concentration dependences of electric conductivity and diffusion permeability of anion-exchange membranes soaked in wine. *Petrol. Chem.* **2017**, *57*, 511–517. [CrossRef]
77. Rybalkina, O.A.; Tsygurina, K.A.; Melnikova, E.D.; Pourcelly, G.; Nikonenko, V.V.; Pismenskaya, N.D. Catalytic effect of ammonia-containing species on water splitting during electrodialysis with ion-exchange membranes. *Electrochim. Acta* **2019**, *299*, 946–962. [CrossRef]
78. Rybalkina, O.; Tsygurina, K.; Melnikova, E.; Mareev, S.; Moroz, I.; Nikonenko, V.; Pismenskaya, N. Partial fluxes of phosphoric acid anions through anion-exchange membranes in the course of NaH_2PO_4 solution electrodialysis. *Int. J. Mol. Sci.* **2019**, *20*, 3593. [CrossRef]

79. Talebi, S.; Chen, G.Q.; Freeman, B.; Suarez, F.; Freckleton, A.; Bathurst, K.; Kentish, S.E. Fouling and in-situ cleaning of ion-exchange membranes during the electrodialysis of fresh acid and sweet whey. *J. Food Eng.* **2019**, *246*, 192–199. [CrossRef]
80. Pintossi, D.; Saakes, M.; Borneman, Z.; Nijmeijer, K. Electrochemical impedance spectroscopy of a reverse electrodialysis stack: A new approach to monitoring fouling and cleaning. *J. Power Sources* **2019**, *444*, 227302. [CrossRef]
81. Zhang, L.; Jia, H.; Wang, J.; Wen, H.; Li, J. Characterization of fouling and concentration polarization in ion exchange membrane by in-situ electrochemical impedance spectroscopy. *J. Membr. Sci.* **2020**, *594*, 117443. [CrossRef]
82. Galama, A.H.; Vermaas, D.A.; Veerman, J.; Saakes, M.; Rijnaarts, H.H.M.; Post, J.W.; Nijmeijer, K. Membrane resistance: The effect of salinity gradients over a cation exchange membrane. *J. Membr. Sci.* **2014**, *467*, 279–291. [CrossRef]
83. Barros, K.S.; Martí-Calatayud, M.C.; Scarazzato, T.; Bernardes, A.M.; Espinosa, D.C.R.; Pérez-Herranz, V. Investigation of ion-exchange membranes by means of chronopotentiometry: A comprehensive review on this highly informative and multipurpose technique. *Adv. Colloid Interfac.* **2021**, *293*, 102439. [CrossRef] [PubMed]
84. Teorell, T. Studies on the "diffusion effect" upon ionic distribution. Some theoretical considerations. *Proc. Natl. Acad. Sci. USA* **1935**, *21*, 152–161. [CrossRef] [PubMed]
85. Meyer, K.H.; Sievers, J.-F. La perméabilité des membranes I. Théorie de la perméabilité ionique. *Helv. Chim. Acta* **1936**, *19*, 649–664. [CrossRef]
86. Gierke, T.; Munn, G.; Wilson, F. Morphology in Nafion perfluorinated membrane products, as determined by wide- and small-angle X-ray studies. *J. Polymer Sci. Part. B.* **1981**, *19*, 1687–1704. [CrossRef]
87. Kreuer, K.D.J. On the development of proton conducting polymer membranes for hydrogen and methanol fuel cells. *J. Membr. Sci.* **2001**, *185*, 29–39. [CrossRef]
88. Geise, G.M.; Paul, D.R.; Freeman, B.D. Fundamental water and salt transport properties ofpolymeric materials. *Prog. Polym. Sci.* **2014**, *39*, 1–42. [CrossRef]
89. Kamcev, J.; Paul, D.R.; Freeman, B.D. Equilibrium ion partitioning between aqueous salt solutions and inhomogeneous ion exchange membranes. *Desalination* **2018**, *446*, 31–41. [CrossRef]
90. Ji, Y.; Luo, H.; Geise, G.M. Specific co-ion sorption and diffusion properties influence membrane permselectivity. *J. Membr. Sci.* **2018**, *563*, 492–504. [CrossRef]
91. Fridman-Bishop, N.; Freger, V. What makes aromatic polyamide membranes superior: New insights into ion transport and membrane structure. *J. Membr. Sci.* **2017**, *540*, 120–128. [CrossRef]
92. Filippov, A.N.; Kononenko, N.A.; Demina, O.A. Diffusion of electrolytes of different natures through the cation-exchange membrane. *Colloid J.* **2017**, *79*, 556. [CrossRef]
93. Gnusin, N.P.; Zabolotskii, V.I.; Nikonenko, V.V.; Meshechkov, A.I. Development of the generalized conductance principle to the description of transfer phenomena in disperse systems under the acting of different forces. *Rus. J. Phys. Chem.* **1980**, *54*, 1518–1522.
94. Golubenko, D.V.; Safronova, E.Y.; Ilyin, A.B.; Shevlyakov, N.V.; Tverskoi, V.A.; Pourcelly, G.; Yaroslavtsev, A.B. Water state and ionic conductivity of grafted ion exchange membranes based on polyethylene and sulfonated polystyrene. *Mendeleev Commun.* **2017**, *27*, 380–381. [CrossRef]
95. Kamcev, J.; Paul, D.R.; Freeman, B.D. Effect of fixed charge group concentration on equilibrium ion sorption in ion exchange membranes. *J. Mater. Chem. A* **2017**, *5*, 4638–4650. [CrossRef]
96. Kamcev, J.; Sujanani, R.; Jang, E.-S.; Yan, N.; Moe, N.; Paul, D.R.; Freeman, B.D. Salt concentration dependence of ionic conductivity in ion exchange membranes. *J. Membr. Sci.* **2018**, *547*, 123. [CrossRef]
97. Khoiruddin, K.; Ariono, D.; Subagjo, S.; Wenten, I.G. Structure and transport properties of polyvinyl chloride-based heterogeneous cation-exchange membrane modified by additive blending and sulfonation. *J. Electroanal. Chem.* **2020**, *873*, 114304. [CrossRef]
98. Niftaliev, S.I.; Kozaderova, O.A.; Kim, K.B. Electroconductance of heterogeneous ion-exchange membranes in aqueous salt solutions. *J. Electroanal. Chem.* **2017**, *794*, 58–63. [CrossRef]
99. Vasil'eva, V.I.; Akberova, E.M.; Kostylev, D.V.; Tzkhai, A.A. Diagnostics of the structural and transport properties of an anion-exchange membrane MA-40 after use in electrodialysis of mineralized natural waters. *Membr. Membr. Technol.* **2019**, *1*, 153–167. [CrossRef]
100. Choy, T.C. *Effective Medium Theory: Principles and Applications*; Oxford Science Publication: Oxford, UK, 1999; p. 200. [CrossRef]
101. Yaroslavtsev, A.B.; Nikonenko, V.V. Ion-exchange membrane materials: Properties, modification, and practical application. *Nanotechnol. Russ.* **2009**, *4*, 137–159. [CrossRef]
102. Larchet, C.; Nouri, S.; Auclair, B.; Dammak, L.; Nikonenko, V. Application of chronopotentiometry to determine the thickness of diffusion layer adjacent to an ion-exchange membrane under natural convection. *Adv. Colloid Interface* **2008**, *139*, 45–61. [CrossRef]
103. Porozhnyy, M.; Huguet, P.; Cretin, M.; Safronova, E.; Nikonenko, V. Mathematical modeling of transport properties of proton-exchange membranes containing immobilized nanoparticles. *Int. J. Hydrog. Energy.* **2016**, *41*, 15605–15614. [CrossRef]
104. Bukhovets, A.; Eliseeva, T.; Oren, Y. Fouling of anion-exchange membranes in electrodialysis of aromatic amino acid solution. *J. Membr. Sci.* **2010**, *364*, 339–343. [CrossRef]
105. Nevakshenova, E.E.; Sarapulova, V.V.; Nikonenko, V.V.; Pismenskaya, N.D. Application of sodium chloride solutions to regeneration of anion-exchange membranes used for improving grape juices and wines. *Membr. Membr. Technol.* **2019**, *1*, 14–22. [CrossRef]

106. Guo, H.; You, F.; Yu, S.; Li, L.; Zhao, D. Mechanisms of chemical cleaning of ion exchange membranes: A case study of plant-scale electrodialysis for oily wastewater treatment. *J. Membr. Sci.* **2015**, *496*, 310–317. [CrossRef]
107. Labbé, D.; Bazinet, L. Effect of membrane type on cation migration during green tea electromigration and equivalent mass transported calculation. *J. Membr. Sci.* **2006**, *275*, 220–228. [CrossRef]
108. Labbé, D.; Araya-Farias, M.; Tremblay, A.; Bazinet, L. Electromigration feasibility of green tea catechins. *J. Membr. Sci.* **2005**, *254*, 101–109. [CrossRef]
109. Jackson, R.S. *Wine Science: Principles and Applications*; Academic Press: Cambridge, MA, USA, 2014; p. 751. ISBN 9780123736468.
110. Ribéreau-Gayon, P.; Glories, Y.; Maujean, A.; Dubourdieu, D. *Handbook of Enology: The Chemistry of Wine, Stabilization and Treatments*; John Wiley & Sons Ltd: Chichester, UK, 2006; p. 451. ISBN 0-470-01037-1.
111. Lee, H.-J.; Hong, M.-K.; Han, S.-D.; Shim, J.; Moon, S.-H. Analysis of fouling potential in the electrodialysis process in the presence of an anionic surfactant foulant. *J. Membr. Sci.* **2008**, *325*, 719–726. [CrossRef]
112. De Jaegher, B.; De Schepper, W.; Verliefde, A.; Nopens, I. Enhancing mechanistic models with neural differential equations to predict electrodialysis fouling. *Sep. Purif. Technol.* **2021**, *259*, 118028. [CrossRef]
113. Berkessa, Y.W.; Lang, Q.; Yan, B.; Kuang, S.; Mao, D.; Shu, L.; Zhang, Y. Anion exchange membrane organic fouling and mitigation in salt valorization process from high salinity textile wastewater by bipolar membrane electrodialysis. *Desalination* **2019**, *465*, 94–103. [CrossRef]
114. Bai, L.; Wang, X.; Nie, Y.; Dong, H.; Zhang, X.; Zhang, S. Study on the recovery of ionic liquids from dilute effluent by electrodialysis method and the fouling of cation-exchange membrane. *Sci. China Chem.* **2013**, *56*, 1811–1816. [CrossRef]
115. Martí-Calatayud, M.C.; Buzzi, D.C.; García-Gabaldón, M.; Ortega, E.; Bernardes, A.M.; Tenório, J.A.S.; Pérez-Herranz, V. Sulfuric acid recovery from acid mine drainage by means of electrodialysis. *Desalination* **2014**, *343*, 120–127. [CrossRef]
116. Strathmann, H. Electrodialysis, a mature technology with a multitude of new applications. *Desalination* **2010**, *264*, 268–288. [CrossRef]
117. Hwang, S.-T.; Kammermeyer, K. *Membranes in Separation*; Wiley: New York, NY, USA, 1975; p. 559, ISBN 047193268X.
118. Zabolotskii, V.I.; El'nikova, L.E.; Shel'deshov, N.V.; Alekseev, A.V. Precision method for measuring ionic transport numbers in ion-exchange membranes. *Sov. Electrochem.* **1988**, *23*, 1516–1519.
119. Nikonenko, V.V.; Mareev, S.A.; Pis'menskaya, N.D.; Uzdenova, A.M.; Kovalenko, A.V.; Urtenov, M.K.; Pourcelly, G. Effect of electroconvection and its use in intensifying the mass transfer in electrodialysis. *Russ. J. Electrochem.* **2017**, *53*, 1122–1144. [CrossRef]
120. Mani, A.; Wang, K.M. electroconvection near electrochemical interfaces: Experiments, modeling, and computation. *Annu. Rev. Fluid Mech.* **2020**, *52*, 509–529. [CrossRef]
121. Kovalenko, A.V.; Wessling, M.; Nikonenko, V.V.; Mareev, S.A.; Moroz, I.A.; Evdochenko, E.; Urtenov, M.K. Space-Charge breakdown phenomenon and spatio-temporal ion concentration and fluid flow patterns in overlimiting current electrodialysis. *J. Membr. Sci.* **2021**, *636*, 119583. [CrossRef]
122. Tanaka, Y. Water dissociation reaction generated in an ion exchange membrane. *J. Membr. Sci.* **2010**, *350*, 347–360. [CrossRef]
123. Andreeva, M.A.; Gil, V.V.; Pismenskaya, N.D.; Dammak, L.; Kononenko, N.A.; Larchet, C.; Grande, D.; Nikonenko, V.V. Mitigation of membrane scaling in electrodialysis by electroconvection enhancement, pH adjustment and pulsed electric field application. *J. Membr. Sci.* **2018**, *549*, 129–140. [CrossRef]
124. Mikhaylin, S.; Nikonenko, V.; Pismenskaya, N.; Pourcelly, G.; Choi, S.; Kwon, H.J.; Han, J.; Bazinet, L. How physico-chemical and surface properties of cation-exchange membrane affect membrane scaling and electroconvective vortices: Influence on performance of electrodialysis with pulsed electric field. *Desalination* **2016**, *393*, 102–114. [CrossRef]
125. Pismenskaya, N.D.; Pokhidnia, E.V.; Pourcelly, G.; Nikonenko, V.V. Can the electrochemical performance of heterogeneous ion-exchange membranes be better than that of homogeneous membranes? *J. Membr. Sci.* **2018**, *566*, 54–68. [CrossRef]
126. Zabolotskii, V.I.; Shel'deshov, N.V.; Gnusin, N.P. Dissociation of water molecules in systems with ion-exchange membranes. *Rus. Chem. Rev.* **1988**, *57*, 801–808. [CrossRef]
127. Mareev, S.A.; Evdochenko, E.; Wessling, M.; Kozaderova, O.A.; Niftaliev, S.I.; Pismenskaya, N.D.; Nikonenko, V.V. A comprehensive mathematical model of water splitting in bipolar membranes: Impact of the spatial distribution of fixed charges and catalyst at bipolar junction. *J. Membr. Sci.* **2020**, *603*, 118010. [CrossRef]
128. Nikonenko, V.; Urtenov, M.; Mareev, S.; Pourcelly, G. Mathematical modeling of the effect of water splitting on ion transfer in the depleted diffusion layer near an ion-exchange membrane. *Membranes* **2020**, *10*, 22. [CrossRef] [PubMed]
129. Mafé, S.; Ramírez, P.; Alcaraz, A. Electric field-assisted proton transfer and water dissociation at the junction of a fixed-charge bipolar membrane. *Chem. Phys. Lett.* **1998**, *294*, 406–412. [CrossRef]
130. Bobreshova, O.; Novikova, L.; Kulintsov, P.; Balavadze, E. Amino acids and water electrotransport through cation-exchange membranes. *Desalination* **2002**, *149*, 363–368. [CrossRef]
131. Aristov, I.V.; Bobreshova, O.V.; Kulintsov, P.I.; Zagorodnykh, L.A. Transfer of amino acids through a membrane/solution interface in the presence of heterogeneous chemical protonation reaction. *Rus. J. Electrochem.* **2001**, *37*, 218–221. [CrossRef]
132. Martí-Calatayud, M.C.; Evdochenko, E.; Bär, J.; Garcia-Gabaldon, M.; Wessling, M.; Perez-Herranz, V. Tracking homogeneous reactions during electrodialysis of organic acids via EIS. *J. Membr. Sci.* **2019**, *595*, 117592. [CrossRef]

133. Pismenskaya, N.D.; Rybalkina, O.A.; Kozmai, A.E.; Tsygurina, K.A.; Melnikova, E.D.; Nikonenko, V.V. Generation of H^+ and OH^- ions in anion-exchange membrane/ampholyte-containing solution systems: A study using electrochemical impedance spectroscopy. *J. Membr. Sci.* **2020**, *601*, 117920. [CrossRef]
134. Tanaka, Y. *Ion Exchange Membranes Fundamentals and Applications*, 2nd ed.; Elsevier Science: Amsterdam, The Netherlands, 2015; p. 522. ISBN 9780444633194.
135. Mishchuk, N.A. Polarization of systems with complex geometry. *Curr. Opin. Colloid Interface Sci.* **2013**, *18*, 137–148. [CrossRef]
136. Rubinstein, I. Electroconvection at an electrically inhomogeneous permselective interface. *Phys. Fluids A* **1991**, *3*, 2301. [CrossRef]
137. Rubinstein, I.; Zaltzman, B. Electro-osmotically induced convection at a permselective membrane. *Phys. Rev. E* **2000**, *62*, 2238. [CrossRef]
138. Nebavskaya, K.A.; Sarapulova, V.V.; Sabbatovskiy, K.G.; Sobolev, V.D.; Pismenskaya, N.D.; Sistat, P.; Cretin, M.; Nikonenko, V.V. Impact of ion exchange membrane surface charge and hydrophobicity on electroconvection at underlimiting and overlimiting currents. *J. Membr. Sci.* **2017**, *523*, 36–44. [CrossRef]
139. Davidson, S.M.; Wessling, M.; Mani, A. On the dynamical regimes of pattern-accelerated electroconvection. *Sci. Rep.* **2016**, *6*, 22505. [CrossRef]
140. Zyryanova, S.; Mareev, S.; Gil, V.; Korzhova, E.; Pismenskaya, N.; Sarapulova, V.; Rybalkina, O.; Boyko, E.; Larchet, C.; Dammak, L.; et al. How electrical heterogeneity parameters of ion-exchange membrane surface affect the mass transfer and water splitting rate in electrodialysis. *Int. J. Mol. Sci.* **2020**, *21*, 973. [CrossRef] [PubMed]
141. Zabolotsky, V.I.; Novak, L.; Kovalenko, A.V.; Nikonenko, V.V.; Urtenov, M.H.; Lebedev, K.A.; But, A.Y. Electroconvection in systems with heterogeneous ion-exchange membranes. *Pet. Chem.* **2017**, *57*, 779–789. [CrossRef]
142. Balster, J.; Yildirim, M.H.; Stamatialis, D.F.; Ibanez, R.; Lammertink, R.G.H.; Jordan, V.; Wessling, M. Morpholgy and microtopology of cation-exchange polymers and the origin of the overlimiting current. *J. Phys. Chem. B* **2007**, *111*, 2152–2165. [CrossRef] [PubMed]
143. Lindstrand, V.; Sundström, G.; Jönsson, A.S. Fouling of electrodialysis membranes by organic substances. *Desalination* **2000**, *128*, 91–102. [CrossRef]
144. Watkins, E.J.; Pfromm, P.H. Capacitance spectroscopy to characterize organic fouling of electrodialysis membranes. *J. Membr. Sci.* **1999**, *162*, 213–218. [CrossRef]
145. Lindstrand, V.; Jönsson, A.-S.; Sundström, G. Organic fouling of electrodialysis membranes with and without applied voltage. *Desalination* **2000**, *130*, 73–84. [CrossRef]
146. Park, J.S.; Choi, J.H.; Yeon, K.H.; Moon, S.H. An approach to fouling characterization of an ion-exchange membrane using current-voltage relation and electrical impedance spectroscopy. *J. Colloid Interf. Sci.* **2006**, *294*, 129–138. [CrossRef]
147. Belloň, T.; Polezhaev, P.; Vobecká, L.; Slouka, Z. Fouling of a heterogeneous anion-exchange membrane and single anion-exchange resin particle by ssDNA manifests differently. *J. Membr. Sci.* **2019**, *572*, 619–631. [CrossRef]
148. Gally, C.R.; Benvenuti, T.; da Trindade, C.D.M.; Rodrigues, M.A.S.; Zoppas-Ferreira, J.; Pérez-Herranz, V.; Bernardes, A.M. Electrodialysis for the tertiary treatment of municipal wastewater: Efficiency of ion removal and ageing of ion exchange membranes. *J. Environ. Chem. Eng.* **2018**, *6*, 5855–5869. [CrossRef]
149. Rudolph, G.; Virtanen, T.; Ferrando, M.; Güell, C.; Lipnizki, F.; Kallioinen, M. A review of in situ real-time monitoring techniques for membrane fouling in the biotechnology, biorefinery and food sectors. *J. Membr. Sci.* **2019**, *588*, 117221. [CrossRef]
150. Lee, H.-J.; Hong, M.-K.; Han, S.-D.; Cho, S.-H.; Moon, S.-H. Fouling of an anion exchange membrane in the electrodialysis desalination process in the presence of organic foulants. *Desalination* **2009**, *238*, 60–69. [CrossRef]
151. Choi, J.-H.; Lee, H.-J.; Moon, S.-H. Effects of electrolytes on the transport phenomena in a cation-exchange membrane. *J. Colloid Interf. Sci.* **2001**, *238*, 188–195. [CrossRef]
152. Zaltzman, B.; Rubinstein, I. Electro-osmotic slip and electroconvective instability. *J. Fluid Mech.* **2007**, *579*, 173–226. [CrossRef]
153. Andersen, M.; Wang, K.; Schiffbauer, J.; Mani, A. Confinement effects on electroconvective instability. *Electrophoresis* **2016**, *38*, 702–711. [CrossRef] [PubMed]
154. Belashova, E.D.; Kharchenko, O.A.; Sarapulova, V.V.; Nikonenko, V.V.; Pismenskaya, N.D. Effect of protolysis reactions on the shape of chronopotentiograms of a homogeneous anion-exchange membrane in NaH_2PO_4 solution. *Petrol. Chem.* **2017**, *57*, 1207–1218. [CrossRef]
155. Choi, J.-H.; Moon, S.-H. Pore size characterization of cation-exchange membranes by chronopotentiometry using homologous a mine ions. *J. Membr. Sci.* **2001**, *191*, 225–236. [CrossRef]
156. Choi, Y.-J.; Kang, M.-S.; Kim, S.-H.; Cho, J.; Moon, S.-H. Characterization of LDPE/polystyrene cation exchange membranes prepared by monomer sorption and UV radiation polymerization. *J. Membr. Sci.* **2003**, *223*, 201–215. [CrossRef]
157. Audinos, R.; Jacquet-Battault, F.; Moutounet, M. Study of the variation in time of the transport properties of electrodialysis permselective membranes fouled by polyphenols of grape must. *J. Chim. Phys. Chim. Biol.* **1985**, *82*, 969. [CrossRef]
158. Maruyama, T.; Katoh, S.; Nakajima, M.; Nabetani, H. Mechanism of bovine serum albumin aggregation during ultrafiltration. *Biotechnol. Bioeng.* **2001**, *75*, 233–238. [CrossRef]
159. Jia, W.; Wu, P. Fast proton conduction in denatured bovine serum albumin coated nafion membranes. *ACS Appl. Mater. Interface* **2018**, *10*, 39768–39776. [CrossRef]
160. Kammerer, J.; Boschet, J.; Kammerer, D.R.; Carle, R. Enrichment and fractionation of major apple flavonoids, phenolic acids and dihydrochalcones using anion exchange resins. *LWT Food Sci. Technol.* **2011**, *44*, 1079–1087. [CrossRef]

161. Rubinstein, I.; Zaltzman, B. Equilibrium electroconvective instability. *Phys. Rev. Lett.* **2015**, *114*, 114502. [CrossRef]
162. Korzhova, E.; Pismenskaya, N.; Lopatin, D.; Baranov, O.; Dammak, L.; Nikonenko, V. Effect of surface hydrophobization on chronopotentiometric behavior of an AMX anion-exchange membrane at overlimiting currents. *J. Membr. Sci.* **2016**, *500*, 161–170. [CrossRef]
163. Newman, J.S. *Electrochemical Systems*; Prentice-Hall: Hoboken, NJ, USA, 1972; p. 432, ISBN-10: 027596356X.
164. Belashova, E.; Mikhaylin, S.; Pismenskaya, N.; Nikonenko, V.; Bazinet, L. Impact of cation-exchange membrane scaling nature on the electrochemical characteristics of membrane system. *Sep. Purif. Technol.* **2017**, *189*, 441–448. [CrossRef]
165. Freijanes, Y.; Barragán, V.M.; Muñoz, S. Chronopotentiometric study of a Nafion membrane in presence of glucose. *J. Membr. Sci.* **2016**, *510*, 79–90. [CrossRef]
166. Shahi, V.K.; Thampy, S.K.; Rangarajan, R. Chronopotentiometric studies on dialytic properties of glycine across ion-exchange membranes. *J. Membr. Sci.* **2002**, *203*, 43–51. [CrossRef]
167. Kang, M.-S.; Cho, S.-H.; Kim, S.-H.; Choi, Y.-J.; Moon, S.-H. Electrodialytic separation characteristics of large molecular organic acid in highly water-swollen cation-exchange membranes. *J. Membr. Sci.* **2003**, *222*, 149–161. [CrossRef]
168. Nagarale, R.K.; Shahi, V.K.; Thampy, S.K.; Rangarajan, R. Studies on electrochemical characterization of polycarbonate and polysulfone based heterogeneous cation-exchange membranes. *React. Funct. Polym.* **2004**, *61*, 131–138. [CrossRef]
169. Scarazzato, T.; Barros, K.S.; Benvenuti, T.; Siqueira Rodrigues, M.A.; Romano Espinosa, D.C.; Moura Bernardes, A.; Amado, F.D.R.; Pérez-Herranz, V. Achievements in electrodialysis process for wastewater and water treatment. *Curr. Trends Future Dev. (Bio-) Membr.* **2020**, *5*, 127–160. [CrossRef]
170. Lerche, D.; Wolf, H. Quantitative characterisation of current-induced diffusion layers at cation-exchange membranes. I. Investigations of temporal and local behaviour of concentration profile at constant current density. *Bioelectrochem. Bioenerg.* **1975**, *2*, 293–303. [CrossRef]
171. Mareev, S.A.; Butylskii, D.Y.; Pismenskaya, N.D.; Nikonenko, V.V. Chronopotentiometry of ion-exchange membranes in the overlimiting current range. Transition time for a finite-length diffusion layer: Modeling and experiment. *J. Membr. Sci.* **2016**, *500*, 171–179. [CrossRef]
172. Titorova, V.D.; Mareev, S.A.; Gorobchenko, A.D.; Gil, V.V.; Nikonenko, V.V.; Sabbatovskii, K.G.; Pismenskaya, N.D. Effect of current-induced co-ion transfer on the shape of chronopotentiograms of cation-exchange membranes. *J. Membr. Sci.* **2021**, *624*, 119036. [CrossRef]
173. Martí-Calatayud, M.C.; Buzzi, D.C.; Garcia-Gabaldon, M.; Bernardes, A.M.; Tenorio, J.A.S.; Perez-Herranz, V. Ion transport through homogeneous and heterogeneous ion-exchange membranes in single salt and multicomponent electrolyte solutions. *J. Membr. Sci.* **2014**, *466*, 45–57. [CrossRef]
174. Barros, K.S.; Martí-Calatayud, M.C.; Perez-Herranz, V.; Espinosa, D.C.R. A three-stage chemical cleaning of ion-exchange membranes used in the treatment by electrodialysis of wastewaters generated in brass electroplating industries. *Desalination* **2020**, *492*, 114628. [CrossRef]
175. Martí-Calatayud, M.C.; Garcia-Gabaldon, M.; Perez-Herranz, V. Mass transfer phenomena during electrodialysis of multivalent ions: Chemical equilibria and overlimiting currents. *Appl. Sci.* **2018**, *8*, 1566. [CrossRef]
176. Martí-Calatayud, M.C.; García-Gabaldón, M.; Pérez-Herranz, V. Effect of the equilibria of multivalent metal sulfates on the transport through cation-exchange membranes at different current regimes. *J. Membr. Sci.* **2013**, *443*, 181–192. [CrossRef]
177. Pismenskaya, N.; Igritskaya, K.; Belova, E.; Nikonenko, V.; Pourcelly, G. Transport properties of ion-exchange membrane systems in LysHCl solutions. *Desalination* **2006**, *200*, 149–151. [CrossRef]
178. Gally, C.; García-Gabaldón, M.; Ortega, E.M.; Bernardes, A.M.; Pérez-Herranz, V. Chronopotentiometric study of the transport of phosphoric acid anions through an anion-exchange membrane under different pH values. *Sep. Purif. Technol.* **2020**, *238*, 116421. [CrossRef]
179. Barros, K.S.; Martí-Calatayud, M.C.; Ortega, E.M.; Perez-Herranz, V.; Espinosa, D.C.R. Chronopotentiometric study on the simultaneous transport of EDTA ionic species and hydroxyl ions through an anion-exchange membrane for electrodialysis applications. *J. Electroanal. Chem.* **2020**, *879*, 114782. [CrossRef]
180. Zerdoumi, R.; Chatta, H.; Mellahi, D.; Oulmi, K.; Ferhat, M.; Ourari, A. Chronopotentiometric evaluation of enhanced counter-ion transport through anion exchange membranes in electromembrane processes. *Desalin. Water Treat.* **2017**, *78*, 34–40. [CrossRef]
181. Park, J.-S.; Choi, J.-H.; Woo, J.-J.; Moon, S.-H. An electrical impedance spectroscopic (EIS) study on transport characteristics of ion-exchange membrane systems. *J. Colloid Interf. Sci.* **2006**, *300*, 655–662. [CrossRef] [PubMed]
182. Zhang, W.; Ma, J.; Wang, P.; Wang, Z.; Shi, F.; Liu, H. Investigations on the interfacial capacitance and the diffusion boundary layer thickness of ion exchange membrane using electrochemical impedance spectroscopy. *J. Membr. Sci.* **2016**, *502*, 37–47. [CrossRef]
183. Golubenko, D.; Karavanova, Y.; Yaroslavtsev, A. Effects of the surface layer structure of the heterogeneous ion-exchange membranes on their impedance. *J. Electroanal. Chem.* **2016**, *777*, 1–7. [CrossRef]
184. Merino-Garcia, I.; Kotoka, F.; Portugal, C.A.M.; Crespo, J.G.; Velizarov, S. Characterization of poly(Acrylic) acid-modified heterogenous anion exchange membranes with improved monovalent permselectivity for RED. *Membranes* **2020**, *10*, 134. [CrossRef]
185. Li, Y.; Shi, S.; Cao, H.; Xu, B.; Zhao, Z.; Cao, R.; Chang, J.; Duan, F.; Wen, H. Anion exchange nanocomposite membranes modified with graphene oxide and polydopamine: Interfacial structure and antifouling applications. *ACS Appl. Nano Mater.* **2020**, *3*, 588–596. [CrossRef]

186. Długołęcki, P.; Ogonowski, P.; Metz, S.J.; Saakes, M.; Nijmeijer, K.; Wessling, M. On the resistances of membrane, diffusion boundary layer and double layer in ion-exchange membrane transport. *J. Membr. Sci.* **2010**, *349*, 369–379. [CrossRef]
187. Sistat, P.; Kozmai, A.; Pismenskaya, N.; Larchet, C.; Pourcelly, G.; Nikonenko, V. Low-frequency impedance of an ion-exchange membrane system. *Electrochim. Acta* **2008**, *53*, 6380–6390. [CrossRef]
188. Kozmai, A.E.; Nikonenko, V.V.; Pismenskaya, N.D.; Mareev, S.A.; Belova, E.I.; Sistat, P. Use of electrochemical impedance spectroscopy for determining the diffusion layer thickness at the surface of ion-exchange membranes. *Petrol. Chem.* **2012**, *52*, 614–624. [CrossRef]
189. Hurwitz, H.D.; Dibiani, R. Experimental and theoretical investigations of steady and transient states in systems of ion exchange bipolar membranes. *J. Membr. Sci.* **2004**, *228*, 17–43. [CrossRef]
190. Kniaginicheva, E.; Pismenskaya, N.; Melnikov, S.; Belashova, E.; Sistat, P.; Cretin, M.; Nikonenko, V. Water splitting at an anion-exchange membrane as studied by impedance spectroscopy. *J. Membr. Sci.* **2015**, *496*, 78–83. [CrossRef]
191. Slouka, Z.; Senapati, S.; Yan, Y.; Chang, H.-C. Charge inversion, water splitting, and vortex suppression due to DNA sorption on ion-selective membranes and their ion-current signatures. *Langmuir* **2013**, *29*, 8275–8283. [CrossRef]
192. Belloň, T.; Slouka, Z. Overlimiting behavior of surface-modified heterogeneous anion-exchange membranes. *J. Membr. Sci.* **2020**, *610*, 118291. [CrossRef]
193. Merino-Garcia, I.; Velizarov, S. New insights into the definition of membrane cleaning strategies to diminish the fouling impact in ion exchange membrane separation processes. *Sep. Purif. Technol.* **2021**, *277*, 119445. [CrossRef]
194. Corbaton-Baguena, M.J.; Alvarez-Blanco, S.; Vincent-Vela, M.C. Cleaning of ultrafiltration membranes fouled with BSA by means of saline solutions. *Sep. Purif. Technol.* **2014**, *125*, 1–10. [CrossRef]
195. Andersen, O.M.; Markham, K.R. *Flavonoids: Chemistry, Biochemistry and Applications*; CRC Press: Boca Raton, FL, USA, 2005; p. 1256. ISBN 9780849320217.
196. Petrus, H.B.; Li, H.; Chen, V.; Norazman, N. Enzymatic cleaning of ultrafiltration membranes fouled by protein mixture solutions. *J. Membr. Sci.* **2008**, *325*, 783–792. [CrossRef]
197. Bilad, M.R.; Baten, M.; Pollet, A.; Courtin, C.; Wouters, J.; Verbiest, T.; Vankelecom, I.F.J. A novel in-situ enzymatic cleaning method for reducing membrane fouling in membrane bioreactors (MBRs). *Indones. J. Sci. Technol.* **2016**, *1*, 1–22. [CrossRef]
198. Bdiri, M.; Bensghaier, A.; Chaabane, L.; Kozmai, A.; Baklouti, L.; Larchet, C. Preliminary study on enzymatic-based cleaning of cation-exchange membranes used in electrodialysis system in red wine production. *Membranes* **2019**, *9*, 114. [CrossRef]
199. Nichka, V.S.; Nikonenko, V.V.; Bazinet, L. Fouling mitigation by optimizing flow rate and pulsed electric field during bipolar membrane electroacidification of caseinate solution. *Membranes* **2021**, *11*, 534. [CrossRef]
200. Merkel, A.; Ashrafi, A.M. An investigation on the application of pulsed electrodialysis reversal in whey desalination. *Int. J. Mol. Sci.* **2019**, *20*, 1918. [CrossRef] [PubMed]
201. Pelletier, S.; Serre, É.; Mikhaylin, S.; Bazinet, L. Optimization of cranberry juice deacidification by electrodialysis with bipolar membrane: Impact of pulsed electric field conditions. *Sep. Purif. Technol.* **2017**, *186*, 106–116. [CrossRef]
202. Khoiruddin, K.; Ariono, D.; Subagjo, S.; Wenten, I.G. Improved anti-organic fouling of polyvinyl chloride-based heterogeneous anion-exchange membrane modified by hydrophilic additives. *J. Water Proc. Eng.* **2021**, *41*, 102007. [CrossRef]
203. Xie, H.; Pan, J.; Wei, B.; Feng, J.; Liao, S.; Li, X.; Yu, Y. Anti-fouling anion exchange membrane for electrodialysis fabricated by in-situ interpenetration of the ionomer to gradient cross-linked network of Ca-Na alginate. *Desalination* **2021**, *505*, 115005. [CrossRef]
204. Zhao, Z.; Li, Y.; Jin, D.; Van der Bruggen, B. Modification of an anion exchange membrane based on rapid mussel-inspired deposition for improved antifouling performance. *Colloid. Surface. A* **2021**, *615*, 126267. [CrossRef]
205. Ariono, D.; Subagjo, D.; Wenten, I.G. Surface modification of ion-exchange membranes: Methods, characteristics, and performance. *J. Appl. Polym. Sci.* **2017**, *134*, 45540. [CrossRef]
206. Jiang, S.; Sun, H.; Wang, H.; Ladewig, B.P.; Yao, Z. A comprehensive review on the synthesis and applications of ion exchange membranes. *Chemosphere* **2021**, *282*, 130817. [CrossRef] [PubMed]
207. Ran, J.; Wu, L.; He, Y.B.; Yang, Z.J.; Wang, Y.M.; Jiang, C.X.; Ge, L.; Bakangura, E.; Xu, T.W. Ion exchange membranes: New developments and applications. *J. Membr. Sci.* **2017**, *522*, 267–291. [CrossRef]
208. Kotoka, F.; Merino-Garcia, I.; Velizarov, S. Surface modifications of anion exchange membranes for an improved reverse electrodialysis process performance: A review. *Membranes* **2020**, *10*, 160. [CrossRef] [PubMed]
209. Larchet, C.; Zabolotsky, V.I.; Pismenskaya, N.D.; Nikonenko, V.V.; Tskhay, A.; Tastanov, K.; Pourcelly, G. Comparison of different ED stack conceptions when applied for drinking water production from brackish waters. *Desalination* **2008**, *222*, 489–496. [CrossRef]
210. Pawlowski, S.; Crespo, J.G.; Velizarov, S. Profiled ion exchange membranes: A comprehensible review. *Int. J. Mol. Sci.* **2019**, *20*, 165. [CrossRef]

 membranes

MDPI

Article

Electrochemical Impedance Spectroscopy of Anion-Exchange Membrane AMX-Sb Fouled by Red Wine Components

Anton Kozmai [1,*], Veronika Sarapulova [1], Mikhail Sharafan [1], Karina Melkonian [2], Tatiana Rusinova [2], Yana Kozmai [2], Natalia Pismenskaya [1], Lasaad Dammak [3] and Victor Nikonenko [1]

[1] Membrane Institute, Kuban State University, 149 Stavropolskaya Street, 350040 Krasnodar, Russia; vsarapulova@gmail.com (V.S.); shafron80@mail.ru (M.S.); n_pismen@mail.ru (N.P.); v_nikonenko@mail.ru (V.N.)

[2] Central Research Laboratory, Kuban State Medical University, 4 Sedina Street, 350063 Krasnodar, Russia; agaron@list.ru (K.M.); rusinova.tv@mail.ru (T.R.); yana.yutskevich@gmail.com (Y.K.)

[3] Institut de Chimie et des Matériaux Paris-Est (ICMPE), UMR 7182 CNRS, Université Paris-Est, 2 Rue Henri Dunant, 94320 Thiais, France; dammak@u-pec.fr

* Correspondence: kozmay@yandex.ru; Tel.: +7-952-8621139

Abstract: The broad possibilities of electrochemical impedance spectroscopy for assessing the capacitance of interphase boundaries; the resistance and thickness of the foulant layer were shown by the example of AMX-Sb membrane contacted with red wine from one side and 0.02 M sodium chloride solution from the other side. This enabled us to determine to what extent foulants affect the electrical resistance of ion-exchange membranes, the ohmic resistance and the thickness of diffusion layers, the intensity of water splitting, and the electroconvection in under- and over-limiting current modes. It was established that short-term (10 h) contact of the AMX-Sb membrane with wine reduces the water-splitting due to the screening of fixed groups on the membrane surface by wine components. On the contrary, biofouling, which develops upon a longer membrane operation, enhances water splitting, due to the formation of a bipolar structure on the AMX-Sb surface. This bipolar structure is composed of a positively charged surface of anion-exchange membrane and negatively charged outer membranes of microorganisms. Using optical microscopy and microbiological analysis, it was found that more intense biofouling is observed on the AMX-Sb surface, that has not been in contacted with wine.

Keywords: electrochemical impedance spectroscopy; anion-exchange membrane; wine; anthocyanins; fouling; biofouling

Citation: Kozmai, A.; Sarapulova, V.; Sharafan, M.; Melkonian, K.; Rusinova, T.; Kozmai, Y.; Pismenskaya, N.; Dammak, L.; Nikonenko, V. Electrochemical Impedance Spectroscopy of Anion-Exchange Membrane AMX-Sb Fouled by Red Wine Components. *Membranes* **2021**, *11*, 2. https://dx.doi.org/10.3390/membranes11010002

Received: 16 November 2020
Accepted: 21 December 2020
Published: 22 December 2020

1. Introduction

Electrodialysis (ED) with ion-exchange membranes (IEM) is increasingly used for tartrate stabilization of wine and the isolation of valuable components (for example, anthocyanins, which are antioxidants and natural dyes) [1] from pulp and other waste of wine production [2–5]. The advantages of ED in comparison with baromembrane methods are in reducing the loss of valuable wine components [6,7] and the ability to effectively desalinate wine materials and juices with simultaneous reagent-free pH correction [8,9].

Note that wine materials contain more than 600 components, including polysaccharides, amino acids, proteins and polyphenols (anthocyanins, proanthocyanins, tannins, etc.) [10]. The interactions of these substances with the materials comprising the IEM and with each other cause fouling [11–13], which reduces the life cycle of membranes and affects energy consumption and other parameters of the ED process [14]. An analysis of works devoted to this problem allows us to conclude that the main mechanisms of fouling are: van der Waals interactions of polyphenols with materials comprising the IEM; $\pi - \pi$ (stacking) interactions [15,16] between aromatic rings of polyphenols and the ion-exchange matrix of IEM (which, in most cases, consists of a copolymer of divinylbenzene and polystyrene). Electrostatic interactions [17–19] take place if the components of wine materials gain an electric charge opposite to the charge of membrane fixed groups. Indeed, in an acidic environment, anthocyanins become cations, and in an alkaline environment they are anions [20]. The colloidal state of organic molecules (formed due to the interactions between polyphenols, polysaccharides, amino acids and proteins) on the surface and in the volume of ion-exchange membranes is supported by weak van der Waals attraction forces, electrostatic repulsion forces and hydrogen bonds [11].

Nitrogen-containing fixed groups of anion-exchange membranes are a nutrient medium for microorganisms [21]. Besides this, wine materials and waste from winemaking are rich in saccharides, amino acids and other substances, the presence of which stimulates biofouling [22]. It should be noted that some microorganisms can be adsorbed at the membrane surface as early as during the first hours of membrane module operation [23]. Piping, storage tanks, pretreatment systems [24] can be a source of microbial contamination. Microorganisms such as those responsible for brewing [25] may find their habitat in these parts of apparatuses.

Electrochemical impedance spectroscopy (EIS) is increasingly being used to study IEM fouling [26–31]. The method of equivalent electrical circuits (EEC) is most often applied for the interpretation of experimental spectra [28–31]. For example, an increase in the ohmic resistance of a membrane, which is determined by processing a high-frequency EIS arc using the EEC method, is the basis for online registration of the effect of fouling on the transport characteristics of membranes in reverse and conventional ED [30,31]. If the electrical resistance (and capacitance) of foulants or modifying films used for fouling prevention and control differs markedly from similar characteristics of the pristine membrane, researchers observe the appearance of additional arc in the high-frequency domain of the impedance spectra [28–31]. In this case, processing the high-frequency domain of impedance spectra using EEC gives information on the resistance, capacitance and thickness of the fouling or modifying layers. It is shown in [32] that an increase in the membrane surface fouling degree by sodium dodecyl sulfate is accompanied by an increase in the high-frequency arc of impedance spectra, as well as a decrease (and, finally, the disappearance) of the low-frequency (diffusion) arc. The authors of [32] believe that the degradation of the low-frequency arc indicates an almost complete cessation of ion migration through the IEM. Park et al. [27] used the EEC method to analyze the manifestation of fouling in the mid-frequency and low-frequency EIS domains, where spectra were obtained under conditions of an applied direct current (as in electrodialysis). They found that the presence of bovine serum albumin (BSA) on the anion-exchange membrane (AEM) surface causes the appearance of an arc in the mid-frequency domain of the impedance spectra and an additional capacitive loop in the low-frequency domain (<100 Hz), which were absent in the case of the pristine membrane. Park et al. [27] suggested that the observed changes in the impedance spectra are caused by the formation of the bipolar AEM/BSA boundary, which contributes to enhanced water splitting. To approximate these spectra, they introduced into the equivalent circuit an additional parallel element of resistance and capacitance to take into account the appearance of a BSA layer on the AEM surface. Besides, they replaced the capacitor, indicator of Warburg impedance, on the inductor, which had to take into account the chemical reaction (water splitting) in a parallel element that described transport phenomena in diffusion layers.

In a number of works [33–35], it was theoretically and experimentally shown that the mid-frequency arc (Gerischer-type impedance) is primarily an indicator of water-splitting at the IEM/solution boundary or at the bipolar interface. The low-frequency arc (Warburg-type impedance), first of all, characterizes the ion diffusion transport in the solution layers adjacent to the IEM and can be used to evaluate the influence of various factors (for example, the development of electroconvection) on the diffusion layer thickness [36,37].

In this work, we applied both the mathematical models of EIS and EEC method to interpret the impedance spectra in the full frequency range. By the example of AMX-Sb anion-exchange membrane contacted with red wine, we will demonstrate the capabilities of the EIS for the first time, for a comprehensive study of the fouling and biofouling effect, not only on the characteristics of AEMs, but also on their behavior under conditions of ED desalination of a NaCl solution.

2. Experiment

2.1. Membranes and Solutions

The homogeneous anion-exchange membrane AMX-Sb (Astom, Tokyo, Japan) was selected as the object of study. This membrane is manufactured by paste method [38]. It contains an inert filler: granules of polyvinyl chloride, whose diameter reaches 60 nm. The matrix of the ion-exchange material of the AMX-Sb consists of a copolymer of styrene and divinylbenzene. The fixed groups are mainly quaternary ammonium bases [38].

The main characteristics of the membrane under study are presented in Table 1.

Table 1. Main physicochemical characteristics of the AMX-Sb membrane (experimental data).

Type	Homogeneous, Strong Base [a]
Thickness in 0.02 M NaCl solution, μm	160 ± 10 [b]
Conductivity in 0.02 M NaCl solution, S m^{-1}	0.28 ± 0.02 [b]
Ion-exchange capacity, meq g^{-1} (swollen membrane)	1.30 ± 0.05 [b,c]
Water content, g H_2O• (g dry membrane)$^{-1}$	0.20 ± 0.05 [b]
Membrane density, g cm^{-3}	1.10 [c]

[a] Manufacturer data [38]. [b] Our measurements. [c] [39].

In the experiments we used: distilled water (electrical conductivity was 0.5 μS cm^{-1}, pH = 5.5, 25 °C), solid NaCl of analitycal grade (JSC Vekton, St. Petersburg, Russia), red dry wine made from the varieties of the Murvedr, Syrah, Grenache grapes (pH = 3.5).

2.2. Methods

2.2.1. Membrane Fouling Procedure

Before the study, all membrane samples underwent standard salt preparation [40] and were equilibrated with 0.02 M NaCl solution. One of these samples was used for comparison. Other samples (indicated by the subscript "w") were placed in a two-compartment flow cell (Figure 1).

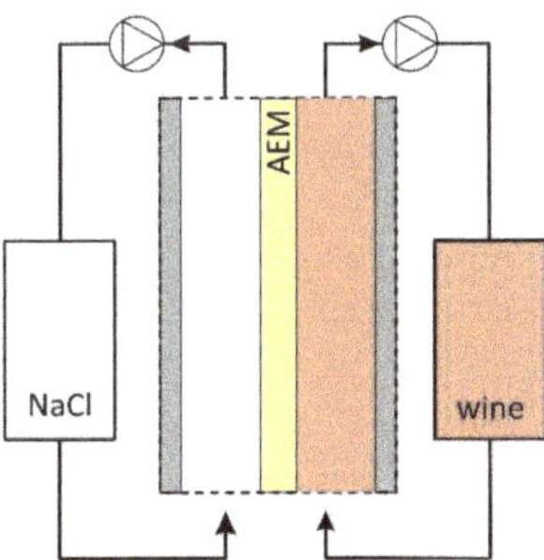

Figure 1. Scheme of the flow-cell used for AEM fouling.

The 0.02 M NaCl solution was circulated through one of the compartments and the red wine through the other. The time of the sample contact with wine in hours is indicated by a subscript. Thus, "AMX-Sb_{w10}" denotes the sample which was in contact with wine during 10 h; "AMX-Sb_{w72}" relates to the sample which was in contact with wine during 72 h. In the course of the experiment, no special measures were taken to prevent the ingress of microorganisms into the wine and NaCl solution circuits in order to simulate conditions favorable for biofouling of the studied membrane. As already mentioned in the

Introduction [11], such conditions often arise during the ED processing of food industry liquid media.

After the completion of the fouling stage, the concentrations of ethyl alcohol and glucose were determined by liquid chromatography [41] and spectrophotometry [42] methods.

2.2.2. Visualization

The visualization of surface and cross-sections of swollen AMX-Sb membranes before and after fouling was carried out using a SOPTOP CX40M optical microscope (Yuyao, China) with a set of 5×, 10×, 20×, 50× and 100× objectives and a digital eyepiece camera.

2.2.3. Microbiological Analysis

After 72 h of contact with wine, the studied sample was removed from the cell (Figure 1) and placed in a sterile Petri dish. Touch smears of the AMX-Sb surface exposed to NaCl solution and of the surface exposed to wine were made on defatted glass slides. The obtained preparations were dried in air, fixed over the flame of an alcohol lamp, and then stained by the Gram-staining method [43] in the following sequence: 2–3 drops (50–75 μL) of carbolic solution of gentian violet for 2 min; 2–3 drops of Lugol's solution for 1 min; discoloration with 96% ethyl alcohol for 30–45 s; rinsing with distilled water; 2–3 drops of an aqueous solution of fuchsin for 2 minutes; rinsing with distilled water.

Gram-positive microorganisms turned blue-violet, and gram-negative ones turned pink-red. Optical images of the stained preparations were performed using a Primo Star microscope, Carl Zeiss, Oberkochen, Germany, with a set of 10×, 100× in the presence of immersion liquid. Similar manipulations were made on the pristine AMX-Sb membrane just after salt pretreatment procedure.

2.2.4. Electrochemical Impedance Spectroscopy

The spectra of the electrochemical impedance were measured in a flow electrochemical cell using the Autolab PGStat-100 electrochemical complex. The desalination compartment was formed by the investigated anion-exchange membrane AMX-Sb (Astom, Japan) and the auxiliary cation-exchange membrane MK-40 (UCC Schekinoazot, Pervomayskiy, Russia). Installation, as well as the method of data-obtaining and processing, are described in detail in [44]. The scheme of the studied membrane system is presented in Figure 2.

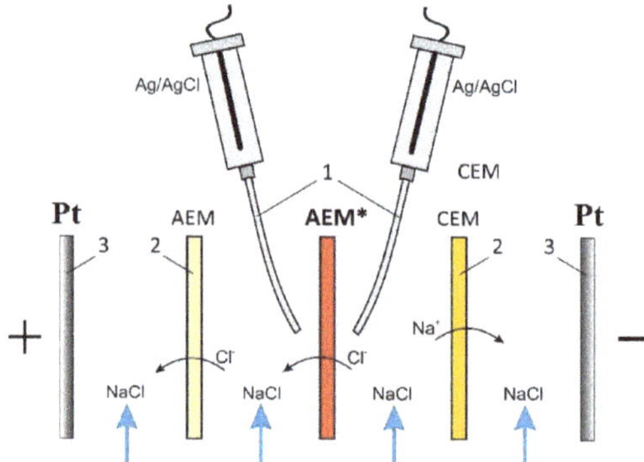

Figure 2. Scheme of the membrane system under study. In figure: AEM* is the anion-exchange membrane under study, the Luggin capillaries (1), two auxiliary membranes, an anion-exchange and a cation-exchange ones (2), platinum polarizing the working and counter electrodes (3).

The intermembrane distance, h, is 6.5 mm; the linear flow velocity of the 0.02 M NaCl solution, V, is 0.4 cm s^{-1}; the area of the polarized section is 2 × 2 cm^2. The AMX-Sb_W was oriented towards the desalination compartment with a surface that was in contact with the wine. Before the measurements, AMX-Sb_W samples were preliminarily soaked in a 0.02 M

NaCl solution for 24 h. The investigations were carried out at a temperature of 20 ± 1 °C and various values of $i/i_{\lim}^{theor}$.

The limiting current density, $i_{\lim}^{theor}$, was calculated from the Lévêque equation obtained in the framework of the convection-diffusion model [45]

$$i_{\lim}^{theor} = 1.47\frac{FDC}{h(T_{Cl} - t_{Cl})}\left(\frac{h^2V}{LD}\right)^{1/3}, \tag{1}$$

where L is the length of the desalination compartment; C is the molar concentration of NaCl at the entrance to the desalination compartment, T_{Cl} and t_{Cl} are the transport numbers of chlorine ion in the membrane and in the solution, respectively; D is the diffusion coefficient of NaCl at infinite dilution; F is the Faraday constant.

The limiting current density, calculated using Equation (1) for the membrane system under study, is equal to 2.9 mA cm^{-2}.

The time for obtaining each electrochemical impedance spectrum is about two hours: at first, the membrane is held for 20 minutes at a given direct current, then equilibrated at each given frequency (from 3×10^{-3} to 1.3×10^5 Hz), starting from the low frequencies.

Spectra are obtained passing from lower current densities, i, to higher ones. The time intervals between the measurements in the absence of direct current (i = 0) are 40 min. The impedance of a membrane and adjoining diffusion boundary layers (DBLs) was determined by subtracting spectra measured with a membrane and without it at corresponding frequencies [36].

A characteristic impedance spectrum of an IEM in a solution of strong electrolyte is shown in Figure 3. It is represented on the complex Argand plane and includes three arcs. The first arc appears at high frequencies (in the range from 10^3 to 1.3×10^5 Hz). Its shape is mainly determined [46,47] by electrical capacitances and ohmic resistances of the layers in the membrane system under study (DBL/membrane/DBL). The width of the high-frequency arc is equal to R^{Ω}, which is the ohmic resistance of the membrane and the adjacent DBLs. The maximal value of the imaginary component of impedance on this arc, $-\mathrm{Im}Z_{\max}$, and the frequency corresponding to this value, $f_{\max}^{\Omega}$, is used to estimate the effective capacitance according to equation [46,48]

$$C = \frac{1}{4\pi f_{\max}^{\Omega}(-\mathrm{Im}Z_{\max})} \tag{2}$$

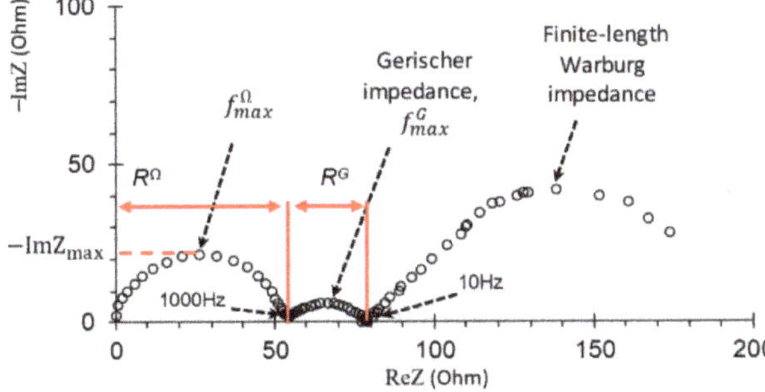

Figure 3. Typical impedance spectrum of a monopolar ion-exchange membrane and adjacent diffusion layers in the over-limiting current mode.

This capacitance includes the capacitances of the electric double layers (EDL) on the membrane/solution boundaries, as well as the capacity caused by the asymmetry of the depleted and enriched DBLs, which occurs when a direct electric current flows. The capacity of the EDL is dominant in the given (10^3–10^5 Hz) frequency range [47].

The mid-frequency (10–10^3 Hz) arc of so-called Gerischer impedance is observed when additional charge carriers appear in a system as a result of the chemical reaction (in case of the considered membrane system, the water-splitting reaction at the membrane/solution boundary) [35,46]. The width of this arc is equal to the effective resistance of the reaction layer, R^G.

The frequency corresponding to the maximal value of imaginary component of the Gerischer impedance spectrum, f_{max}^G, is used to calculate the effective water-splitting reaction constant on the membrane/solution boundary [35]

$$\chi = \frac{2\pi f_{max}^G}{\sqrt{3}} \tag{3}$$

The low-frequency (0.003–10 Hz) arc of the spectrum (finite-length Warburg impedance) characterizes the diffusion and electroconvective transport of ions in DBL adjacent to a membrane. The DBL thickness can be found from the difference in values of the real component of the low-frequency arc for the lowest and the highest frequencies [37].

3. Results and Discussion

3.1. Optical Images

The results of swollen pristine and fouled AMX-Sb membrane surface and cross-section optical microscopy are shown in Figure 4.

As can be seen from Figure 4b, after 10 h of contact with wine, the surface and cross-section of the AMX-Sb_{w10} sample acquire a pale ruby color. This color is an indicator of the penetration of anthocyanins into the membrane volume [12,13]. On the surface of AMX-Sb_{w10}, structures are visualized, the distribution of which has an "island" character. These structures are mainly composed of high-molecular-weight components of wine [13].

After 72 h of contact with wine (Figure 4c), the color of the surface and cross-section of the membrane turns red-brown, indicating that the volume of AMX-Sb_{w72} sample is enriched with tannins and/or anthocyanin-tannin adducts (for example, catechins), which are yellow and brown, respectively [49]. Colored aggregates of wine components and microorganisms almost completely cover the surface of AMX-Sb_{w72}, which was exposed to the wine compartment (Figure 4c). On the opposite surface of the membrane, which was facing the NaCl compartment, a translucent film is visualized. As will be shown in Section 3.2, this film is formed by microorganisms.

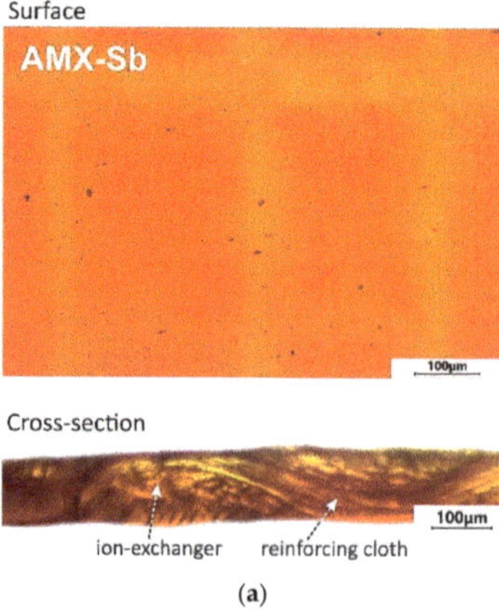

(a)

Figure 4. *Cont.*

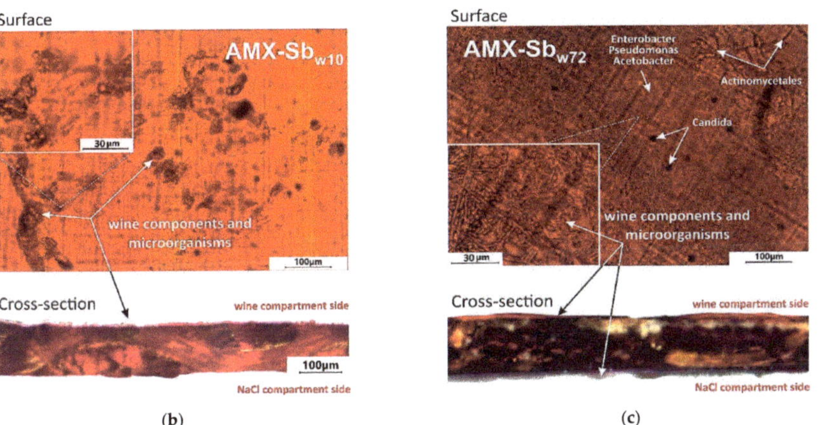

Figure 4. Optical images of the swollen AMX-Sb (**a**), AMX-Sb$_{w10}$ (**b**) and AMX-Sb$_{w72}$ (**c**) samples surface and cross-section.

3.2. Results of Microbiological Analysis

The pristine AMX-Sb sample surfaces immediately after salt pretreatment do not contain microorganisms. Figure 5 shows the results of microbiological analysis of the surface microflora of AMX-Sb$_{w72}$. On the surface facing the NaCl solution compartment (Figure 5a,b), gram-negative, non-spore-forming aerobic rod-shaped bacteria were found, morphologically similar to representatives of the genera Enterobacter, Pseudomonas and Acetobacter. In addition, gram-positive bacteria Actinomycetales, which have the ability to form branching pseudomycelium, and single cells of microscopic fungi of the genus Candida, which are stained in dark purple, are visualized. The same representatives of microflora, but in much smaller quantities, were found on the surface facing the wine compartment (Figure 5c,d).

The source of microflora is most likely the elements of the ED system (hoses, etc.), as is often the case in real production. Less rapid development of microflora from the side of wine compartment is caused by the presence of ethyl alcohol, which suppresses its reproduction [50,51]. The more intensive growth of microflora on the AMX-Sb surface facing the NaCl solution compartment is apparently due to the absence of ethyl alcohol in this solution, as well as (as our measurements show) a high concentration of glucose and other nutrients at the AMX-Sb/NaCl interface, that are transported across the membrane from the wine compartment. In the future, we plan to obtain more detailed information about this phenomenon, as well as about the structure of foulant on the membrane surface, using optical coherence tomography, OCT, and fluorescence microscopy method.

The obtained results are important for the practice of ED tartrate stabilization of wine and ED separation of anthocyanins from winemaking waste. First, they confirm the legitimacy of the established electrodialysis practice, where membrane stacks in the food industry are sterilized every 10–12 h [11,52,53]. Indeed, during this time, biofouling does not have time to significantly affect the membrane transport characteristics. Secondly, these results show that it makes sense to pay more attention to the periodic sterilization of the concentration compartment (circuit), and not just the circuits of electrodialyzer with wine or other food media, as is most often done now.

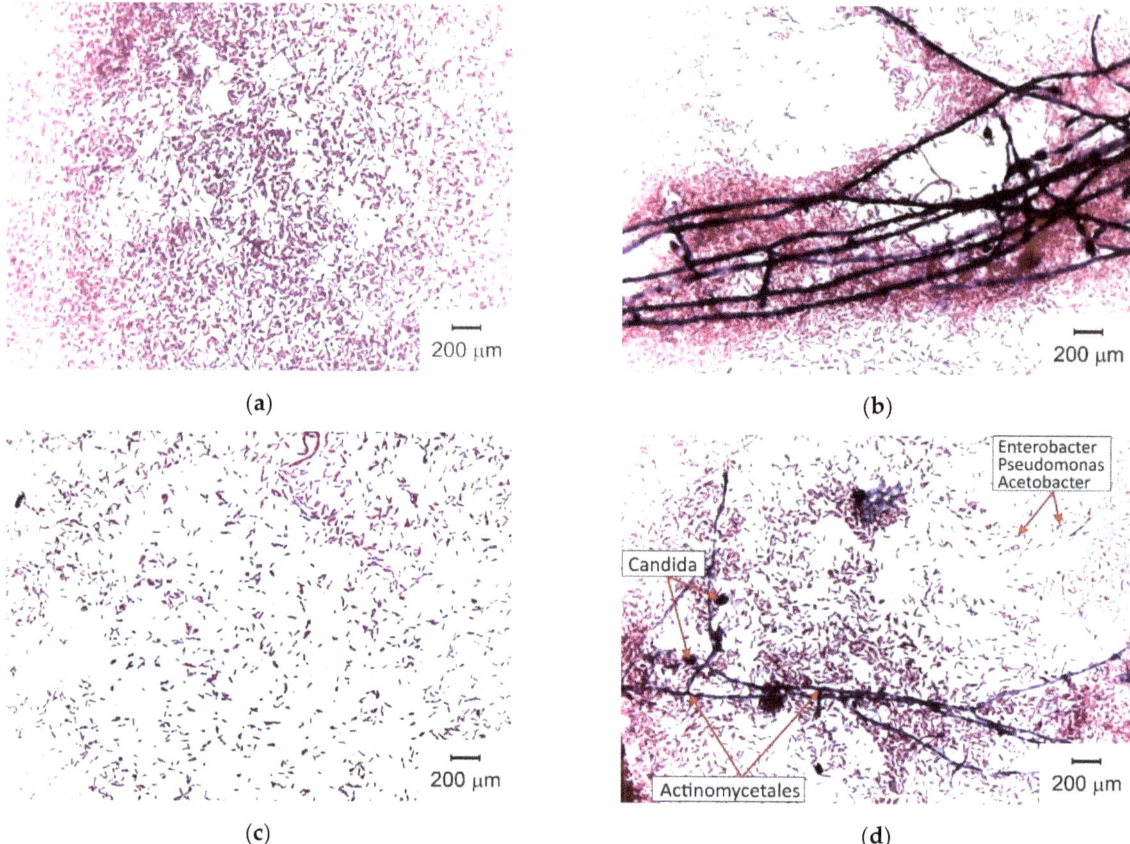

Figure 5. Optical images of the inoculation from touch smears taken from AMX-Sb surfaces facing the NaCl compartment (**a**,**b**) and the wine compartment (**c**,**d**).

3.3. Electrochemical Impedance Spectra

Using electrochemical impedance spectroscopy (EIS), let us analyze to what extent the changes in the volume and surface of membranes after contact with wine affect their behavior in the superimposed electric field (DC bias).

Figure 6 shows the EIS of the AMX-Sb (Figure 6a) and AMX-Sb_{w10} (Figure 6b) obtained in under- and over-limiting current modes. The shape of the spectra for the pristine membrane and the sample contacted with wine are significantly different. These differences are observed in all three (high-, mid- and low-frequency) domains.

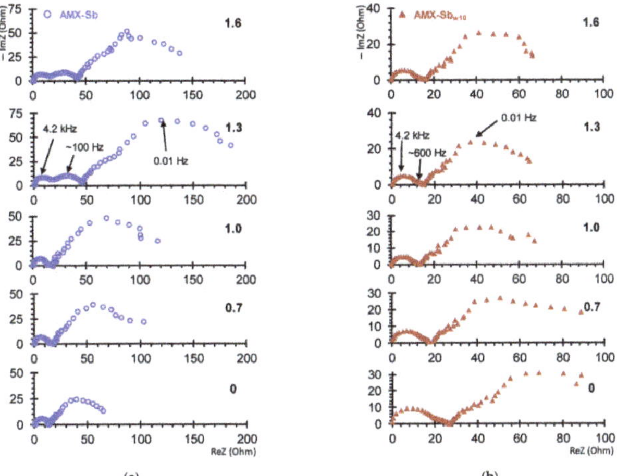

Figure 6. Electrochemical impedance spectra of AMX-Sb (**a**) and AMX-Sb$_{w10}$ (**b**) samples. The numbers near the curves denote the values of $i/i_{\rm lim}^{theor}$.

3.3.1. The High-Frequency EIS Domain

Consider first the high-frequency (10 Hz–1.3 × 10^5 Hz) domain. In the case of the AMX-Sb membrane, the spectra in this domain take the form of a semicircle (Figures 6a and 7c). They can be approximated using the *RC* element (Figure 7a), where the resistance, *R*, is equivalent to the ohmic resistance of the membrane, its interphase boundaries and adjacent DBLs [47]; the effective electrical capacitance, *C*, is mainly controlled by the capacity of EDL on the interphase boundaries [47,54].

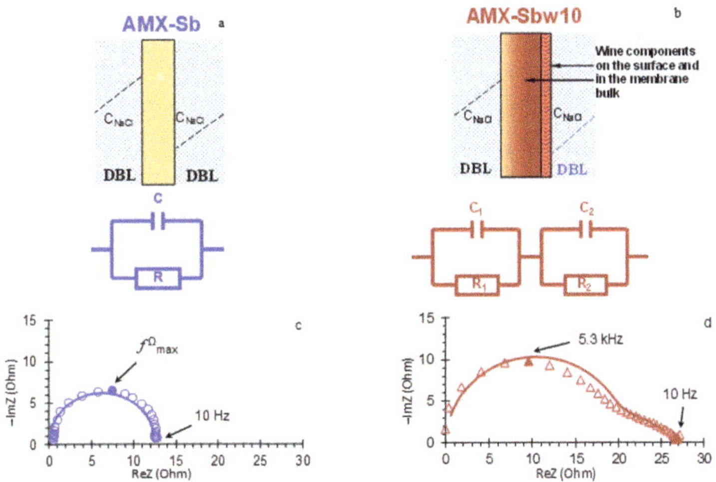

Figure 7. Electrical equivalent circuits for systems containing pristine AMX-Sb (**a**), AMX-Sb$_{w10}$ (**b**) and adjacent DBLs. A high-frequency (10 Hz–1.3 × 10^5 Hz) arcs of AMX-Sb (**c**) and AMX-Sb$_{w10}$ (**d**) impedance spectra, obtained at $i/i_{\rm lim}^{theor}$= 0: experimental values are indicated by dots, approximations using EEC are indicated by solid lines. Explanations are given in the text.

The difference between this equivalent circuit and those described in a number of works, for example, in [32,46,55,56], lies in the fact that it does not include the resistance of a solution located between the outer boundaries of the DBLs (where the concentration of the solution is equal to the concentration of the feed solution) and the measuring electrodes. This component is excluded from spectra at the stage of the experimental data-processing.

The impedance spectra of the AMX-Sb$_{w10}$ in a similar frequency domain has a specific shape (Figures 6b and 7d). Such a form of spectra, as a rule, is recorded in the presence of two layers, which are characterized by significantly different transport time constants [47,57,58]. A schematic representation of such a membrane system and its equivalent circuit is shown in Figure 7b. This circuit consists of two series-connected RC elements.

The first one, R_1C_1, describes the ohmic resistance and effective electrical capacitance of the membrane, its interphase boundaries and DBLs, which have changed as a result of fouling by the components of wine.

The second element, R_2C_2, characterizes the layer of wine components on the membrane surface. The example of experimental data approximation using these equivalent circuits is shown in Figure 7c,d. Figure 8 generalizes the results of such approximations for all the current densities applied.

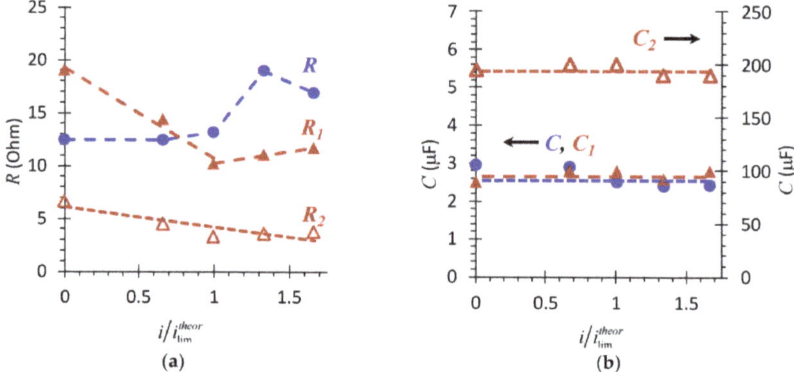

Figure 8. Dependence of the resistances (**a**) and effective electrical capacitances (**b**) of the AMX-Sb (R, C) and AMX-Sb$_{w10}$ (R_1, C_1 and R_2, C_2) samples upon the direct current density normalized to the limiting current, calculated using Equation (1).

The dependence of the resistance of the system with the pristine membrane (Figure 8a) upon the direct current density has a complicated form. In the underlimiting current modes, ($0 < i/i_{lim}^{theor} < 1$), R slowly increases with increasing current (Figure 8a). At currents close to the limiting value, the value of R increases sharply. This dependence in the underlimiting and close to the limiting currents is mainly determined by changes in the resistance of the DBLs, primarily the depleted one [47]. Indeed, in the range $0 < i/i_{lim}^{theor} < 1$, there are no reasons for a change in the resistance of the system with AMX-Sb with increasing current. Therefore, the registered dependence of R on i/i_{lim}^{theor} is due to the concentration profiles formation (schematically indicated by the dashed line in Figure 7a,b) under conditions where the increase in the resistance of the depleted DBL is close to the decrease in the resistance of the enriched DBL.

At current densities close to i/i_{lim}^{theor}, the resistance of the depleted DBL greatly increases because of the emerging deficiency of charge carriers near the surface of the membrane. In the overlimiting current modes, water splitting begins at the membrane/solution boundary. This is evidenced, in particular, by the appearance of the pronounced Gerischer impedance (Figure 6a) on the impedance spectra of the pristine membrane. Protons, which enter the adjacent to the AEM solution, as well as the development of electroconvection, reduce the resistance of the depleted DBL [59]. The decrease in the values of R can also be promoted by competitive transfer through the membrane of hydroxyl ions: according to [60], the electrical resistance of the AEM (which is close to the membranes under study characteristics) decreases by a factor of 2 with the replacement of Cl^- counter ions by OH^- ions.

The calculated value of the effective capacitance, $C = 0.62$ μF cm^{-2} (2.5 μF), (Figure 8b), for the AMX-Sb membrane is consistent with the results obtained in work [56] for the AMX membrane. Our estimates show that the value of this capacitance remains practically unchanged with increasing current density (Figure 8b). Taking into account the fact that there is no contribution of the geometric capacitance of the DBLs in the investigated frequency domain [47], one can conclude that the charge of the AMX-Sb surface remains constant during the experiment. This means that the conditions where the EIS is obtained do not cause changes in the concentration or composition of the fixed groups at the membrane/solution interface, which are sometimes observed in the over-limiting current modes [61].

Measured in the absence of a direct electric current ($i = 0$), the total ($R_1 + R_2$) resistance of the AMX-Sb$_{w10}$ after contact with wine is two times higher (Figure 7d) than that of the pristine membrane (Figure 7c). The contribution of the resistance of wine components layer on the AMX-Sb$_{w10}$ surface, R_2, to the total resistance is no more than 15%. However, its specific electric conductivity is two orders of magnitude lower than that of the membrane bulk, given that the thickness of this layer is only 2–3 μm (Figure 4b). The dependences R_1, R_2 on the direct current density (Figure 8a) differs markedly from those observed for the pristine AMX-Sb membrane. In the range of $0 < i/i_{lim}{}^{theor} \leq 1$, the value of R_1 for AMX-Sb$_{w10}$ sample decreases rapidly. At currents close to the limiting value and higher, $i/i_{lim}{}^{theor} \geq 1$, the values of R_1 become lower than the values of R obtained for the pristine membrane. In the overlimiting modes, R_1 gradually increases, and R_2 gradually decreases with increasing current density.

A decrease in R_2, most likely, is due to a decrease in the thickness of the layer of wine components on the surface of the AMX-Sb$_{w10}$ with an increase in time of the membrane stay under current. Apparently, in the superimposed electric field, anthocyanins, which contain positively charged chromophore groups (flavylium cations) and are retained in the layer formed only due to hydrogen bonds and van der Waals forces, move to a negatively charged cathode. The determining factor in reducing the thickness of the layer is not so much the current strength as the duration of the membrane stay under electric field.

The reduction in R_1 in the under-limiting current modes and its smoother growth in the over-limiting modes in comparison to R, most likely, are due to the partial destruction of the complex colloidal structures that are formed by wine components in the pores of the AMX-Sb$_{w10}$. The cause of their destruction, apparently, is the salting-out effect [62]. A similar effect is observed, for example, in protein solutions after the addition of certain electrolytes [63]. Such electrolyte in our case is NaCl. Note that some of the wine components (polyphenols, saccharides) appear in the 0.02 M NaCl solution at the stage of AMX-Sb$_{w10}$ and AMX-Sb$_{w72}$ preparation for electrochemical studies. Apparently, the application of an electric field contributes to the removal of the products of colloidal structures destruction. In addition, the high electric field strength suppressed the attachment of microorganisms to the AEMs surface [64]. As a result, the mobility of counterions in the membrane increases, and its resistance decreases.

As for the effective electric capacitances C_1 and C_2 (Figure 8b, Equation (2)), their values do not depend on the current density, just as in the case of the pristine membrane. Moreover, the value of C_1 for the AMX-Sb$_{w10}$ is only slightly higher, while C_2 is two orders of magnitude higher than C calculated for the pristine membrane. The observed growth in C_2 seems to be associated with a significant increase in the roughness factor and in the real area of the AMX-Sb$_{w10}$/solution interface due to appearance of spatial colloidal structures with distributed positive and negative charges [12]. It should be noted that some of wine components remain on the surface of AMX-Sb$_{w10}$ after a sufficiently long time (more than 20 h) of the membrane operation under current. This is evidenced by the non-zero values of R_2 (Figure 8a) and high values of C_2 for $i/i_{lim}{}^{theor} = 1.6$ (Figure 8b).

3.3.2. The Middle-Frequency EIS Domain

Apparently, foulants partially screen the catalytically active fixed groups on the surface of the membrane. As a result, the surface of AMX-Sb$_{w10}$ loses the ability of water-splitting. This is evidenced by the absence of the Gerischer impedance arc on the impedance spectra of the AMX-Sb$_{w10}$ (Figures 6b and 9a), whereas in case of AMX-Sb (Figures 6a and 9b), the arc appears at $i/i_{lim}{}^{theor} > 1$ and develops with an increase in current in the frequency range from 10 to 10^3 Hz. For the AMX-Sb, the values of the effective water-splitting reaction constants, χ (Figure 9b), found from the frequency corresponding to the maximal value of imaginary component on the Gerischer impedance spectrum (Figure 9a), are in the range from 250 s^{-1} ($i/i_{lim}{}^{theor}$ = 1.0) to 700 s^{-1} ($i/i_{lim}{}^{theor}$ = 1.6). These values have good correlation with those found for similar membranes [35]. At the same time, the values of χ for the AMX-Sb$_{w10}$ sample tend to zero (Figure 9b).

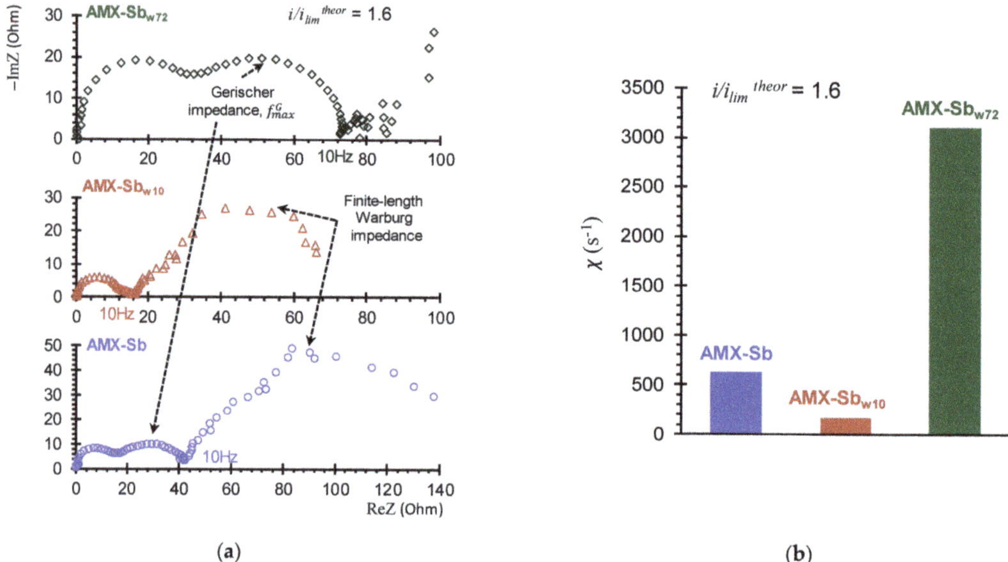

Figure 9. Electrochemical impedance spectra of the AMX-Sb, AMX-Sb$_{w10}$, AMX-Sb$_{w72}$ (**a**) and the values of the effective water splitting constants (**b**) found from the frequencies of the maximum points on the Gerischer arc (Equation (3)).

A longer (72 h) contact of the membrane with wine leads to an enhancement in water splitting: the Gerischer impedance arcs for AMX-Sb$_{w72}$ sample increase noticeably as compared to AMX-Sb and AMX-Sb$_{w10}$ (Figure 9a). The value of χ reaches 3100 s^{-1} (Figure 9b), eight times greater than those found for the AMX-Sb$_{w10}$. The reason for the enhancement in water-splitting is, most probably, the appearance on the surface of AMX-Sb$_{w72}$ of microorganisms (Section 3.2) that do not have time to proliferate on the surfaces of AMX-Sb and AMX-Sb$_{w10}$ (Figure 4a,b).

It is known [65] that the bacterial outer membrane mainly consists of phosphatidylethanolamine, phosphatidylglycerol and diphosphatidylglycerol. Moreover, the deprotonation of phosphorus groups in the region of neutral pH values gives bacteria and other microorganisms a negative charge, which ensures their preferential adsorption on the surface of AEMs with positively charged fixed amino groups [66]. Therefore, biofouling leads to the formation of bipolar boundaries that facilitate the generation of H^+ and OH^- ions under the action of an electric current (Figure 10a). T. Belloň et al. [67,68] observed a similar phenomenon after ssDNA sorption by the membrane surface.

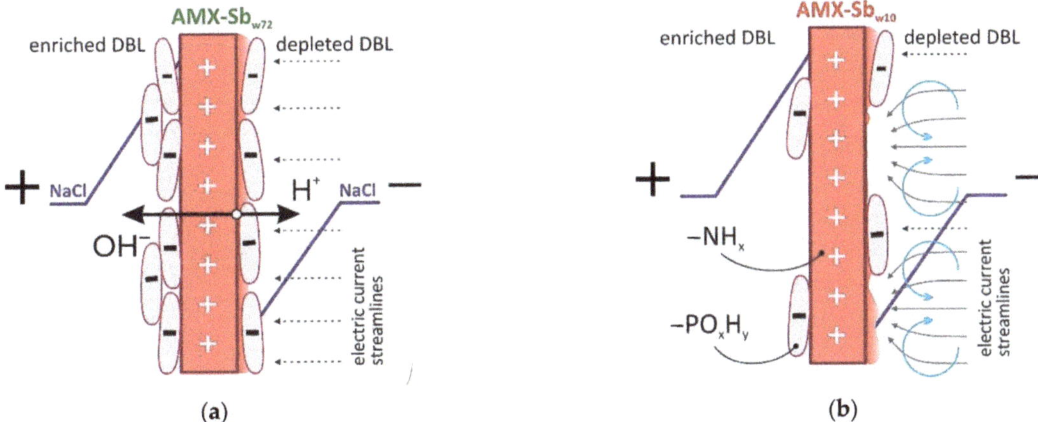

Figure 10. Schematic representation of the effect of foulant nature and its distribution over the membrane surface upon water-splitting (**a**) and electrovonvection (**b**).

The EIS was carried out in conditions where the membrane surface contacted with the wine was facing the desalination compartment of the electrodialysis cell. A high electric field strength, which is necessary for water splitting [69], could arise exactly at the bipolar boundary of this surface. This means that more intense biofouling on the receiving surface of AMX-Sb$_{w72}$ could hinder the transport of salt ions through this membrane, i.e., promote an increase in the ohmic resistance of a fouled sample (Section 3.1), but did not have any effect on water-splitting.

3.3.3. The Low-Frequency EIS Domain

Analysis of impedance spectra in the low-frequency (10 Hz–3 $\times$ 10^{-3} Hz) domain was carried out in order to assess the effect of changes in membrane surface properties after contact with wine on DBL thickness. In the case of AMX-Sb$_{w72}$, the impedance spectra in this frequency domain have a significant dispersion (Figure 9a), which is caused by the effect of intensive water splitting. This is why the dependence of the depleted DBL thickness on the i/i_{lim}^{theor}, found from difference in values of the real component of Warburg impedance at high and low frequencies [37], was obtained for AMX-Sb and AMX-Sb$_{w10}$ only.

Figure 11 shows that in both cases (the pristine membrane and the membrane after contact with wine), there is a decrease in the DBL thickness with increasing current density. This decrease is observed at currents close to the limiting one and higher, and is more significant for the AMX-Sb$_{w10}$. Our previous studies [59] allow one to conclude that under the conditions of the experiment (see Section 2), the reason for decrease in the depleted DBL thickness is electroconvection. Note that in the case of the AMX-Sb$_{w10}$, the intensification of electroconvection takes place against the background of a decrease in the surface charge and its hydrophilization compared to the AMX-Sb (contact angels for AMX-Sb, AMX-Sb$_{w10}$ and AMX-Sb$_{w72}$ are equal to 58 $\pm$ 2, 50 $\pm$ 2 and 45 $\pm$ 2, respectively [12]). These conditions are directly opposite to those that contribute to the development of electroconvection [44,59]. Apparently, the determining factor for the enhancement of electroconvection in case of the AMX-Sb$_{w10}$ is the growth in the inhomogeneity of the electric field, caused by the isle-type localization of the more hydrophilic but less conductive components of wine (Figure 10a) on a sufficiently hydrophobic pristine surface as well as the reduction in water splitting.

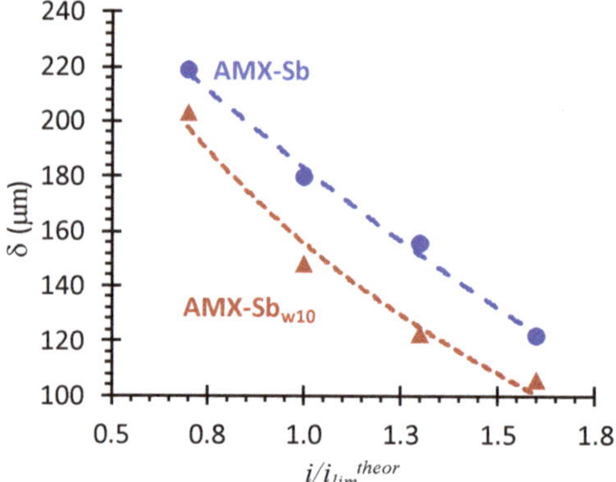

Figure 11. Dependence of the DBL thickness (δ) upon current density in systems with AMX-Sb (circles) and AMX-Sb_{w10} (triangles) membranes.

4. Conclusions

The analysis of the electrochemical impedance spectra obtained in the frequency range from 3×10^{-3} to 1.3×10^{5} Hz at zero-current density, as well as in under- and over-limiting current modes, is extremely informative for understanding the mechanisms of membrane fouling and identifying their consequences for the behavior of membranes in the applied electric field.

Analysis of the AMX-Sb_{w10} impedance spectra shows that the contribution of the wine components layer (on the membrane surface) resistance to the total resistance does not exceed 15%. However, its specific electric conductivity is two orders of magnitude lower than that of the membrane bulk, given that the thickness of this layer is only 2–3 μm.

With the increase in the duration of the AMX-Sb_{w10} operation in electrodialysis desalination of 0.02 M NaCl solution, the resistance of the membrane volume and the layer of wine components on its surface decreases. This is due to the partial destruction of the complex colloidal structures that are formed by wine components. The cause of their destruction, apparently, is the salting-out effect.

The value of electric capacitance of the foulant layer on the surface of AMX-Sb_{w10} is two orders of magnitude higher than that calculated for the pristine membrane. The observed growth seems to be associated with a significant increase in the roughness factor and in the real area of the AMX-Sb_{w10}/solution interface due to the appearance of a spatial colloidal structures with distributed positive and negative charges.

These changes in the structure and chemical composition of the AMX-Sb_{w10} membrane surface lead to a reduction in water-splitting and an enhancement of electroconvection in the over-limiting current modes in comparison to the pristine membrane. Despite the noticeable hydrophilization of the AMX-Sb_{w10} surface as compared to AMX-Sb, electroconvection develops. This causes the decrease in the thickness of the depleted diffusion layer. The observed effect is apparently due to the isle-type distribution of anthocyanin-containing substances along the undulating surface of the investigated membrane.

The subsequent contact of the membrane with wine (72 h) leads to the formation of a fairly uniform layer, the mixture of wine components and microorganisms, on the AMX-Sb_{w72} membrane surfaces. The biofouling of AMX-Sb_{w72}, led to an increase in water-splitting and the reduction in electroconvection in over-limiting current modes. The fact that biofouling and its negative effect on the behavior of the membrane system begins to manifest itself no earlier than 10 h later, allows for concluding that the cleaning-in-place procedure used in industry will be most effective when applied on a daily basis.

We hope that the demonstrated application of impedance spectroscopy will stimulate other researchers to use this method more widely, not only in the study of fouling, but also as a support for the design of membrane cleaning protocols.

Author Contributions: Conceptualization, N.P. and V.N.; methodology, V.S., K.M. and L.D.; formal analysis, A.K. and M.S.; investigation, A.K., V.S., T.R. and Y.K.; writing—original draft preparation, A.K. and N.P.; writing—review and editing, V.S. and L.D. All authors have approved the final article. All authors have read and agreed to the published version of the manuscript.

Funding: This research was funded by Kuban Scientific Foundation, Project MFI-20.1/130.

Acknowledgments: The authors thank the Center for Collective Use of the Kuban State University "Diagnostics of the structure and properties of nanomaterials" for the equipment provided.

Conflicts of Interest: The authors declare no conflict of interest.

References

1. Teixeira, A.; Baenas, N.; Dominguez-Perles, R.; Barros, A.; Rosa, E.; Moreno, D.; Garcia-Viguera, C. Natural bioactive compounds from winery by-products as health promoters: A review. *Int. J. Mol. Sci.* **2014**, *15*, 15638–15678. [CrossRef]
2. El Rayess, Y.; Mietton-Peuchot, M. Membrane technologies in wine industry: An overview. *Crit. Rev. Food Sci. Nutr.* **2016**, *56*, 2005. [CrossRef]
3. Romanov, A.M.; Zelentsov, V.I. Use of electrodialysis for the production of grape-based soft and alcoholic drinks. *Surf. Eng. Appl. Elect.* **2007**, *43*, 279–286. [CrossRef]
4. Serre, E.; Rozoy, E.; Pedneault, K.; Lacour, S.; Bazinet, L. Deacidification of cranberry juice by electrodialysis: Impact of membrane types and configurations on acid migration and juice physicochemical characteristics. *Sep. Purif. Technol.* **2016**, *163*, 228–237. [CrossRef]
5. De Pinho, M.N. Membrane processes in must and wine industries. In *Membrane Technology*; Peinemann, K.-V., Nunes, S.P., Giorno, L., Eds.; Wiley-VCH: Weinheim, Germany, 2010; Volume 3, pp. 105–118.
6. Riponi, C.; Nauleau, F.; Amati, A.; Arfelli, G.; Castellari, M. Electrodialysis. 2. Tartrate stabilization of wines by electrodialysis. *Rev. Fr. Oenol.* **1992**, *137*, 59–63.
7. Low, L.; O'Neill, B.; Ford, C.; Godden, J.; Gishen, M.; Colby, C. Economic evaluation of alternative technologies for tartrate stabilisation of wines. *Int. J. Food Sci. Technol.* **2008**, *43*, 1202–1216. [CrossRef]
8. Gonçalves, F.; Fernandes, C.; dos Santos, P.C.; de Pinho, M.N. Wine tartaric stabilization by electrodialysis and its assessment by the saturation temperature. *J. Food Eng.* **2003**, *59*, 229–235. [CrossRef]
9. Rozoy, E.; Boudesocque, L.; Bazinet, L. Deacidification of cranberry juice by electrodialysis with bipolar membranes. *J. Agric. Food Chem.* **2015**, *63*, 642–651. [CrossRef] [PubMed]
10. Jackson, R.S. *Wine Science: Principles and Applications*, 4th ed.; Academic Press: Cambridge, MA, USA, 2014; p. 751.
11. Mikhaylin, S.; Bazinet, L. Fouling on ion-exchange membranes: Classification, characterization and strategies of prevention and control. *Adv. Colloid Interface Sci.* **2016**, *229*, 34–56. [CrossRef] [PubMed]
12. Sarapulova, V.; Nevakshenova, E.; Nebavskaya, X.; Kozmai, A.; Aleshkina, D.; Pourcelly, G.; Nikonenko, V.; Pismenskaya, N. Characterization of bulk and surface properties of anion-exchange membranes in initial stages of fouling by red wine. *J. Membr. Sci.* **2018**, *559*, 170–182. [CrossRef]
13. Bdiri, M.; Perreault, V.; Mikhaylin, S.; Larchet, C.; Hellal, F.; Bazinet, L.; Dammak, L. Identification of phenolic compounds and their fouling mechanisms in ion-exchange membranes used at an industrial scale for wine tartaric stabilization by electrodialysis. *Sep. Purif. Technol.* **2020**, *233*, 115995. [CrossRef]
14. Cifuentes-Araya, N.; Pourcelly, G.; Bazinet, L. How pulse modes affect proton-barriers and anion-exchange membrane mineral fouling during consecutive electrodialysis treatments. *J. Colloid Interface Sci.* **2013**, *392*, 396–406. [CrossRef] [PubMed]
15. Ghafari, M.; Cui, Y.; Alali, A.; Atkinson, J.D. Phenol adsorption and desorption with physically and chemically tailored porous polymers: Mechanistic variability associated with hyper-cross-linking and amination. *J. Hazard. Mater.* **2019**, *361*, 162–168. [CrossRef] [PubMed]
16. Kammerer, J.; Boschet, J.; Kammerer, D.R.; Carle, R. Enrichment and fractionation of major apple flavonoids, phenolic acids and dihydrochalcones using anion exchange resins. *LWT Food Sci. Technol.* **2011**, *44*, 1079–1087. [CrossRef]
17. Caetano, M.; Valderrama, C.; Farran, A.; Cortina, J.L. Phenol removal from aqueous solution by adsorption and ion exchange mechanisms onto polymeric resins. *J. Colloid Interface Sci.* **2009**, *338*, 402–409. [CrossRef]
18. Zhang, K.; Yang, S.-T. Effect of pH on fumaric acid adsorption onto IRA900 ion exchange resin. *Sep. Sci. Technol.* **2015**, *50*, 56–63. [CrossRef]
19. Hashim, H.; Wan Ahmad, W.Y.; Zubairi, S.I.; Maskat, M.Y. Effect of pH on adsorption of organic acids and phenolic compounds by amberlite IRA 67 resin. *J. Teknol.* **2019**, *81*, 69–81. [CrossRef]
20. Ribéreau-Gayon, P.; Glories, Y.; Maujean, A.; Dubourdieu, D. *Handbook of Enology: The Chemistry of Wine, Stabilization and Treatments*, 2nd ed.; John Wiley & Sons Ltd.: Chichester, UK, 2006; p. 450.

21. Ping, Q.; Cohen, B.; Dosoretz, C.; He, Z. Long-term investigation of fouling of cation and anion exchange membranes in microbial desalination cells. *Desalination* **2013**, *325*, 48–55. [CrossRef]
22. Drews, A. Membrane fouling in membrane bioreactors-characterisation, contradictions, cause and cures. *J. Membr. Sci.* **2010**, *363*. [CrossRef]
23. Baker, J.S.; Dudley, L.Y. Biofouling in membrane systems—A review. *Desalination* **1998**, *118*, 81–89. [CrossRef]
24. Wingender, J.; Flemming, H.C. Biofilms in drinking water and their role as reservoir for pathogens. *Int. J. Hyg. Environ. Health* **2011**, *214*, 417–423. [CrossRef] [PubMed]
25. Multon, J.L.; Flanzy, C. *Oenologie: Fondements Scientifiques et Technologiques*; TEC&DOC Lavoisier: Paris, France, 1998.
26. Park, J.S.; Chilcott, T.C.; Coster, H.G.L.; Moon, S.H. Characterization of BSA-fouling of ion-exchange membrane systems using a subtraction technique for lumped data. *J. Membr. Sci.* **2005**, *246*, 137–144. [CrossRef]
27. Park, J.S.; Choi, J.H.; Yeon, K.H.; Moon, S.H. An approach to fouling characterization of an ion-exchange membrane using current-voltage relation and electrical impedance spectroscopy. *J. Colloid Interface Sci.* **2006**, *294*, 129–138. [CrossRef] [PubMed]
28. Merino-Garcia, I.; Kotoka, F.; Portugal, C.A.M.; Crespo, J.G.; Velizarov, S. Characterization of poly(Acrylic) acid-modified heterogenous anion exchange membranes with improved monovalent permselectivity for RED. *Membranes* **2020**, *10*, 134. [CrossRef] [PubMed]
29. Li, Y.; Shi, S.; Cao, H.; Xu, B.; Zhao, Z.; Cao, R.; Chang, J.; Duan, F.; Wen, H. Anion exchange nanocomposite membranes modified with graphene oxide and polydopamine: Interfacial structure and antifouling applications. *ACS Appl. Nano Mater.* **2020**, *3*, 588–596. [CrossRef]
30. Pintossi, D.; Saakes, M.; Borneman, Z.; Nijmeijer, K. Electrochemical impedance spectroscopy of a reverse electrodialysis stack: A new approach to monitoring fouling and cleaning. *J. Power Sources* **2019**, *444*, 227302. [CrossRef]
31. Zhang, L.; Jia, H.; Wang, J.; Wen, H.; Li, J. Characterization of fouling and concentration polarization in ion exchange membrane by in-situ electrochemical impedance spectroscopy. *J. Membr. Sci.* **2020**, *594*, 117443. [CrossRef]
32. Zhao, Z.; Shi, S.; Cao, H.; Li, Y. Electrochemical impedance spectroscopy and surface properties characterization of anion-exchange membrane fouled by sodium dodecyl sulfate. *J. Membr. Sci.* **2017**, *530*, 220–231. [CrossRef]
33. Hurwitz, H.D.; Dibiani, R. Experimental and theoretical investigations of steady and transient states in systems of ion exchange bipolar membranes. *J. Membr. Sci.* **2004**, *228*, 17–43. [CrossRef]
34. Zabolotskii, V.; Sheldeshov, N.; Melnikov, S. Heterogeneous bipolar membranes and their application in electrodialysis. *Desalination* **2014**, *342*, 183–203. [CrossRef]
35. Kniaginicheva, E.; Pismenskaya, N.; Melnikov, S.; Belashova, E.; Sistat, P.; Cretin, M.; Nikonenko, V. Water splitting at an anion-exchange membrane as studied by impedance spectroscopy. *J. Membr. Sci.* **2015**, *496*, 78–83. [CrossRef]
36. Sistat, P.; Kozmai, A.; Pismenskaya, N.; Larchet, C.; Pourcelly, G.; Nikonenko, V. Low-frequency impedance of an ion-exchange membrane system. *Electrochim. Acta* **2008**, *53*, 6380. [CrossRef]
37. Kozmai, A.E.; Nikonenko, V.V.; Pismenskaya, N.D.; Mareev, S.A.; Belova, E.I.; Sistat, P. Use of electrochemical impedance spectroscopy for determining the diffusion layer thickness at the surface of ion-exchange membranes. *Petrol. Chem.* **2012**, *52*, 614. [CrossRef]
38. Astom Corporation. Available online: http://www.astom-corp.jp/en/product/10.html (accessed on 14 December 2020).
39. Le, X.T. Permselectivity and microstructure of anion exchange membranes. *J. Colloid Interface Sci.* **2008**, *325*, 215–222. [CrossRef] [PubMed]
40. Berezina, N.P.; Kononenko, N.A.; Dyomina, O.A.; Gnusin, N.P. Characterization of ion-exchange membrane materials: Properties vs. structure. *Adv. Colloid Interface Sci.* **2008**, *139*, 3–28. [CrossRef] [PubMed]
41. Calull, M.; Marce, R.M.; Borrull, F. Determination of carboxylic acids, sugars, glycerol and ethanol in wine and grape must by ion-exchange high-performance liquid chromatography with refractive index detection. *J. Chromatogr. A* **1992**, *590*, 215–222. [CrossRef]
42. Hussain, Z.; Ali, A.; Khan, K.M.; Perveen, S.; Mabood, F. A novel Spectrophotometric method for the trace analysis of glucose. *J. Pharm. Res.* **2011**, *4*, 4731–4733.
43. Harrigan, W.F.; McCance, M.E. *Laboratory Methods in Microbiology*; Academic press: Cambridge, MA, USA, 2014; p. 374.
44. Belashova, E.D.; Melnik, N.A.; Pismenskaya, N.D.; Shevtsova, K.A.; Nebavsky, A.V.; Lebedev, K.A.; Nikonenko, V.V. Overlimiting mass transfer through cation-exchange membranes modified by Nafion film and carbon nanotubes. *Electrochim. Acta* **2012**, *59*, 412. [CrossRef]
45. Newman, J.S. *Electrochemical Systems*; Prentice Hall: Englewood Cliffs, NJ, USA, 1973; p. 309.
46. Barsukov, Y.; Macdonald, J.R. *Impedance Spectroscopy: Theory, Experiment, and Applications*, 2nd ed.; Wiley: New York, NY, USA, 2005; p. 616.
47. Nikonenko, V.; Kozmai, A. Electrical equivalent circuit of an ion-exchange membrane system. *Electrochim. Acta* **2011**, *56*, 1262–1269. [CrossRef]
48. Lasia, A. *Electrochemical Impedance Spectroscopy and Its Applications*; Springer: New York, NY, USA, 2014; p. 369.
49. Simoes Costa, A.M.; Costa Sobral, M.M.; Delgadillo, I.; Cerdeira, A.; Rudnitskaya, A. Astringency quantification in wine: Comparison of the electronic tongue and FT-MIR spectroscopy. *Sens. Actuators B Chem.* **2015**, *207*, 1095–1103. [CrossRef]
50. Luong, J.H.T. Kinetics of ethanol inhibition in alcohol fermentation. *Biotechnol. Bioeng.* **1985**, *27*, 280–285. [CrossRef] [PubMed]
51. Cray, J.A.; Stevenson, A.; Ball, P.; Bankar, S.B.; Eleutherio, E.C.; Ezeji, T.C.; Hallsworth, J.E. Chaotropicity: A key factor in product tolerance of biofuel-producing microorganisms. *Curr. Opin. Biotechnol.* **2015**, *33*, 228–259. [CrossRef] [PubMed]
52. Porcelli, N.; Judd, S. Chemical cleaning of potable water membranes: A review. *Sep. Purif. Technol.* **2010**, *71*, 137–143. [CrossRef]

53. Dammak, L.; Larchet, C.; Grande, D. Ageing of ion-exchange membranes in oxidant solutions. *Sep. Purif. Technol.* **2009**, *69*, 43–47. [CrossRef]
54. Rubinstein, I.; Zaltzman, B.; Futerman, A.; Gitis, V.; Nikonenko, V. Reexamination of electrodiffusion time scales. *Phys. Rev. E* **2009**, *79*, 021506. [CrossRef]
55. Abdu, S.; Martí-Calatayud, M.-C.; Wong, J.E.; García-Gabaldón, M.; Wessling, M. Layer-by-layer modification of cation-exchange membranes controls ion selectivity and water splitting. *ACS Appl. Mater. Inter.* **2014**, *6*, 1843. [CrossRef]
56. Długołęcki, P.; Ogonowski, P.; Metz, S.J.; Saakes, M.; Nijmeijer, K.; Wessling, M. On the resistances of membrane, diffusion boundary layer and double layer in ion-exchange membrane transport. *J. Membr. Sci.* **2010**, *349*, 369–379. [CrossRef]
57. Macdonald, J.R. *Impedance Spectroscopy: Emphasizing Solid Materials and Systems*; Wiley-Interscience: New York, NY, USA, 1987.
58. Femmer, R.; Marti-Calatayud, M.C.; Wessling, M. Mechanistic modeling of the dielectric impedance of layered membrane architectures. *J. Membr. Sci.* **2016**, *520*, 29–36. [CrossRef]
59. Nikonenko, V.V.; Mareev, S.A.; Pis'menskaya, N.D.; Uzdenova, A.M.; Kovalenko, A.V.; Urtenov, M.K.; Pourcelly, G. Effect of electroconvection and its use in intensifying the mass transfer in electrodialysis (Review). *Russ. J. Electrochem.* **2017**, *53*, 1122–1144. [CrossRef]
60. Pismenskaya, N.; Nikonenko, V.; Auclair, B.; Pourcelly, G. Transport of weak-electrolyte anions through anion-exchange membranes. Current-voltage characteristics. *J. Membr. Sci.* **2001**, *189*, 129–140. [CrossRef]
61. Choi, J.-H.; Moon, S.-H. Structural change of ion-exchange membrane surfaces under high electric field and its effect on membrane properties. *J. Colloid Interface Sci.* **2003**, *265*, 93–100. [CrossRef]
62. Tadros, T. *Encyclopedia of Colloid and Interface Science*; Springer: Berlin, Germany, 2013; p. 1449.
63. Piazza, R. Protein interactions and association: An open challenge for colloid science. *Curr. Opin. Colloid Interface Sci.* **2004**, *8*, 515–522. [CrossRef]
64. Vaselbehagh, M.; Karkhanechi, H.; Takagi, R.; Matsuyama, H. Biofouling phenomena on anion exchange membranes under the reverse electrodialysis process. *J. Membr. Sci.* **2017**, *530*, 232–239. [CrossRef]
65. Strahl, H.; Errington, J. Bacterial membranes: Structure, domains, and function. *Annu. Rev. Microbiol.* **2017**, *71*, 519–538. [CrossRef]
66. Herzberg, M.; Pandit, S.; Mauter, M.S.; Oren, Y. Bacterial biofilm formation on ion exchange membranes. *J. Membr. Sci.* **2020**, *596*, 117564. [CrossRef]
67. Slouka, Z.; Senapati, S.; Yan, Y.; Chang, H.-C. Charge inversion, water splitting, and vortex suppression due to DNA sorption on ion-selective membranes and their ion-current signatures. *Langmuir* **2013**, *29*, 8275–8283. [CrossRef]
68. Belloň, T.; Polezhaev, P.; Vobecká, L.; Slouka, Z. Fouling of a heterogeneous anion-exchange membrane and single anion-exchange resin particle by ssDNA manifests differently. *J. Membr. Sci.* **2019**, *572*, 619–631. [CrossRef]
69. Simons, R. Water splitting in ion exchange membranes. *Electrochim. Acta* **1985**, *30*, 275–282. [CrossRef]

Article

Assessment of the Performance of Electrodialysis in the Removal of the Most Potent Odor-Active Compounds of Herring Milt Hydrolysate: Focus on Ion-Exchange Membrane Fouling and Water Dissociation as Limiting Process Conditions

Sarah Todeschini [1,2,3], Véronique Perreault [1,2,3], Charles Goulet [4], Mélanie Bouchard [5], Pascal Dubé [5], Yvan Boutin [3,6] and Laurent Bazinet [1,2,3,*]

1. Department of Food Sciences, Université Laval, Québec, QC G1V 0A6, Canada; sarah.todeschini.1@ulaval.ca (S.T.); veronique.perreault.5@ulaval.ca (V.P.)
2. Laboratoire de Transformation Alimentaire et Procédés ElectroMembranaires (LTAPEM, Laboratory of Food Processing and ElectroMembrane Processes), Université Laval, Québec, QC G1V 0A6, Canada
3. Institute of Nutrition and Functional Foods (INAF), Université Laval, Québec, QC G1V 0A6, Canada; yvan.boutin@tbt.qc.ca
4. Department of Phytology, Université Laval, Québec, QC G1V 0A6, Canada; charles.goulet@fsaa.ulaval.ca
5. Investissement Québec-Centre de Recherche Industrielle du Québec (CRIQ, Quebec Investment–Industrial Research Center of Quebec), Québec, QC G1P 4C7, Canada; melanie.bouchard@invest-quebec.com (M.B.); pascal.dube@invest-quebec.com (P.D.)
6. Centre Collégial de Transfert de Technologie en Biotechnologie (TransBIOTech, College Center for Technology Transfer in Biotechnology), Lévis, QC G6V 6Z9, Canada

* Correspondence: laurent.bazinet@fsaa.ulaval.ca; Tel.: +418-656-2131 (ext. 407445); Fax: +418-656-3353

Received: 25 May 2020; Accepted: 16 June 2020; Published: 20 June 2020

Abstract: Herring milt hydrolysate (HMH), like many fish products, presents the drawback to be associated with off-flavors. As odor is an important criterion, an effective deodorization method targeting the volatile compounds responsible for off-flavors needs to be developed. The potential of electrodialysis (ED) to remove the 15 volatile compounds identified, in the first part of this work, for their main contribution to the odor of HMH, as well as trimethylamine, dimethylamine and trimethylamine oxide, was assessed by testing the impact of both hydrolysate pH (4 and 7) and current conditions (no current vs. current applied). The ED performance was compared with that of a deaerator by assessing three hydrolysate pH values (4, 7 and 10). The initial pH of HMH had a huge impact on the targeted compounds, while ED had no effect. The fouling formation, resulting from electrostatic and hydrophobic interactions between HMH constituents and ion-exchange membranes (IEM); the occurrence of water dissociation on IEM interfaces, due to the reaching of the limiting current density; and the presence of water dissociation catalyzers were considered as the major limiting process conditions. The deaerator treatment on hydrolysate at pH 7 and its alkalization until pH 10 led to the best removal of odorant compounds.

Keywords: electrodialysis; deaerator; herring milt hydrolysate; deodorization; off-flavors; trimethylamine; fouling; water dissociation

---

1. Introduction

Due to the important increase in fish consumption, by-products generated by industries have known an important increase over the last twenty years [1]. One major way of valorizating these by-products is based on their hydrolysis through the use of chemicals or enzymes. The resulting hydrolysates have gained more and more attention from the food, nutraceutical and cosmetics sectors, since they could be of biological interest, having antioxidant, anticancer, antidiabetic, antihypertensive or antimicrobial activities [2–7]. Recently, the valorization of herring milt hydrolysate (HMH) has been much more considered, since it contains high value-added molecules, such as peptides, amino acids, nucleic acids and vitamins, and since it has already proven its relevance regarding metabolic syndrome illnesses [1,8]. However, in spite of these promising aspects, HMH, similarly to other fish hydrolysates, presents the substantial drawback of being associated with unpleasant smells [9]. Since odor is an important criterion for consumer acceptance of a product, there is a real need to overcome this main issue [10].

Unpleasant odors in fishery products and by-products such as hydrolysates can be, in general, ascribed to the contribution of several compounds [9,11]. Nevertheless, all these compounds present the similarities of being volatile, low-molecular-weight and hydrophobic molecules [12]. Among these compounds, trimethylamine (TMA), a tertiary amine characterized by a strong fishy odor, is the most well-known agent involved in the off-flavors of marine products [13]. With other amines like dimethylamine (DMA), TMA is considered to be a major indicator of fish spoilage [14]. TMA and DMA originate from the breakdown of trimethylamine oxide (TMAO) due to bacterial and enzymatic reactions, respectively, as well as environmental conditions [15]. In addition to nitrogen-containing compounds, other volatile compounds belonging to various chemical groups, such as aldehydes, ketones or alcohols, have already been proven to be associated with off-flavors in fishery materials [9,11,16]. These latter compounds originate from enzymatic reactions, lipid oxidation, microbial actions and environmental or thermic stresses [10]. Nonetheless, as no study has been carried out so far regarding the odor of HMH, there is no evidence concerning the contribution of these compounds in this type of matrix, and this needs to be further investigated.

Different strategies are currently available to remove off-flavors. Depending on their principles, they can be divided into three groups of deodorization methods: biological, chemical and physical methods. Firstly, biological methods are based on the use of microorganisms. However, their action spectrum is often reduced, as they target specific compounds, and their mechanisms remain inaccurate [9,17,18]. Chemical methods involve processes like ozonation or the use of antioxidants. The ozonation process presents the main drawback of giving rise to side reactions [19], while the use of antioxidants would be more considered as a preventive measure than a curative one [9,20]. Finally, physical methods include various processes, such as extraction or adsorption on microporous materials. Extraction methods can involve high temperatures that may damage thermolabile compounds [21], while the action of adsorption methods is limited [9]. Thus, disadvantages seem to be associated with all the available methods. Surprisingly, other physical processes, based on the use of selective membranes such as pressure-driven and electromembrane processes, have never been investigated for deodorization purposes so far. Therefore, developing a new deodorization method based on the use of membranes would be an innovative and challenging idea.

It is of importance to note that pressure-driven processes have already been reported for the recovery of valuable components from cooking effluents of seafood processing industries. In that case, pressure-driven technologies allowed to obtain two distinct fractions: one rich in low-sized compounds, including volatile compounds, and another one containing purified water [22–24]. Nevertheless, in a deodorization context, the aim is only to isolate volatile compounds. Therefore, the use of pressure-driven technologies does not look appropriate for this problem. Concerning electromembrane processes, despite the fact they have never been used for deodorization purposes, the impact of a demineralization process performed by conventional electrodialysis (ED) on volatile compounds was already considered by Cros et al. (2005) and Chindapan et al. (2011). Interestingly, these studies showed that ED could lead to a decrease in the odor intensity of the treated product resulting from a decrease in the abundance of certain volatile compounds [25,26]. More specifically, Cros et al. (2005) observed a decrease in (Z)-4-heptenal, 2,3-butanedione, 3-octen-2-one, limonene, phenol and 1-propanol, while Chindapan et al. (2011) experienced a decrease in TMA, 2,6-dimethylpyrazine, various carboxylic acids and phenol, as well [25,26]. According to these results, ED would be a plausible effective process to lower the concentration of these specific molecules. To assess the feasibility of ED on HMH, it appears to be even more necessary to identify the volatile compounds contributing to its odor. Furthermore, since ED is based on the selective migration of ionic species through ion-exchange membranes (IEM) using an electric field as driving force [27], it is particularly interesting to mention that two of the major compounds associated with off-flavors in seafood products, TMA and DMA, are positively charged when the pH is lower than their pKa values, respectively equal to 9.80 and 10.70 [28], making these cations likely to migrate during an ED process.

In this context, the aim of the present study was to (1) identify the main volatile compounds of HMH contributing to its odor, (2) challenge ED as a new potential deodorization method by comparison with a deaerator and (3) understand the potential fouling mechanisms. Even if the purpose of a deaerator is more to remove air from liquid samples, as compounds responsible for off-odors are volatile, they are removed at the same time as gases, making this device suitable as a control for this type of problem.

2. Materials and Methods

2.1. Materials

2.1.1. Chemicals

Sodium sulphate (Na_2SO_4), potassium chloride (KCl), sodium chloride (NaCl), methylene chloride and nonyl acetate were purchased from Fisher (Montréal, QC, Canada). Hydrochloric acid (HCl) and sodium hydroxide (NaOH) were obtained from VWR (Montréal, QC, Canada). TMAO, TMA and DMA standards came from Sigma-Aldrich (St-Louis, MO, USA). Helium and nitrogen were obtained from Praxair (Mississauga, ON, Canada). Oxygerm, Blizzard and Extrem solutions were purchased from Sani Marc (Victoriaville, QC, Canada).

2.1.2. HMH

HMH powder, the composition of which is listed on Table 1, was provided by Ocean NutraSciences (Matane, QC, Canada). The hydrolysate was stored at −30 °C under vacuum and protected from light before its use.

Table 1. Chemical composition of the herring milt hydrolysate (HMH) according to the manufacturer specifications.

Compounds	Composition of Dry Powder (%)
Proteins	40.0 ± 5.0
Arginine	16.0 ± 2.0
Leucine	2.0 ± 0.3
Lysine	1.8 ± 0.3
Isoleucine	1.1 ± 0.2
Histidine	0.4 ± 0.2
Methionine	<0.1
Nucleic acids	26.0 ± 5.0
Lipids	14.0 ± 4.0
Phospholipids	4.5 ± 1.5
Omega-3 (n-3) acids: Eicosapentaenoic acid (EPA) and docosahexaenoic acid (DHA)	4.0 ± 1.0
Minerals	12.5 ± 2.5
Phosphorus	3.2 ± 0.6
Sodium	1.2 ± 0.1
Potassium	0.7 ± 0.1

2.2. Methods

2.2.1. Protocols

Two protocols were specifically designed for this study. The first aimed to identify the volatile compounds of the HMH and their contribution to its odor, while the second assessed the potential of ED to remove these compounds.

Volatile Compounds Analysis

Volatile Compound Extraction

Volatile compounds were extracted according to the procedure described by Tremblay et al. (2020) [29]. Briefly, 6 g of solid samples dissolved in 24 mL of deionized water were absorbed by cotton balls. The soaked cotton balls were then introduced into glass tubes in which air circulated for 1h at room temperature. Volatile compounds were collected on a divinylbenzene column (HayeSep® Q 80/100, Bandera, TX, USA). After that, they were eluted with 150 μL of an elution solvent. The elution solvent consisted in dichloromethane containing nonyl acetate as an internal standard (3.71×10^{-5} M). The extracts were stored at −80 °C before undergoing further analyses.

Volatile Compound Identification

Volatile compounds were identified using a gas chromatography-mass spectrometry (GC-MS) system consisting of a 7890B GC (Agilent Technologies, Santa Clara, CA, USA) equipped with a 5977B mass selective detector (MSD) with a high efficiency source (Agilent Technologies, Santa Clara, CA, USA). The extract (1.8 μL) was injected in splitless mode into a capillary column (DB-5MS Ultra Inert, 30 m length × 250 μm id, 1 μm thickness) (Agilent Technologies, Santa Clara, CA, USA). The flow rate of the carrier gas (helium) was 1.3 mL/min. The temperature of the oven was programmed according to the following steps: from 35 °C (temperature hold for 1.46 min) to 47 °C at 6 °C/min, and to 250 °C (temperature hold for 3 min) at 10 °C/min. MSD conditions were as follows: source temperature, 230 °C; quad temperature, 150 °C; ionization energy, 70 eV; mass range, 30 to 250 arbitrary unit of mass (a.m.u); scan rate, 6,1 scans/s. Data obtained were processed with MassHunter Qualitative Analysis software version B.07.00 (Agilent Technologies, Santa Clara, CA, USA). Volatile compounds were identified by comparing their mass spectra with those available in the NIST 17.L mass spectral database (National Institute of Standards and Technology, Gaithersburg, MD, USA), and were reported when the match degree exceeded 60%.

Most Potent Odor-Active Volatile Compound Determination

The most potent odor-active volatile compounds of the HMH were determined using gas chromatography-olfactometry (GC-O) and the frequency detection method. The GC-O system consisted of a 7890B GC (Agilent Technologies, Santa Clara, CA, USA) equipped with a 5977B MSD with a high efficiency source (Agilent Technologies, Santa Clara, CA, USA) and an olfactory detection port ODP3 (Gerstel, Linthicum, MD, USA) supplied with humidified air to avoid the drying of the mucous membrane in the nasal cavity. The GC effluent was split in the ratio 1: 2 between the MSD and the olfactory detection port. The extract (1.8 μL) was analyzed according to the same procedure as those described previously. A panel made of 13 judges (11 females and 2 males between 23 and 50 years old) took part in the olfactometric experiment. They all previously followed a training session to become familiar with the GC-O system. The sniffing of the HMH extract took place during a session of 26.76 min. The judges were asked to assess the intensity of each odorant on a scale of 1–4 (1 = very weak odor intensity; 4 = very strong odor intensity) and, at the same time, to qualify the perceived odor as bad, good or ok (the latter being if the perceived odor was acceptable without being all that pleasant). Data obtained from GC-MS were processed with MassHunter Qualitative Analysis software version B.07.00 (Agilent Technologies, Santa Clara, CA, USA), while those obtained from olfactometry were treated by Gerstel ODP dataviewer software version 1.0.2.8 (Linthicum, MD, USA).

Compounds detected by at least 5 of the 13 judges were considered as the most potent odor-active volatile compounds.

Analysis of the TMAO, TMA and DMA Contents

As the GC-MS method previously described was not sensitive enough to detect TMAO, TMA and DMA, another procedure was used. Samples were prepared as follows: 2 mL standards or samples pipetted into 20-mL vials were alkalized to pH 10 using 0.01, 0.1 or 5 M NaOH. The molarity of the NaOH solution used for this purpose depended on the initial pH of the analyzed samples in order to avoid any dilution effect. TMAO, TMA and DMA contents were then determined using headspace (HS) gas chromatography equipped with a nitrogen-phosphorus detector (NPD). Briefly, samples were equilibrated for 10 min in a 7697A headspace sampler (Agilent Technologies, Santa Clara, CA, USA) at 90 °C. Then 1-mL sample was injected into a 7890B GC (Agilent Technologies, Santa Clara, CA, USA) equipped with a NPD. The loop and transfer line were at 110 °C and 115 °C, respectively. Volatile amines were separated using a capillary column (CP-Volamine, 30 m length x 0.32 mm id) (Agilent Technologies, Santa Clara, CA, USA). GC conditions were as follows: split injection, 1:5; injector temperature, 250 °C; helium carrier gas flow, 3 mL/min; detector temperature, 300 °C; oven programmed from 50 °C (4 min) to 200 °C at 25 °C/min. Data were processed with Open Lab CDS EZChrom software version A.03.02.023 (Agilent Technologies, Santa Clara, CA, USA).

Deodorization by ED

ED Cell and Configuration

The ED cell used was a MP type cell with an effective area of 100 cm^2, manufactured by Electrocell AB (Taby, Sweden), including 4 Neosepta AMX-FG anionic membranes (AEM, Astom Ltd., Tokyo, Japan) and 3 Neosepta CMX-FG cationic membranes (CEM, Astom Ltd., Tokyo, Japan) (Figure 1). This arrangement defined three closed loops containing the 20 g/L Na_2SO_4 electrolyte solution (3 L), the 2 g/L KCl solution (3 L) for the potential recovery of volatile compounds and the dry HMH dissolved to 4% of proteins in water overnight at 4 °C under nitrogen, protected from light; the initial pH of 7.3 was adjusted to the desired value (3 L). Each loop was connected to an external tank allowing a continuous recirculation of the solutions. Each tank was closed in order to avoid any loss of volatile compounds due to their volatility. All the solutions were circulated at a flow rate of 3 L/min (1.5 psi) using three centrifugal pumps. The cathode and anode used were respectively a 316 stainless steel electrode and a dimensionally-stable electrode (DSA-O_2). A 0-100 V power supply was employed to generate the potential difference between the electrodes.

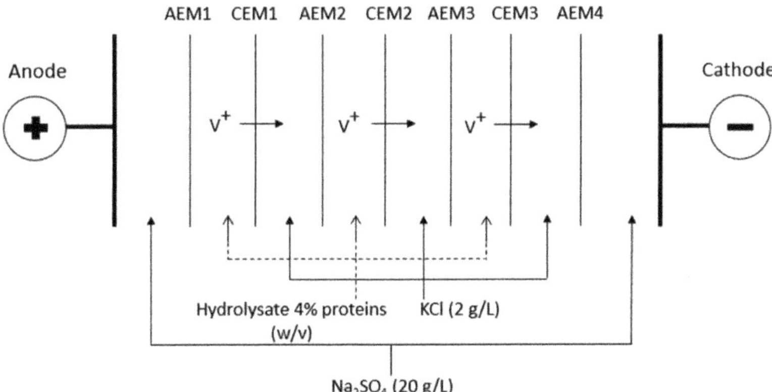

Figure 1. Electrodialysis cell configuration; V^+: cationic volatile compounds, AEM: anion-exchange membrane, CEM: cation-exchange membrane.

ED Parameters

The ED experiment was conducted for 240 min. This treatment duration was set in order to meet two main requirements: allowing the potential migration of volatile compounds without risking the degradation of polyunsaturated fatty acids present in the HMH. Also, for these reasons, in addition to microbiological concerns, temperature was controlled around 10 °C during all the ED treatments.

ED treatments were performed under different constant voltage conditions corresponding to no current application (a residual electrode potential difference of 0.7 V) and current application (voltage value of 10 V) to assess the impact of current on the volatile compound content of the HMH. Also, to assess if TMA and DMA were able to migrate in their cationic forms, the ED treatments were conducted on HMH solution preliminarily acidified at pH 4 or pH 7 with, respectively, 14 and 7 mL of 6 M HCl. Indeed, while pH is inferior to the pKa values of these two amines, namely 9.80 for TMA and 10.70 for DMA [28], they became positively charged. The pH of hydrolysate was maintained either at 4 or at 7 with, respectively, a total addition of 25 and 10 mL of 6 M HCl during the ED run.

45-mL samples of hydrolysate and KCl recovery solutions were collected at the initial and final times. These samples were stored at −30 °C and protected from light before being further analyzed in terms of volatile compounds and ash contents. Each condition was conducted in triplicate and in a random order. Following each run, the ED cell was rinsed with NaCl 2% (w/v) twice. The stack was cleaned with 0.1 N HCl and 0.1 N NaOH solutions every four runs. All the membranes were characterized in terms of average thickness and conductivity before any experiment and then after each run. Different membranes were used for the different conditions. However, the same membranes were used for the three replicates of the same condition.

Comparison with a Deaerator

To assess the performance of ED as a new potential deodorization method, deaerator assays were carried out. The device used was a deaerator model ERV2 (Koruma, Neuenburg, Germany). Based on preliminary tests, HMH (3L) was treated for 30 min in closed circuit under 100 Torr. In this case, treatments were performed on hydrolysate solution prepared according to the same procedure as those for ED runs. The only difference was that, in addition to pH 4 and 7, experiments on hydrolysate solution alkalized to pH 10 with 6 M NaOH were conducted as well. Indeed, in the case of deaerator, it did not matter to charge TMA and DMA molecules, as this device only uses the volatility state of compounds. Similarly to the ED experiments, 45-mL samples of hydrolysate were collected at the initial and final time. They were then stored at −30 °C and protected from light before being further analyzed in terms of volatile compounds. Experiments were conducted in triplicate for each studied pH value. Before use, the deaerator was decontaminated with Oxygerm solution, involving a complex blend of organic acids and powerful oxidizers. After each run, the deaerator was cleaned with Blizzard and Extrem solutions that are, respectively, a foaming chlorinated alkaline degreaser and a blend of wetting, sequestering agents and caustic soda.

2.2.2. Analyses

pH

The pH of the HMH and KCl recovery solutions was measured using a pH-meter model SP20 (Thermo Orion, West Chester, PA, USA) equipped with a VWR Symphony epoxy gel combination pH electrode (Montreal, QC, Canada). During the first hour of the ED run, pH values were recorded every 5 min and then every 15 min.

Conductivity

The conductivity of hydrolysate and KCl recovery solutions was measured using a YSI conductivity meter (Model 3100) equipped with a YSI immersion probe model 3252, (cell constant K = 1 cm^{-1}) (Yellow Springs Instrument Co., Yellow Springs, OH, United States). Similarly to the pH measurements, the conductivity values were recorded every 5 min during the first hour of the ED run, and then every 15 min. The demineralization rate (DR, in %) of the HMH and the mineralization rate (MR, in %) of the corresponding KCl recovery solution were determined according to Equations (1) and (2), respectively, where k_t refers to the solution conductivity at time t, and k_0 refers to the initial solution conductivity [30].

$$DR = \left(1 - \frac{k_t}{k_0}\right) \times 100 \tag{1}$$

$$MR = \left(1 - \frac{k_0}{k_t}\right) \times 100 \tag{2}$$

Ash Content

Ash content (in %) was determined using a method adapted from the Association of Official Analytical Chemists [31]. Briefly, 10 mL samples of HMH and KCl recovery solutions at initial and final times were weighted before being dried overnight at 105 °C in an oven (VWR Gravity Convection Oven, Radnor, PA, USA). Dried samples were then reduced to ashes in a furnace at 550 °C until they turned white. Samples were weighted after cooling, and the ash content was determined according to Equation (3), where m refers to the measured weight.

$$Ash\ content = \left(\frac{m_{crucible+ashes} - m_{crucible}}{m_{crucible+sample} - m_{crucible}}\right) \times 100 \tag{3}$$

The DR (in %) and the MR (in %) were also determined based on the ash content values according to Equations (4) and (5).

$$DR = \left(1 - \frac{ash\ content\ at\ time\ t}{ash\ content\ at\ initial\ time}\right) \times 100 \tag{4}$$

$$MR = \left(1 - \frac{ash\ content\ at\ initial\ time}{ash\ content\ at\ time\ t}\right) \times 100 \tag{5}$$

Global System Resistance

The global system resistance (R, in Ω) was calculated based on the Ohm's law (R = U/I). As voltage (U, in V) value was maintained constant during all of the ED runs, only current intensity (I, in A) values were recorded, which were directly obtained from the power supply.

Membrane Thickness

Membrane thickness was measured before and after each ED run in 0.5 M NaCl solution, according to the same procedure as Lemay et al. (2019) [32], using an electronic digital micrometer equipped with a 10-mm-diameter flat contact point from Marathon watch company LTD (Richmond Hill, ON, Canada). Six measurements, taking place at different locations on the membrane's surface, were used to obtain the average membrane thickness.

Membrane Electrical Conductivity

Membrane electrical conductivity was measured before and after each ED run in 0.5 M NaCl solution, according to the procedure described by Lemay et al. (2019) [32], using a YSI conductivity meter model 3100 (Yellow Springs Instrument Co., Yellow Springs, OH, USA) equipped with a specially designed clip from the Laboratoire des Matériaux Echangeurs d'Ions (Université Paris XII, Créteil, Val de Marne, France).

Volatile Compound Content

Volatile compounds determined as being the most potent odor-active compounds of the studied HMH, according to the GC-O procedure described previously, were used as deodorization indicators. Volatile compounds of HMH and KCl recovery solutions at initial and final times were obtained and analyzed according to the same extraction and GC-MS procedures described previously. The only difference was that extraction was performed on 45-mL liquid samples instead of solid samples. The area under each peak corresponding to the most potent odor-active volatile compounds was considered to assess their abundance in the different samples.

The TMAO, TMA and DMA contents of HMH and KCl recovery solutions were also determined according to the procedure described previously. The quantification was carried out with a calibration curve of known amounts of TMAO, TMA and DMA standards (from 25 ppm to 755 ppm for TMAO; from 2.5 ppm to 10 ppm for TMA and DMA).

Statistical Analyses

Analyses of variance (ANOVA) were performed, using SAS software version 9.4 for Windows, on data concerning ED parameters, as well as those regarding volatile compounds (SAS Institute Inc., Cary, NC, USA). A Tukey test ($\alpha = 0.05$ as probability level) was used to compare the different treatments.

3. Results and Discussion

3.1. Volatile Compound Analysis

3.1.1. Overall Content in Volatile Compounds

The overall content of the HMH was determined using GC-MS, while HS-GC-NPD was used to verify the presence of TMAO, TMA and DMA. The GC-MS procedure led to the identification of a total of 86 compounds, as listed in Table 2. Although a minimal match degree of 60% was initially selected, all of the volatile compounds were identified with a score ranging from 72.29% to 98.7%. The identified volatile compounds mainly belonged to eight groups, based on their chemical structures. More specifically, volatile compounds included 27 aldehydes, 17 ketones, 12 alcohols, 8 alkenes, 8 nitrogenous compounds, 5 alkanes, 2 furans and 2 esters, while the 5 other compounds were miscellaneous representatives of other chemical groups. Aldehydes and ketones were the most abundant compounds of the hydrolysate. Aldehyde compounds included saturated and unsaturated compounds. The unsaturated aldehydes involved alkenals, alkadienals and aromatic aldehydes. Aldehyde compounds generally originate from lipid oxidation [11,33]. This would be consistent with the composition of HMH, as it results from a rich source of polyunsaturated fatty acids, as shown in Table 1. Aldehydes can also derive from the Strecker degradation of amino acids [34]. Strecker degradation refers to the reaction involving a dicarbonylated compound and an amino acid [35]. Concerning the ketones content, it may also be the result of lipid and/or amino acid degradation [34]. Alkanes, alkenes and alcohols, also identified in large numbers in the HMH, are mainly known to directly derive from lipid oxidation [35–37]. Then, among the nitrogenous compounds content, 8 compounds were identified by GC-MS, while TMAO, TMA and DMA, also listed in Table 2, were detected by HS-GC-NPD. On the one hand, among the nitrogenous compounds identified by GC-MS, 2 pyrazines and 2 thiazoles were detected. More precisely, concerning the pyrazine content, methylpyrazine and 2,5-dimethylpyrazine were identified. Pyrazines are generally formed during Maillard or pyrolysis reactions in heat-processed foods [36,38]. However, despite the fact that high temperatures promote pyrazine formation, mild temperatures are also sufficient to lead to the formation of these compounds [13]. In the present study, since the HMH was obtained through enzymatic hydrolysis under minor heating, followed by a drying step by atomization involving higher temperatures, the two pyrazines identified might have been formed during one of these steps. Pyrazines can be biosynthesized by microorganisms, as well [34,36]. The thiazole content of HMH involved 2 compounds, 2-acetylthiazole and benzothiazole. Thiazole are sulfur-containing compounds that are supposed to originate from Strecker degradation [35]. Surprisingly, while Cha and Cadwaller (1995a) studied the volatile components of various fish and crustacean pastes, including herring paste and shrimp pastes, they identified 2-acetylthiazole only in shrimp paste, while benzothiazole was only identified in anchovy and hair tailed viscera pastes. Nonetheless, they were not detected in herring paste [39]. Regarding the four other nitrogenous compounds identified by GC-MS, three of them, namely 1-methyl-1H-tetrazole, 3-methyl-butanenitrile and 3-methyl oxime butanal, were not evidenced in fishery materials. These compounds may be characteristic of the volatile content of HMH. However, N-nitrosodimethylamine was already known to be generated from TMA and DMA [40]. On the other hand, the HS-GC-NPD procedure allowed the detection of TMAO, TMA and DMA (Table 2). In general, these compounds play a key role in the volatile content of fishery products. TMAO is a well-known osmoregulator, mainly found in marine fish, which aims to counteract protein destabilization [14]. The transformation of TMAO leads to the formation of TMA and DMA. More specifically, TMA can be formed from TMAO through bacterial degradation, while the formation of DMA is mainly attributed to the action of the endogenous TMAO aldolase enzyme [14,41]. Moreover, while TMAO is in presence of reducing agents, these molecules would be generated as well. Finally, compounds belonging to furan and ester groups were also found in HMH. More specifically, two furan compounds, 2-methylfuran and 5-isopropyl-3,3-dimethyl-2-methylene-2,3-dihydrofuran, were identified. Furans are known to originate from lipid oxidation [35]. Another mechanism of the formation of furan compounds involves Maillard and Strecker reactions [38]. Concerning the ester compounds identified, cis-cyclohexane-1,4-dimethanol diacetate and 2,2,4-trimethyl-1,3-pentanediol diisobutyrate, they may have originated from the esterification reaction occurring between alcohols and carboxylic acids generated through microbial or enzymatic lipid degradation [39]. It is of interest to note that Cha and Cadwallader (1995a) identified several ester compounds in herring paste, but none of them corresponded to those identified in this study. In addition, in herring paste, a high content of esters was observed. This should probably be linked to the fermentation process taking place in this type of products [39]. Also, it is interesting to note that, in this study, no carboxylic acid was identified. Whilst this was consistent with the study of Cha and Cadwaller (1995a) dealing with herring paste, this was not in line with the study of Aro et al. (2002), in which few carboxylic acids were identified in herring raw material [39,42]. Carboxylic acids are known to be lipid oxidation products, or to originate from amino acid degradation [42]. The absence of carboxylic acids might be a property of the volatile content of HMH. Globally, despite the few differences noted and specified previously, most of the compounds identified in HMH have already been reported in several marine products of different origins: fishing and aquaculture products [33,36,38,42,43], fish oil [21], fish sauces and fish pastes [25,34,39,44–47], and mollusk hydrolysates [9,48].

Table 2. Volatile compounds identified in HMH.

Classification	Compounds	Chemical Formula	Retention Time (Min)
Aldehydes	Butanal	C_4H_8O	6.050
	(E)-2-Butenal	C_4H_6O	7.086
	3-Methylbutanal	$C_5H_{10}O$	7.200
	2-Methylbutanal	$C_5H_{10}O$	7.381
	Pentanal	$C_5H_{10}O$	8.009
	2-Methyl-2-pentenal	$C_6H_{10}O$	8.777
	(E)-2-Methyl-2-butenal	C_5H_8O	8.968
	(E)-2-Pentenal	C_5H_8O	9.201
	Hexanal	$C_6H_{12}O$	10.121
	2-Ethyl-trans-2-butenal	$C_6H_{10}O$	10.786
	Furfural	$C_5H_4O_2$	10.845
	2-Hexenal	$C_6H_{10}O$	11.235
	2-Ethyl-2-pentenal	$C_7H_{12}O$	11.669
	(Z)-4-Heptenal	$C_7H_{12}O$	12.114
	Heptanal	$C_7H_{14}O$	12.156
	Methional	C_4H_8OS	12.338
	(E,E)-2,4-Hexadienal	C_6H_8O	12.391
	Benzaldehyde	C_7H_6O	13.552
	(E,E)-2,4-Heptadienal	$C_7H_{10}O$	13.978
	Benzeneacetaldehyde	C_8H_8O	14.984
	4-(1-methylethyl)-1-cyclohexene-1-carboxaldehyde	$C_{10}H_{16}O$	16.025
	2-Isopropyl-5-oxohexanal	$C_9H_{16}O_2$	16.200
	(E,Z)-2,6-Nonadienal	$C_9H_{14}O$	16.642
	2-Phenylpropenal	C_9H_8O	18.887
	4-Ethyl-benzaldehyde	$C_9H_{10}O$	17.027
	Decanal	$C_{10}H_{20}O$	17.411
	Lilac aldehyde D	$C_{10}H_{16}O_2$	19.394

Table 2. *Cont.*

Classification	Compounds	Chemical Formula	Retention Time (Min)
Ketones	2,3-Pentanedione	$C_5H_8O_2$	7.918
	2,3-Hexanedione	$C_6H_{10}O_2$	9.751
	2-Hexanone	$C_6H_{12}O$	9.854
	2-Heptanone	$C_7H_{14}O$	11.887
	6-Methyl-2-heptanone	$C_8H_{16}O$	13.115
	(Z)-6-Octen-2-one	$C_8H_{14}O$	13.680
	2-Octanone	$C_8H_{16}O$	13.772
	5-Methyl-3-hepten-2-one	$C_8H_{14}O$	13.873
	3,5-Octadien-2-one	$C_8H_{12}O$	15.215
	Acetophenone	C_8H_8O	15.396
	2-Nonanone	$C_9H_{18}O$	15.526
	(E,E)-3,5-octadien-2-one	$C_8H_{12}O$	15.650
	2-Decanone	$C_{10}H_{20}O$	17.155
	2-(2-nitro-2-propenyl)-cyclohexanone	$C_9H_{13}NO_3$	17.890
	2-Undecanone	$C_{11}H_{22}O$	18.679
	7-Methylene-6 (or 8)-methyl-bicyclo [3.3.0]octan-2-one	$C_{10}H_{14}O$	18.918
	2,6-Di-tert-butyl-4-hydroxy-4-methylcyclohexa-2,5-dien-1-one	$C_{15}H_{24}O_2$	21.084
Alcohols	2-Methylbutan-2-ol	$C_5H_{12}O$	6.882
	1-Penten-3-ol	$C_5H_{10}O$	7.682
	1-Pentanol	$C_5H_{12}O$	9.358
	(Z)-2-Penten-1-ol	$C_5H_{10}O$	9.412
	1-Hexen-3-ol	$C_6H_{12}O$	9.649
	(Z)-3-Octen-2-ol	$C_8H_{16}O$	10.494
	2-Octyn-1-ol	$C_8H_{14}O$	13.361
	(E)-2-Octen-1-ol	$C_8H_{16}O$	14.045
	2,7-Octadien-1-ol	$C_8H_{14}O$	15.170
	6,6-Dimethyl-bicyclo[3 .1.1]hept-2-ene-2-methanol	$C_{10}H_{16}O$	17.641
	5-(methylenecyclopropyl)-1-pentanol	$C_9H_{16}O$	19.758
	2,6-bis(1,1-dimethylethyl)-1,4-benzenediol	$C_{14}H_{22}O_2$	21.317
Alkenes	2,3-Dimethyl-2-butene	C_6H_{12}	6.227
	(E)-3-Methyl-2-pentene	C_6H_{12}	6.471
	1,4-Cyclohexadiene	C_6H_8	8.307
	Toluene	C_7H_8	9.573
	(Z,Z)-3,5-Octadiene	C_8H_{14}	10.337
	3,5,5-Trimethyl-2-hexene	C_9H_{18}	13.502
	1,3-bis(1,1-dimethylethyl)-benzene	$C_{14}H_{22}$	18.192
	Butylated Hydroxytoluene	$C_{15}H_{24}O$	21.730
Nitrogenous compounds	Dimethylamine *	C_2H_7N	4.350
	Trimethylamine *	C_3H_9N	4.581
	Trimethylamine oxide *	C_3H_9NO	4.883
	1-Methyl-1H-tetrazole	$C_2H_4N_4$	7.739
	3-Methyl-butanenitrile	C_5H_9N	8.716
	N-Nitrosodimethylamine	$C_2H_6N_2O$	8.862
	Methylpyrazine	$C_5H_6N_2$	10.696
	3-Methyl oxime butanal	$C_5H_{11}NO$	10.992
	2,5-Dimethylpyrazine	$C_6H_8N_2$	12.449
	2-Acetylthiazole	C_5H_5NOS	14.513
	Benzothiazole	C_7H_5NS	18.255
Alkanes	Cyclohexane	C_6H_{12}	6.704
	Hexylidene cyclopropane	C_9H_{16}	9.912
	4-Methyloctane	C_9H_{20}	11.355
	Dodecane	$C_{12}H_{26}$	14.903
	(1.alpha.,3.alpha.,5.alpha.)-1,5-diethenyl-3-methyl-2-methylene-cyclohexane	$C_{12}H_{18}$	20.135
Furans	2-Methylfuran	C_5H_6O	9.155
	5-Isopropyl-3,3-dimethyl-2-methylene-2,3-dihydrofuran	$C_{10}H_{16}O$	16.740
Esters	Cis-cyclohexane-1,4-dimethanol diacetate	$C_{12}H_{20}O_4$	18.508
	2,2,4-trimethyl-1,3-pentanediol diisobutyrate	$C_{16}H_{30}O_4$	22.692
Others	Trimethyloxirane	$C_5H_{10}O$	6.5140
	Di-t-butylacetylene	$C_{10}H_{18}$	12.546
	(Z)-3-undecen-1-yne	$C_{11}H_{18}$	15.796
	2-(hexyn-1-yl)-3-methoxymethylene oxirane	$C_{10}H_{14}O_2$	18.042
	Acetic acid, nonyl ester	$C_{11}H_{22}O_2$	18.820

* Compounds identified by HS-GC-NPD while all the others were identified by GC-MS.

3.1.2. Most Potent Odor-Active Volatile Compounds

To better understand the contribution of each compound to the overall aroma of the studied HMH, the frequency detection method was used. As shown in Table 3, a total of 15 compounds were perceived by at least 5 of the 13 judges, and would be considered to be the most potent odorants of the HMH. 11 of these 15 odorants were aldehyde compounds. In a general way, aldehydes impact significantly the overall aroma, due to their lower odor thresholds compared to other chemical groups. This means that only a small concentration of these compounds is necessary to make them perceptible [33]. Among the aldehyde compounds, 3-methylbutanal was perceived by all the panelists, while 2-methylbutanal was perceived by 6 of the 13 judges. 3-methylbutanal and 2-methylbutanal are generally generated from the Strecker reaction of leucine and isoleucine, respectively [21,45]. Both of these saturated aldehydes are characterized by burnt, malted, dark chocolate smells. 3-methylbutanal was identified as being the major aldehyde contributing to the odor of white herring [49]. Pentanal, hexanal, heptanal and octanal were sniffed by at least 6 of the 13 judges. These compounds are derived from lipid oxidation. On the one hand, pentanal and hexanal are respectively responsible for pungent and green smells, while green, rancid and fishy notes are ascribed to heptanal [21,34,36,38]. On the other hand, octanal is described as having citrus and orange peel smells, but also as having fatty and fishy smells [34,36,38]. The smell of octanal may probably depend on its concentration and its possible interactions with other compounds present in the matrix of interest. An interesting point was noticed by Liu et al. (2017) concerning hexanal. They noted that hexanal boosted the global fishy odor when it acted synergistically with other compounds, while its odor could not really be considered fishy when it stood alone [50]. Also, it is noteworthy that hexanal was already found in herring paste and in white herring [39,49]. (Z)-4-Heptenal was perceived by 7 of the 13 judges. This alkenal would be formed during lipid oxidation, and it is characterized by a fishy off-flavor [11,38]. In a general way, alkanal and alkenals were known to contribute to the fatty-oily, slightly rancid odor of marine products [36]. Methional was sniffed by 11 of the 13 judges, while benzaldehyde was sniffed by all the panel members. On the one hand, methional is an aldehyde sulfur-containing compound characterized by cooked potato notes, whose formation is associated to the Strecker degradation of methionine. Interestingly, it was the only sulfur-containing compound perceived by the panel. Surprisingly, other sulfur-compounds, such as methanethiol, dimethyl disulfide and dimethyl trisulfide, well-known for the considerable role they play in the odor of fishery products, were not identified in this study, while the two other sulfur-compounds, 2-acetylthiazole and benzothiazole, present in HMH were not sniffed by the panel [49,51]. Methional was already identified as an odorant compound in white herring [49]. On the other hand, benzaldehyde is an aromatic aldehyde characterized by almond, nutty or fruity aromas [38,46]. The two alkadienals, (E,E)-2,4-heptadienal and (E,Z)-2,6-nonadienal, were perceived by, respectively, five and 12 of the 13 judges. They both originate from lipid oxidation [35]. Aidos et al. (2002) found that (E,E)-2,4-heptadienal was representative of the oxidative status of herring oil [52]. (E,E)-2,4-heptadienal is characterized by fatty and fishy odors, while (E,Z)-2,6-nonadienal is reported to have a cucumber-like smell [21,34,38]. Nevertheless, it was reported that (E,Z)-2,6-nonadienal promotes a fishy off-flavor while it stands with other compounds [52]. Another group of chemical compounds that was well perceived by the panel was ketone compounds. More precisely, 2,3-pentadione, (Z)-6-octen-2-one and 2-nonanone were sniffed by at least nine of the 13 judges. The presence of ketones can, similarly to aldehyde compounds, be attributed to lipid and/or amino acid degradation. Iglesias et al. (2009) mentioned that 2,3-pentanedione could be even used as an indicator of lipid oxidation in chilled fish muscle [51]. 2,3-Pentadione is responsible for butter and fruity notes, while 2-nonanone is characterized by grass and green notes [9,34,49]. However, no data was found in the literature concerning the odor of (Z)-6-octen-2-one, and it seemed that this compound was never identified in marine products. Three hypotheses could be made to explain the perception of (Z)-6-octen-2-one in this study. The first one is that, as the hydrolysate studied in that case is an HMH, this compound may be a specific compound of this type of matrix. Nevertheless, as no study has been carried out so far regarding the volatile compounds of fish milt, to the best of our knowledge, it was not possible to confirm this first hypothesis. A second possible explanation could be that a wrong identification occurred. In fact, there is another ketone with a similar structure, 1-octen-3-one, that is highly reported in the literature for its fishy smell, but it was not identified in the case of this study [11]. Finally, the last explanation could be that the judges continued to perceive the odor of benzaldehyde, which was eluted just before (Z)-6-octen-2-one. Indeed, some studies already highlighted that the use of GC-O was not an exact science, and some mistakes could happen due to the fact that the odor detector of this method is the human nose [38,53]. Since (Z)-6-octen-2-one was identified by GC-MS with a match degree of 92%, and since no information was available in the literature regarding the duration of the smell of benzaldehyde, the hypothesis that seemed to be the most plausible was that (Z)-6-octen-2-one was a specific odor-active compound of HMH. It is of interest to note that (E,E)-3,5-octadien-2-one, a ketone compound identified in this study but not perceived by the panel, is well recognized to contribute to fatty-fishy odors in fishery materials [39]. Finally, 1-Methyl-1H-tetrazole was also a compound detected by at least five of the 13 judges. 1-Methyl-1H-tetrazole is a cyclic nitrogen-containing compound. Similarly to (Z)-6-octen-2-one, no information was found in the literature concerning the odor and the presence of this compound in seafood materials. Moreover, the judge responses did not allow us to clearly distinguish if the smell of 1-methyl-1H-tetrazole was negative or positive. Therefore, it could be supposed that 1-methyl-1H-tetrazole was a specific odor-active compound of HMH. Except for (Z)-6-octen-2-one and 1-Methyl-1H-tetrazole, all the compounds perceived by the panel were already reported for their contribution to the odor of various fishery materials [21,25,34,36,38,44,45,47]. Interestingly, no compound belonging to the alcohol group was detected by the judges, particularly the alcohol compound 1-penten-3-ol. In fact, this compound, well-known to contribute to the fishy smell, was identified in HMH. However, it was not perceived by the judges as a contributor to the overall aroma of this hydrolysate. This could be explained by the fact that alcohols generally do not have a huge contribution to the overall aroma of food products due to their high odor detection thresholds [21,33,36,47]. This also explained the fact that no alkane and alkene compounds were perceived by the panel members.

Table 3. Most potent odor-active volatile compounds of HMH.

RT (Min)	Compounds	No. of Judges [a]	Average Intensity [b]	No. of Judges Having Qualified the Odor as Bad or Good [c]		Odorant Properties [d]	
				Bad	Good	Descriptors	Odor Threshold (μg/kg)
7.200	3-Methylbutanal	13	4.0	5	6	Burnt, malted, dark chocolate	0.2–2.0
7.381	2-Methylbutanal	6	3.0	1	4	Burnt, malted, dark chocolate	1.0–3.0
7.739	1-Methyl-1H-tetrazole	9	2.0	5	3	-	-
7.918	2,3-Pentanedione	12	3.0	0	10	Butter, fruity	15–5505.6
8.009	Pentanal	7	2.0	3	4	Pungent	12–42
10.121	Hexanal	10	2.0	1	7	Green	5.0
12.114	(Z)-4-heptenal	7	3.0	5	2	Fishy	0.04–0.8
12.156	Heptanal	6	3.0	3	2	Green, rancid, fishy	0.7–2.9
12.338	Methional	11	3.0	5	5	Cooked potatoes	0.2
13.552	Benzaldehyde	13	3.0	9	3	Almond, nutty, fruity	41.7–4600
13.680	(Z)-6-octen-2-one	9	3.0	7	0		
13.978	(E,E)-2,4-heptadienal	5	2.0	3	2	Fatty, fishy	10–15.4
14.045	Octanal	7	3.0	1	6	Citrus, orange, fatty, fishy	0.6
15.526	2-Nonanone	12	4.0	8	3	Grass, green	200
16.642	(E,Z)-2,6-nonadienal	12	3.0	4	6	Cucumber-like	1

[a] Number of judges (out of thirteen) who have perceived an odor. [b] The average intensity given by the judges (out of thirteen) who perceived an odor. The average intensity was rounded to the nearest whole number i.e., an average intensity between 2.0 and 2.5 was rounded to 2.0, and an average intensity between 2.5 and 3.0 was rounded to 3.0. [c] The judges were asked to qualify the odor as bad or good. Sometimes the total number was inferior to those related to the number of judges having perceived an odor, since some of them simply considered the odor to be OK. [d] Odorant properties were indicated only when they were available. They were gathered from the following literature and online database [9,11,16,21,33–36,45–47,54,55]; (http://www.odour.org.uk; http://www.flavornet.org).

3.2. *Deodorization by ED*

3.2.1. ED Parameters

pH

The evolution of the pH of both HMH and KCl recovery solutions during ED treatments in the four different conditions is shown in Figure 2. Firstly, independently of the conditions of the current and pH tested, the pH of hydrolysate solution during the four different ED treatment conditions was steady at pH 4 or 7. However, concerning the KCl recovery solution, its pH varied differently according to the current conditions applied ($p < 0.05$), the pH of the hydrolysate ($p < 0.05$), and a combination of current conditions and pH ($p < 0.05$). Hence, for the hydrolysate at pH 4 without current, the pH of the KCl recovery decreased from 7.13 ± 0.06 to 5.60 ± 0.60 ($p < 0.05$), while those with current rapidly decreased from 7.13 ± 0.08 to 3.57 ± 0.08 ($p < 0.05$) during the first twenty minutes, and then continued to steadily drop to reach the final value of 2.42 ± 0.03. In the same time, regarding the treatment of hydrolysate at pH 7 without current, the pH of the KCl recovery solution remained steady at 7.06 ± 0.03 during all of the experiment, while with current, the pH of the KCl recovery increased rapidly from 7.09 ± 0.07 to 8.80 ± 0.65 ($p < 0.05$) during the first twenty minutes, and then continued to steadily rise to reach the plateau value of 9.50 ± 0.38.

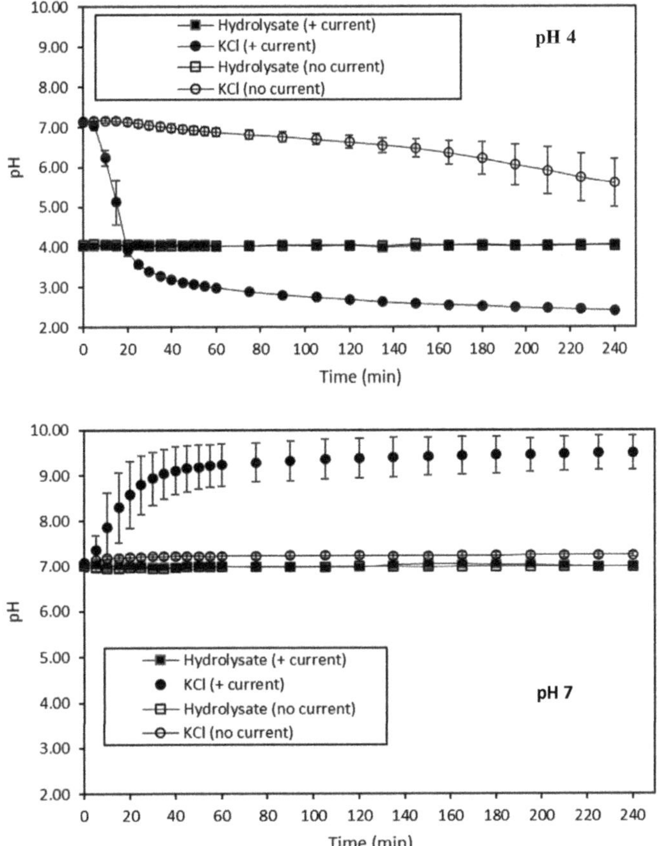

Figure 2. pH evolution in HMH solutions at pH 4 and 7, and in the corresponding KCl recovery solutions treated with and without current during ED treatments of 240 min.

The steady value of the pH observed for the hydrolysate solution during the four ED treatments was consistent with the fact that the pH of this solution was constantly adjusted to the desired value. The slight decrease in pH noticed for the KCl recovery solution of the hydrolysate at pH 4 without current would suggest the potential diffusion of acid species through the CEMs. Indeed, as HCl was added to acidify the hydrolysate at pH 4, H^+ coming from the dissociation of HCl molecules diffused to the KCl solution. However, the rapid decrease in pH during the first twenty minutes of the experiment of the KCl recovery solution with current at the same pH could indicate that water dissociation took place at an early stage. Indeed, even if HCl molecules were initially added to lower and then to maintain the pH of the hydrolysate at pH 4, the fact that the pH of the corresponding KCl solution rapidly dropped below that of the hydrolysate implied that the electromigration of H^+ acid species coming from the dissociation of HCl molecules would not have been sufficient to obtain such a decrease. For this reason, the occurrence of water dissociation could be considered. Water dissociation leads to the formation of protons and hydroxyl ions at the IEM (AEM or CEM) diluate interfaces, bringing about a pH variation [56,57]. This phenomenon generally occurs when the electrolyte concentration near the diluate side of the membranes becomes close to zero, and is due to the reaching of the limiting current density (LCD). As a consequence, the usual mass transfer of ionic species is hampered, and water dissociation takes place [27]. In that case, the decrease in pH value observed in the KCl recovery solution could suggest that the dissociation of water was more important at the CEM interfaces than at the AEM ones [56]. Regarding the KCl solution for the hydrolysate at pH 7 without current, where no variation of pH was observed, on the contrary to pH 4, no diffusion of H^+ occurred. However, in that case, the quantity of HCl required to maintain the pH value of the hydrolysate at pH 7 was lower than that which was necessary to maintain the pH value at 4. This might explain the difference observed between pH 4 and 7 conditions. Finally, regarding the KCl recovery solution of hydrolysate at pH 7 with current, the drastic pH increase during the first twenty minutes was due to a rapid migration of basic species. Similarly to the ED experiment conducted on hydrolysate at pH 4 with current, water dissociation took place [32,58]. Nonetheless, in the present case, the increase in pH in the KCl recovery solution was related to the migration of OH^- basic species. This implied that the dissociation of water was more intensive at the AEM interfaces than at the CEM ones [32,56,58].

Conductivity

The evolution of the conductivity of both HMH and KCl recovery solutions during the ED treatments in the four different conditions is presented in Figure 3. Firstly, the conductivity of the hydrolysate solution was mainly impacted by the current conditions ($p < 0.05$). The same trend was observed for the KCl recovery solution. Hence, when a current was applied on hydrolysate at pH 4, a decrease in the hydrolysate conductivity, from 4.57 ± 0.20 mS/cm to 3.40 ± 0.13 mS/cm, and an increase in the KCl solution conductivity, from 3.63 ± 0.07 mS/cm to 6.95 ± 0.13 mS/cm, were observed ($p < 0.05$). However, when the hydrolysate at pH 4 was treated without current, no change in conductivity of both the hydrolysate and the KCl recovery solutions was observed ($p > 0.05$). Regarding the ED treatment with current on hydrolysate at pH 7, a decrease in the conductivity of the hydrolysate, from 2.79 ± 0.06 mS/cm to 1.91 ± 0.11 mS/cm, and an increase in the conductivity of the KCl solution, from 3.69 ± 0.06 mS/cm to 5.10 ± 0.15 mS/cm, were also observed ($p < 0.05$), as for pH 4, but in a lower way. Finally, without current on hydrolysate at pH 7, no change was observed in the conductivity of both the hydrolysate and KCl recovery solutions, as for pH 4 ($p > 0.05$).

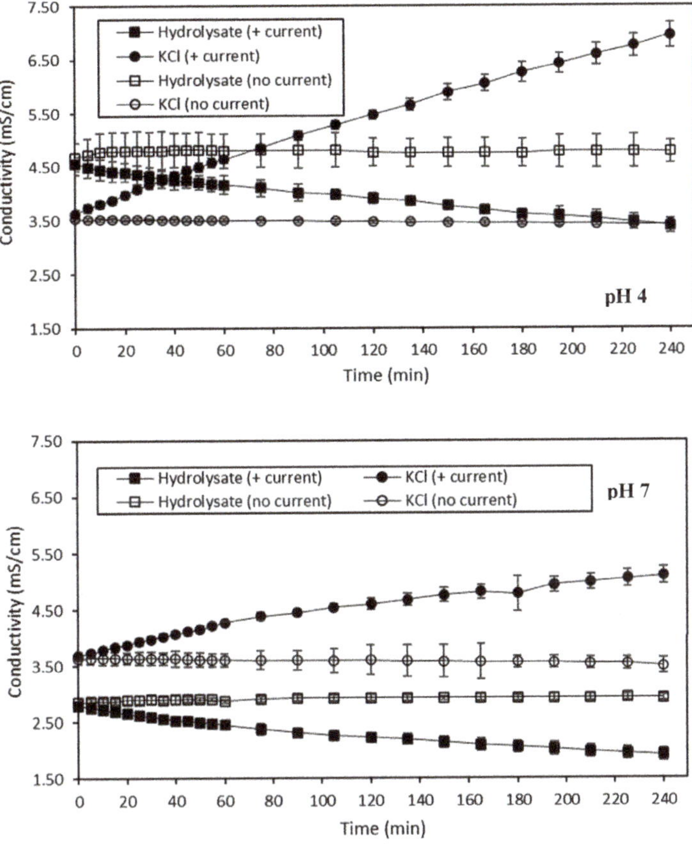

Figure 3. Conductivity evolution in HMH solutions at pH 4 and 7, and in the corresponding KCl recovery solutions, treated with and without current during ED treatments of 240 min.

The decrease in conductivity observed for hydrolysate at pH 4 and 7 during the ED treatments conducted with current was representative of their demineralization, resulting in final respective DRs of 25.62 ± 0.92% and 31.45 ± 2.49%. Compared to other demineralizations performed by ED using a similar configuration, these two DRs were quite low. Indeed, in their studies, Dufton et al. (2018) and Lemay et al. (2019) reached a final DR of almost 70% for acid and sweet whey, respectively [58,59]. However, as it was already mentioned, the fact that, in this present study, the pH of the hydrolysate was constantly adjusted to the desired pH value with HCl hindered the demineralization process. In addition, in accordance with the pH evolution observed, some water dissociation took place rapidly after the first twenty minutes of both of these treatments, thus counteracting the efficiency of the demineralization. Nevertheless, the conductivity of hydrolysate at pH 4 was surprisingly quite high, to totally justify the possible occurrence of water dissociation suggested by the pH evolution. Indeed, Dufton et al. (2018) noted the occurrence of water dissociation at an acid whey conductivity close to 3.0 mS/cm [58], whereas, in the case of HMH at pH 4, the water dissociation would have begun at a conductivity close to 4.0 mS/cm. On the contrary, the fact that the conductivity of the hydrolysate at pH 7 was lower than 2.5 mS/cm made the occurrence of water dissociation due to the limited availability of ionic species for electric current transport even more plausible [27,60]. Regarding the two KCl solutions of the ED treatments performed with current, the increase in their conductivity values was correlated to the demineralization of the corresponding hydrolysate solutions, and thus to their mineralization, resulting in final MRs of 48.67 ± 0.50% and 27.65 ± 1.74% for the KCl recovery solution corresponding to the treatment of hydrolysate at pH 4 and 7, respectively. As a comparison, Dufton et al. (2018) obtained an MR of 74% for the recovery compartment of their study dealing with the demineralization of acid whey [58]. The lower MRs obtained in the case of this study were in line with the lower DRs of the hydrolysate discussed previously. Moreover, the conductivity of the KCl solution for hydrolysate at pH 4 with current increased linearly. This suggests the continuous migration of ionic species, among them H^+ species, into the KCl compartment. However, it was less obvious to qualify the evolution of KCl conductivity as linear for the ED treatment conducted on hydrolysate at pH 7 with current, indicating that the H^+ migration would not be as continuous in that case. Finally, the absence of changes in the conductivity of the hydrolysate solutions at pH 4 and 7, and the corresponding KCl recovery solutions for the ED treatments conducted without current, showed that no demineralization and no mineralization occurred. This was in line with the fact that no current was applied. Concerning ED on the hydrolysate at pH 4 without current, the diffusion of H^+ suggested by the pH evolution was not perceived in terms of conductivity in that case. This observation was not consistent with the fact that H^+ species are known to impact conductivity [61]. However, the observed drop of pH was approximately of one unit, meaning that this variation was caused by around 0.000001 M of H^+ and, normally, the conductivity contribution of this ionic species would not have been perceived in 0.02678 M of 2 g/L KCl. The fact that such pH variation was visible suggests that the membrane integrity could have been altered by a potential fouling.

Ash Content

The ash content of HMH and KCl recovery solutions was analyzed, and is presented in Tables 4 and 5. The ash content of the hydrolysate solution was only impacted by the current conditions applied ($p < 0.05$). The same trend was observed for the KCl recovery solution. At the initial time, the ash content of hydrolysate at pH 4 was 0.478 ± 0.005%. Hence, after ED with current, the ash content of the hydrolysate at pH 4 decreased to 0.400 ± 0.048% ($p < 0.05$), while no change was observed regarding the ash content of hydrolysate at pH 4 after ED without current ($p > 0.05$). Concerning the hydrolysate at pH 7, its ash content was 0.484 ± 0.007% at the initial time. This value was similar to those obtained for hydrolysate at pH 4 at the initial time ($p > 0.05$). After ED of the hydrolysate at pH 7 with current, the ash content decreased to 0.427 ± 0.020% ($p < 0.05$). The final ash contents of both hydrolysates at pH 4 and 7 after ED with current applied were similar ($p > 0.05$). However, ED carried out without current did not lead to any change in the ash content of the hydrolysate at pH 7 at the final time compared to the initial time ($p > 0.05$). With regard to the KCl recovery solution, its initial ash content was 0.166 ± 0.007%. After ED with current on the hydrolysate at pH 4, the ash content of the KCl recovery solution rose to 0.240 ± 0.022% ($p < 0.05$). At the same time, the ash content of the KCl recovery solution for ED with current on hydrolysate at pH 7 increased as well, compared to the initial time, since the final value obtained in that case was 0.230 ± 0.005% ($p < 0.05$). The ash content of both KCl recovery solutions after ED with current on the hydrolysate at pH 4 and pH 7 was similar ($p > 0.05$). Finally, no change was observed in the ash content of KCl recovery solutions after ED of the hydrolysate at pH 4 and 7 without current, compared to initial time ($p > 0.05$).

Table 4. Ash content of the HMH at the initial time and after the different ED treatments (mean ± standard deviation).

	pH 4			**pH 7**		
	Initial Time	**Final Time (ED + Current)**	**Final Time (ED No Current)**	**Initial Time**	**Final Time (ED + Current)**	**Final Time (ED No Current)**
Ash content (%)	0.478 ± 0.005 ^{a}	0.400 ± 0.018 ^{c}	0.460 ± 0.020 ^{ab}	0.484 ± 0.007 ^{a}	0.427 ± 0.020 ^{bc}	0.462 ± 0.017 ^{ab}

Values within the same row with different letters (a–c) are significantly different $p < 0.05$ (Tukey test).

Table 5. Ash content of the KCl recovery solution at initial time and after the different ED treatments (mean ± standard deviation).

		pH 4		pH 7	
	Initial Time	Final Time (ED + Current)	Final time (ED No Current)	Final Time (ED + Current)	Final Time (ED No Current)
Ash content (%)	0.166 ± 0.007 [b]	0.240 ± 0.022 [a]	0.140 ± 0.023 [b]	0.230 ± 0.005 [a]	0.167 ± 0.016 [b]

Values within the same row with different letters (a–b) are significantly different, $p < 0.05$ (Tukey test).

The ash content of the hydrolysate at pH 4 and the hydrolysate at pH 7 at the initial time was surprisingly similar. Indeed, as the acidification of the hydrolysate to pH 4 required more HCl than the acidification to pH 7, it would have been more logical to obtain a higher value of ash content for the hydrolysate at pH 4 than for the hydrolysate at pH 7, due to the contribution of Cl^- species formed by the dissociation of HCl molecules. Nevertheless, the fact that the difference in HCl volume added to the initial hydrolysate to reach these two pH values was only in the order of a few milliliters might explain this result. The lower ash content of the hydrolysate at pH 4 and 7 after ED with current was related to their demineralization. This observation was in line with the decrease in the conductivity of these two solutions, as described previously. Based on the initial and final ash content values, the DRs obtained for the hydrolysate at pH 4 and for the hydrolysate at pH 7 was 16.71 ± 4.59% and 11.62 ± 4.71%, respectively. However, these two DR values were not consistent with those of 25.62 ± 0.92% for the hydrolysate at pH 4 and those of 31.45 ± 2.49% for the hydrolysate at pH 7, determined by means of the conductivity measurements. This discrepancy between the DR values would confirm the occurrence of water dissociation during these two ED treatments. Indeed, it was already observed that the calculation of DR based on conductivity values could be biased in the case of water dissociation, since this phenomenon leads to the formation of ionic species, namely H^+ and OH^-, impacting the conductivity but not the ash content [61]. Regarding the corresponding KCl solutions, their increase in ash content was correlated to their mineralization. This was consistent with the increase in their conductivity mentioned before. For the KCl recovery solutions, based on their ash content values, the corresponding MRs were 29.27 ± 3.40% and 28.87 ± 3.24% for the hydrolysate at pH 4 and 7, respectively. However, the MRs obtained in the case of the conductivity measurements were 48.67 ± 0.50% and 27.65 ± 1.74% for the KCl recovery solutions after ED of the hydrolysate at pH 4 and 7, respectively. If the MRs of the KCl recovery solution of the hydrolysate at pH 7 were similar in both cases, at pH 4 they were highly different depending on the equation used to calculate them. This could suggest that a higher migration of H^+ species took place in the KCl recovery solution during ED of the hydrolysate at pH 4 with current. As explained previously, the MR of the KCl recovery solution was overestimated, based on conductivity values [61]. This could also explain why the conductivity evolved linearly for the KCl solution during ED with current of the hydrolysate at pH 4, but not for the KCl with current at pH 7. Finally, the fact that the ash content of the hydrolysate at pH 4 and 7 after the ED treatments without current remained unchanged compared to initial time was in line with the absence of demineralization, since no current was applied. This also explained the fact that no change was observed in the ash content of the corresponding KCl solutions.

Global System Resistance

The global system resistance evolutions of ED treatments conducted at pH 4 and 7 with current are shown in Figure 4. Regarding the treatment conducted at pH 4, the global system resistance increased significantly from 25.00 ± 0.00 Ω to 88.89 ± 9.62 Ω ($p < 0.05$), while those concerning the treatment carried out at pH 7 increased significantly from 33.33 ± 0.00 Ω to 125.00 ± 0.00 Ω ($p < 0.05$). Moreover, for both treatments, the increase in global system resistance was even more visible after the first twenty minutes of treatments.

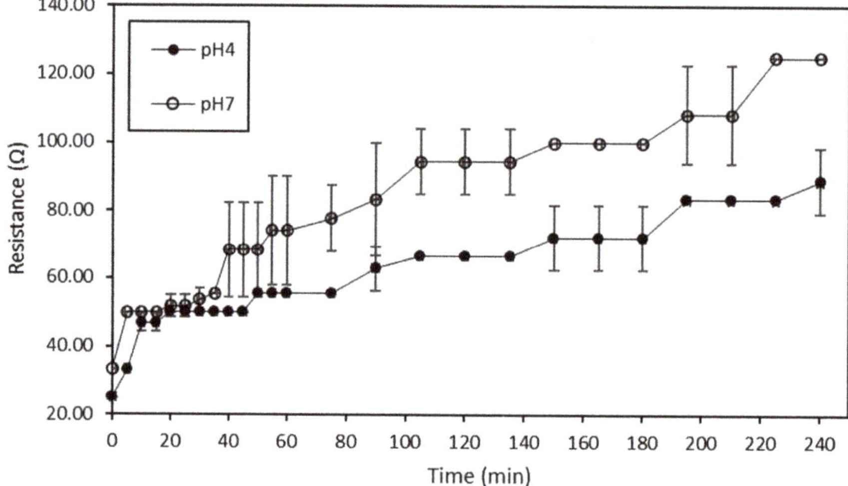

Figure 4. Global system resistance evolution during ED treatments of 240 min at pH 4 and pH 7, with current.

The ED treatments performed on the hydrolysate at pH 4 and 7 presented a 3.5-fold increase and a 3.75-fold increase in global system resistance, respectively. Dufton et al. (2018) experienced a similar increase in global system resistance during the demineralization of acid whey by ED [58]. Such an increase in global system resistance could not only be due to the demineralization process but also to the occurrence of water dissociation as a cause or as a consequence of potential membrane mineral and/or protein fouling, as previously observed by Dufton et al. (2018) [58]. This is corroborated by the fact that this increase in global system resistance for both ED treatments took place at the same time as the changes in pH evolution mentioned previously. Another interesting point to mention was that, during ED performed on hydrolysate at pH 7, the intensity dropped considerably after the first twenty minutes (data not shown). The decrease in intensity was representative of a lack of ionic species to carry the electric current, and was thus consistent with the increase in global system resistance that was even more visible after the first twenty minutes of treatment. This observation was in line as well with the lower conductivity of the hydrolysate at pH 7 noted rapidly after the beginning of the treatment, suggesting that the LCD was reached. This could explain why the global system resistance was, in that case, so high. However, the fact that the global system resistance of ED conducted on the hydrolysate at pH 4 was lower could indicate that, at that stage, it was still not possible to clearly identify whether the LCD was reached and thus explain the water dissociation.

Membrane Thickness

The evolution of the membrane thickness over the different ED treatments is shown in Figures 5 and 6. None of the membranes evidenced an increase in thickness over ED, with and without current, performed on the hydrolysate at pH 4 ($p > 0.05$) (Figure 5). Regarding the ED treatments carried out on the hydrolysate at pH 7, both AEM1 and CEM3 showed an increase in thickness ($p < 0.05$) for the experiment without current, while only AEM4 presented an increase in thickness ($p < 0.05$) for the experiment with current (Figure 6).

Membrane thickness is an indicator of membrane integrity, and more particularly of membrane fouling [58]. This could indicate that no fouling occurred during ED treatments on the hydrolysate at pH 4, showing the efficiency of the NaCl rinsing after each run. This was consistent with the visual observations of the membranes after each run as well. However, even if a few membranes evidenced an increase in thickness after the ED treatments on the hydrolysate at pH 7, the final membrane thickness values were still representative of those reported in the literature. Indeed, Lemay et al. (2019) reported average values of 0.142 ± 0.006 mm and 0.141 ± 0.005 mm for AEMs and CEMs, respectively [32]. Therefore, based on membrane thickness evolution, no fouling phenomena seemed to happen at that stage regarding the treatments on the hydrolysate at pH 7.

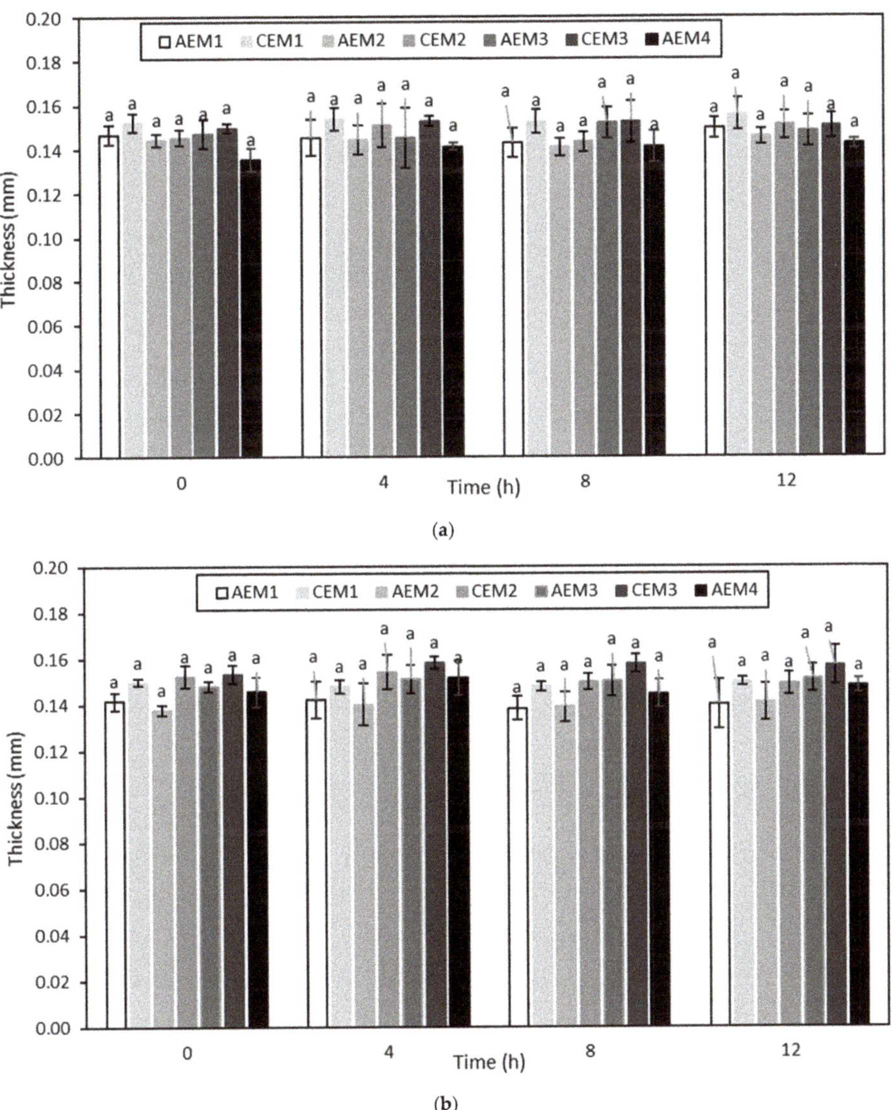

Figure 5. Membrane thickness before and after each 4-h ED treatment conducted at pH 4, (**a**) without current and (**b**) with current. Values with different letters corresponding to the same membranes are significantly different $p < 0.05$ (Tukey test).

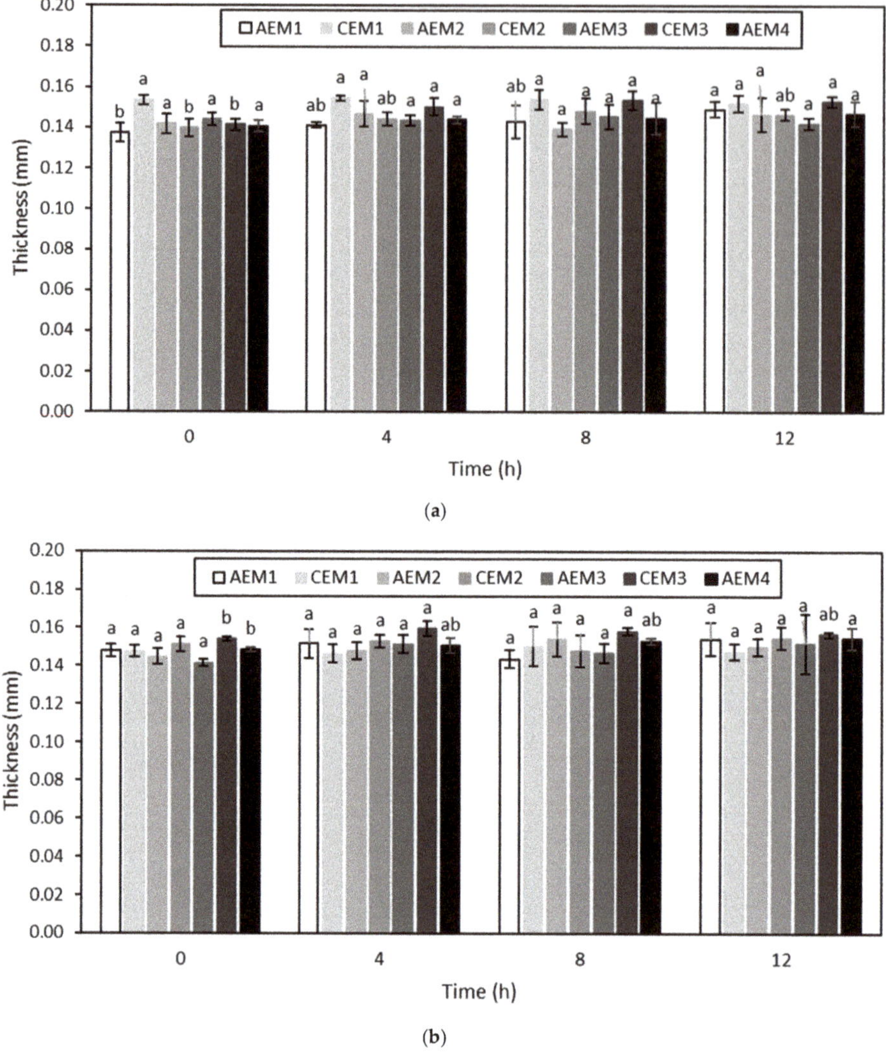

(a)

(b)

Figure 6. Membrane thickness before and after each 4-h ED treatment conducted at pH 7, (**a**) without current and (**b**) with current. Values with different letters corresponding to the same membranes are significantly different, $p < 0.05$ (Tukey test).

Membrane Conductivity

The membrane conductivity evolution over the different ED treatments is presented in Figures 7 and 8. Regarding ED conducted on the hydrolysate at pH 4, AEM2 and all the CEMs for the experiment conducted without current, and all the CEMs only for the experiment conducted with current, evidenced a decrease in their conductivity over time ($p < 0.05$) (Figure 7). Concerning ED on the hydrolysate at pH 7, AEM3, AEM4 and all the CEMs for the experiment without current, and all the AEMs except AEM4 and all the CEMs for the experiment using current, experienced a decrease in their conductivity over time ($p < 0.05$) (Figure 8).

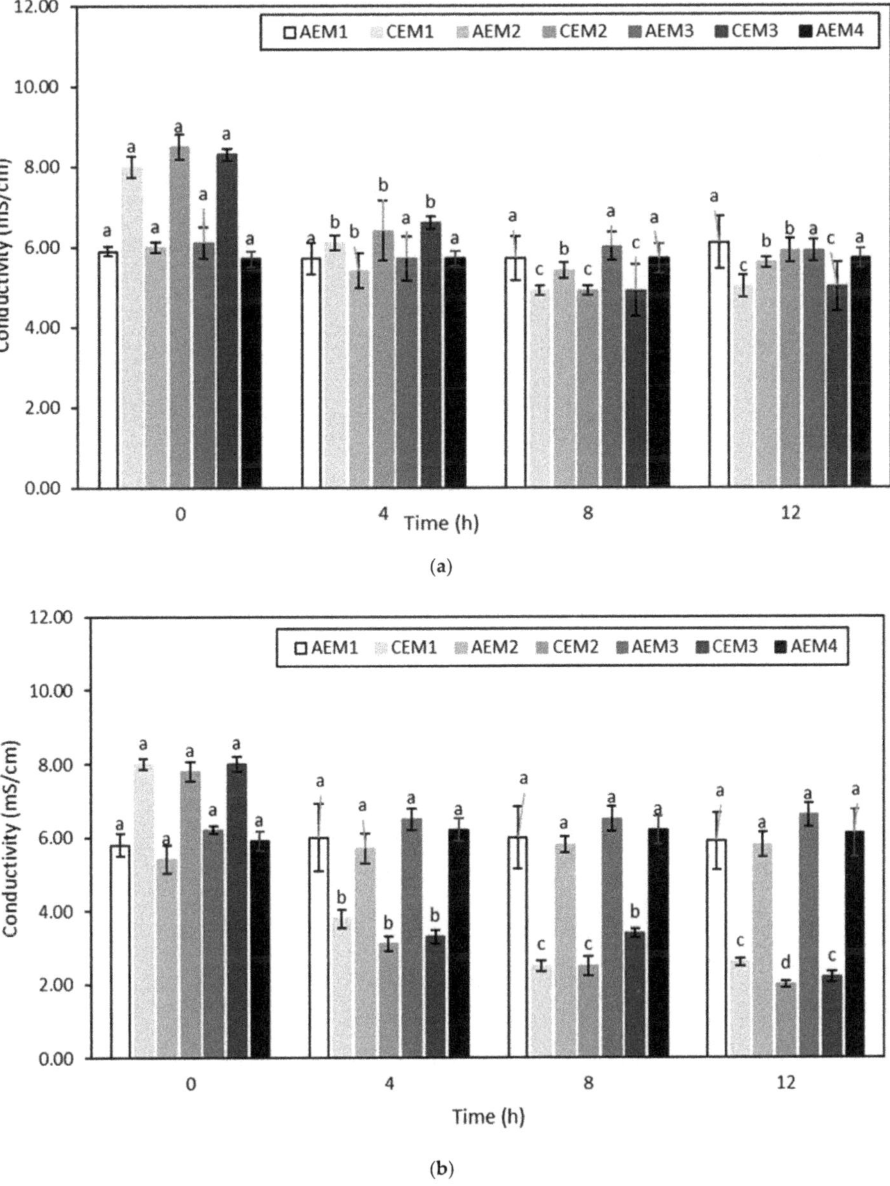

(a)

(b)

Figure 7. Membrane conductivity before and after each 4-h ED treatments conducted at pH 4, (**a**) without current and (**b**) with current. Values with different letters corresponding to the same membranes are significantly different $p < 0.05$ (Tukey test).

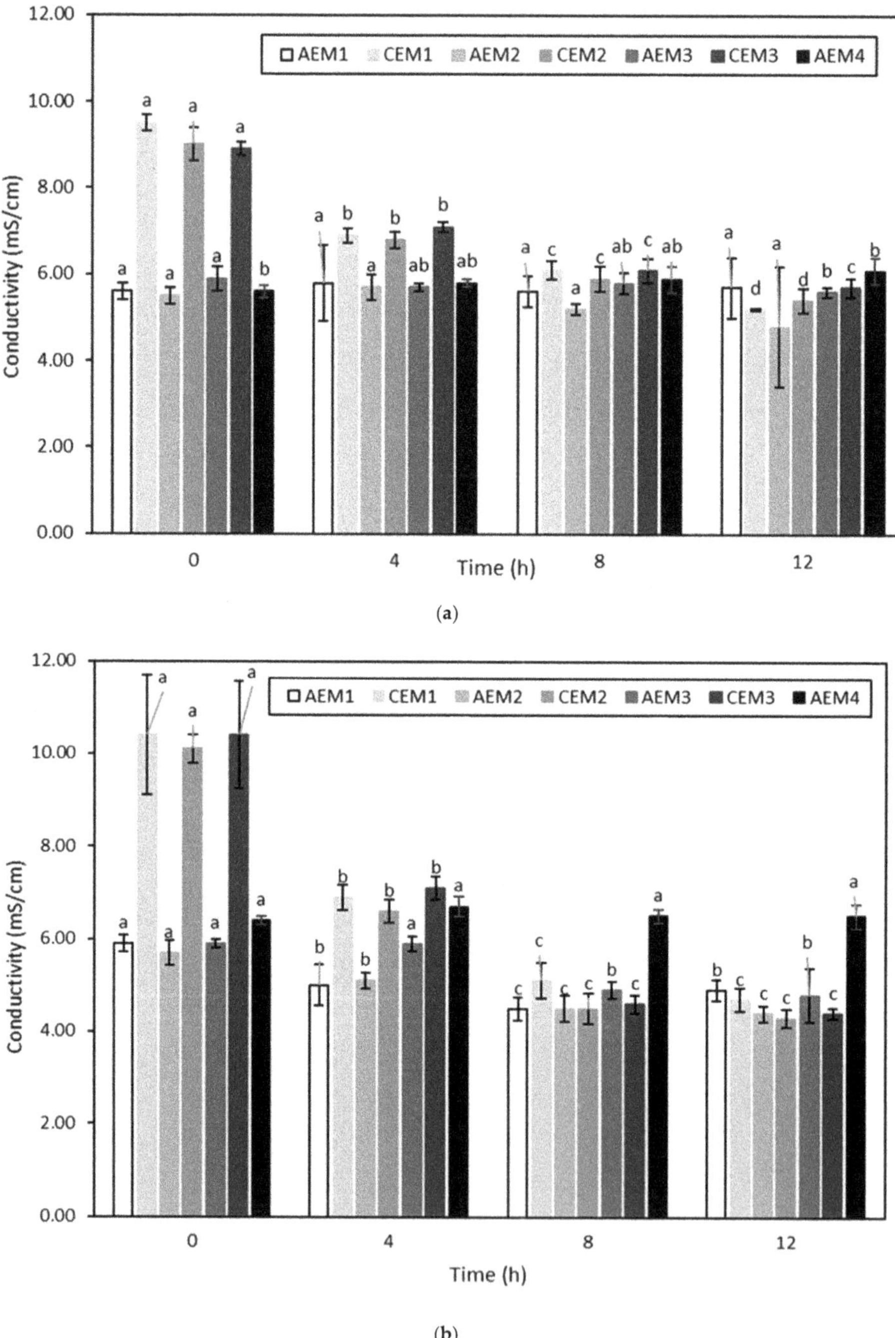

Figure 8. Membrane conductivity before and after each 4-h ED treatments conducted at pH 7, (**a**) without current and (**b**) with current. Values with different letters corresponding to the same membranes are significantly different $p < 0.05$ (Tukey test).

Similarly to membrane thickness evolution, the evolution of membrane conductivity can be considered to be an indicator of membrane integrity [58]. Firstly, the conductivity of both CEMs and AEMs before any run was comparable to the following values reported in the literature: 5.197 ± 0.257 ms/cm and 8.960 ± 0.442 mS/cm for AEMs and CEMs, respectively [32]. Then, all the four ED conditions evidenced a change in membrane conductivity as a function of time. In their study, Lemay et al. (2019) also noted a decrease in membrane conductivity after sweet whey demineralization [32]. In that case, the observed drop was attributed to the substitution of the counterions present in the initial membranes by divalent ionic species of the sweet whey having lower conductivity values, resulting in a decrease in membrane conductivity [32]. Nonetheless, in the present study, the fact that membranes evidenced a decrease in conductivity not only after the ED treatments conducted with current but also after those performed without current may suggest that another explanation could be involved. More specifically, regarding the ED treatments carried out on hydrolysate at pH 4, a decrease in membrane conductivity was already noted after the first run. This could indicate that, independently of the current conditions, ionic compounds present in the hydrolysate solution interacted with the membranes. Indeed, previous works have already showed that charged compounds such as peptides and amino acids could interact with the boundary layers of membranes [56]. As the decrease in membrane conductivity was mainly observed for CEMs, this means that the compounds involved were cationic and interacted with the negatively charged sulfonic groups present in the CEMs. More specifically, Persico et al. (2017) showed in their study the ability of peptides containing histidine (pKa of 6.0), lysine (pKa ~10.5) and arginine (pKa ~12.5) residues, in addition to their amine group at the N-term (pKa ~9.8), to interact electrostatically with the negatively charged sulfonic groups of CEMs due to the positive charges they carry, even when no current was applied [62]. Free arginine could even be considered to be a major agent responsible for the fouling of CEMs [63]. Since HMH is mainly composed of amino acids containing materials (Table 1) including a high amount of arginine in both bound and free forms [1], the important decrease observed in the CEM conductivity could be explained by the neutralization of the fixed membrane charges by these molecules, among them arginine amino acid. At that stage, it is worth it to mention that, after each run, NaCl rinsing was performed. The aim of such rinsing was to reduce the electrostatic interactions occurring between the ionogenic constituents of the membranes and amino acids containing materials from the hydrolysate due to the high ionic strength of the salt rinsing solution [64]. However, the fact that fouling was observed even after this rinsing indicates that other interactions than electrostatic ones could be involved between the ionogenic groups of membranes and components from the hydrolysate. This suggests that hydrophobic interactions could be involved between volatile compounds and membrane constituents as well, as that type of interaction was not impacted by NaCl rinsing [64]. This hypothesis was already formulated in the studies of Cros et al. (2005) and Chindapan et al. (2011), in which they ascribed the decrease in the abundance of certain volatile compounds to hydrophobic interactions occurring with membrane components [25,26]. In addition, the observed decrease in membrane conductivity could be explained by another phenomenon. Indeed, it was already shown that, depending on the nature of the groups from both the membranes and the matrix present at the membrane interface, the catalysis of water dissociation during an ED treatment could take place even if the LCD was not reached [56,65]. More precisely, at the CEM interface, the catalytic action of carboxylic acid present at the C-term of the peptides was already proven [56]. Therefore, the decrease in pH observed in the KCl recovery solution corresponding to the ED treatment of the hydrolysate at pH 4 with current, and the increase in its conductivity, could be effectively due to the occurrence of water dissociation resulting from the action of cationic catalysts compounds, such as peptides or free amino acids involving a carboxylic group, at the boundary layers of the CEMs. On the contrary, for ED of the hydrolysate at pH 4 without current, since a decrease in the CEM conductivity was evidenced, but with no huge pH change of the KCl solution, this could suggest that the interaction of cationic compounds took place, resulting in membrane fouling without bringing about water dissociation due to the absence of current. Regarding ED on the hydrolysate at pH 7, since both AEMs and CEMs presented a decrease in conductivity for both current conditions applied, this may suggest not only that cationic compounds interacted with CEMs but also anionic ones with ammonium groups of AEMs. This observation was consistent with the study of Persico et al. (2016), in which it was observed that fouling of AEMs was absent at acidic pH, while it tended to be more important at a pH close to neutral [64]. This was due to the fact that, at a pH close to neutral, negative residues of carboxylic acid present at the C-term (pKa ~2.1) of peptides, or at the side-chains of aspartic and glutamic acids (pKa ~4.0), were able to interact electrostatically with the positively charged ammonium groups of AEMs [64]. As both aspartic and glutamic acids are present in HMH, in both bound and free forms [1], they could have been responsible for the fouling observed on the AEMs due to their carboxylic acid residues leading to the observed decrease in conductivity. Nevertheless, as noted for CEMs, the fact that fouling was evidenced even after the NaCl rinsing could suggest that other interactions than electrostatic ones could be involved. Moreover, similarly to what was mentioned previously as well, water dissociation phenomena could have taken place at the interface of AEMs, due to the catalytic action of both compounds present in the AEMs and in the hydrolysate [56,65,66]. More precisely, the conversion of the initial quaternary ammonium groups ($-N^+(CH_3)_3$) present in the AEMs into tertiary ($=N(CH_3)_2$) or secondary amine ($\equiv N(CH_3)$) is a factor accelerating water dissociation due to the lone electron pair carried by the newly formed groups [56,66]. Weak-acid anions [65], such as glutamic and aspartic residues present in the studied hydrolysate, can catalytically accelerate the water dissociation as well. Furthermore, this hydrolysate also contains nucleic acids that carry a negative charge due to the presence of phosphoric acid in its anion form. Therefore, nucleic acids could have also established interactions with AEMs, or could have played a role in the catalysis of water dissociation. At this stage, it is worth to mention that the occurrence of water dissociation due to the generation of H^+ and OH^- could be a factor promoting membrane fouling by increasing the interaction with amino-containing materials at the membrane interface [67]. In addition, the fact that both types of IEM were fouled for the condition at pH 7 could explain why the global system resistance increased more significantly than those during the ED treatment conducted on the hydrolysate at pH 4. Another interesting point to note is that the standard deviations corresponding to the global system resistance of the ED treatment of the hydrolysate at pH 7 were quite important compared to those displayed for the ED treatment of the hydrolysate at pH 4. This would probably be due to the progressive decrease in conductivity observed for both types of IEM. It is worth to mention that at that stage, the potential interactions occurring between constituents of the studied hydrolysate, including volatile compounds and the IEM, seemed to be the most plausible explanation. Therefore, the occurrence of water dissociation observed during the ED treatment of the hydrolysate at pH 4 with current could be due more to the action of catalysts than to the reaching of the LCD.

3.2.2. Hydrolysate and Recovery Solutions Analyses

Volatile Compound Analysis

Most Potent Odor-Active Compounds

The abundance of compounds determined to be the most potent odorants of the HMH over the different ED and deaerator treatments is shown in Table 6. Firstly, concerning the composition of the hydrolysate at the initial time, the results indicate that pH had an important impact on the volatile compounds' abundance. It appeared that the abundance of the majority of the most potent odor-active molecules significantly decreased while pH increased from 4 to 10. For example, the abundance of 3-methylbutanal, 2,3-pentanedione, pentanal, hexanal, (Z)-4-heptenal, heptanal, methional, (Z)-6-octen-2-one, (E,E)-2,4-heptadienal, octanal, 2-nonanone and (E,Z)-2,6-nonadienal dropped by at least 50% between the hydrolysate at pH 4 and the hydrolysate at pH 10 at the initial time ($p < 0.05$). 1-methyl-1H-tetrazole was the only compound that did not follow this trend. In fact, it was identified in the hydrolysate at pH 7, but it was not present in the hydrolysate at pH 4 and pH 10. In general, pH is known to be a major factor influencing the content of volatile compounds. Indeed, volatile compounds are able to interact with molecules like lipids through hydrophobic interactions, and amino acids constituents such as proteins, peptides and free amino acids through covalent irreversible bindings, in addition to hydrophobic and ionic interactions [68–70]. Among these different interactions, those taking place between volatile compounds and amino-acid containing compounds are the most impacted by pH, as this factor modifies the conformation and charge of proteins, peptides and free amino acids, and thus the ability of binding of volatile compounds [69]. Based on this fact, two hypotheses can be made concerning the general decrease in volatile compounds observed while pH increased. The first could be that alkaline pH might be responsible for the breaking of interactions taking place between volatile compounds and amino groups. As the targeted compounds are volatile, breaking these interactions could promote their loss. On the contrary, the second could be that the lower abundance of volatile compounds observed in the HMH at pH 10 at the initial time may be representative of a higher interaction with amino acid materials. As no study has been carried out regarding the impact of pH on the retention of volatile compounds by amino acid constituents from HMH materials so far, it was not obvious to clearly validate one hypothesis rather than the other. Nevertheless, some studies with similar purposes were already conducted on milk proteins [71,72], as well as on animal tissues proteins and peptides [69,73]. These studies showed that there was a general trend of amino acid containing molecules, such as peptides, to retain volatile compounds to a higher extent while pH increased. Several explanations are involved, depending on the proteins. For example, the milk protein β-lactoglobulin is reported to bind a larger proportion of volatile compounds at pH 9 than at pH 3. The increase in retention ability is, in this case, explained by better access to the hydrophobic amino acid residues of β-lactoglobulin due to conformation changes occurring under alkaline conditions [70,72]. Interestingly, leucine, a hydrophobic amino acid, is present in both β-lactoglobulin and HMH. Leucine is even present in high proportion in the latter (Table 1). It could be suggested that leucine might participate in the retention of volatile compounds, and while pH increases, the loss of the proton H^+ on the amine group could promote this phenomenon. Independently of pH, Meynier et al. (2004) observed the unavailability of lysine and histidine of milk proteins in the presence of aldehydes, suggesting a potential interaction occurring between these amino acids and volatile compounds. It was proposed to explain this loss that the carbonyl group of aldehydes could react with the primary amine of lysine either by Michael addition or by Schiff base formation. Concerning histidine, it was suggested that aldehyde and, preferentially, alkenal could react with the imidazole ring of histidine [74]. Since among the 15 compounds identified as being the most potent odor-active of HMH, 11 are aldehydes and one of them is an alkenal, namely (Z)-4-heptenal, and since lysine and histidine are both present in this product (Table 1), it could be possible that these interactions occur between these volatile compounds and these amino acids. Histidine is also a constituent of carnosine, a dipeptide found in animal tissues, and whose ability to retain volatile compounds is also known to increase while pH increases [69]. In that case, as pH affects the retention of volatile compounds, it could be proposed, similarly to leucine, that the loss of the proton H^+ on the imidazole ring occurring under alkaline conditions could promote the interaction between histidine and aldehydes. Interestingly, HMH contains a high amount of arginine (Table 1), which is an amino acid with an amine-containing side-chain similar to lysine and histidine. Based on this, it should be proposed that arginine was also involved in the interactions, explaining partially the decrease in volatile compound abundance observed. Therefore, the hypothesis that seemed to be the most plausible regarding the decrease in the abundance of volatile compounds at pH 10 compared to pH 4 and pH 7 would be those implying a higher degree of interaction occurring between volatile compounds and amino acid containing compounds, such as peptides present in the hydrolysate. In that case, the fact that 1-methyl-1H-tetrazole was not detected in the hydrolysate at pH 4 and 10 might suggest that this compound could have more interactions at these pH values than at pH 7, allowing its detection at this pH value only. Also, it is of interest to mention that, while volatile compounds are bound to other components, both their release and perception are hindered [69]. This means that HMH should be globally less odorous at pH 10 than at pH 4 and pH 7.

Table 6. Abundance of the most-potent odor active compounds of HMH ($\times 10^7$ Arbitrary Unit (A.U)) at the initial time and after the different treatments (mean ± standard deviation).

	pH 4				pH 7				pH 10	
	Initial Time	**Final Time (ED + Current)**	**Final Time (ED No Current)**	**Final Time Deaerator**	**Initial Time**	**Final Time (ED + Current)**	**Final Time (ED No Current)**	**Final Time Deaerator**	**Initial Time**	**Final Time Deaerator**
3-Methylbutanal	10.3 ± 2.14 ^{a}	7.61 ± 0.52 ^{ab}	7.67 ± 1.00 ^{ab}	3.36 ± 0.15 ^{cd}	6.47 ± 1.61 ^{b}	5.53 ± 0.40 ^{bc}	4.89 ± 0.50 ^{bc}	1.26 ± 0.05 ^{d}	4.88 ± 0.12 ^{bc}	5.08 ± 0.80 ^{bc}
2-Methylbutanal	4.15 ± 2.06 ^{a}	3.87 ± 0.46 ^{a}	3.32 ± 0.69 ^{ab}	1.00 ± 0.05 ^{bc}	3.67 ± 1.03 ^{a}	2.66 ± 0.54 ^{abc}	2.42 ± 0.55 ^{abc}	0.831 ± 0.094 ^{c}	2.45 ± 0.14 ^{abc}	2.13 ± 0.61 ^{abc}
1-Methyl-1H-tetrazole	0.00 ± 0.00 ^{b}	0.00 ± 0.00 ^{b}	0.00 ± 0.00 ^{b}	0.00 ± 0.00 ^{b}	3.80 ± 0.92 ^{a}	3.94 ± 0.55 ^{a}	2.95 ± 0.73 ^{a}	0.00 ± 0.00 ^{b}	0.00 ± 0.00 ^{b}	0.00 ± 0.00 ^{b}
2,3-Pentanedione	8.90 ± 1.66 ^{a}	7.03 ± 0.63 ^{ab}	7.13 ± 0.79 ^{ab}	5.02 ± 0.72 ^{bc}	3.05 ± 0.94 ^{cd}	2.94 ± 0.22 ^{cd}	2.53 ± 0.22 ^{d}	1.60 ± 0.19 ^{de}	0.356 ± 0.072 ^{e}	0.237 ± 0.018 ^{e}
Pentanal	3.76 ± 0.10 ^{a}	3.35 ± 0.28 ^{ab}	3.64 ± 0.23 ^{a}	1.31 ± 0.02 ^{cd}	2.79 ± 1.75 ^{abc}	3.37 ± 0.40 ^{ab}	2.77 ± 0.40 ^{abc}	0.00 ± 0.00 ^{d}	1.67 ± 0.11 ^{bcd}	1.18 ± 0.14 ^{cd}
Hexanal	6.94 ± 1.54 ^{a}	4.63 ± 1.88 ^{abc}	5.85 ± 0.90 ^{ab}	3.66 ± 0.08 ^{bc}	5.30 ± 1.34 ^{ab}	4.69 ± 0.27 ^{abc}	3.78 ± 0.66 ^{bc}	2.47 ± 0.15 ^{c}	2.14 ± 0.27 ^{c}	2.18 ± 0.55 ^{c}
(Z)-4-heptenal	4.66 ± 2.59 ^{a}	4.40 ± 0.32 ^{a}	4.75 ± 0.94 ^{a}	4.11 ± 0.26 ^{ab}	3.19 ± 1.25 ^{abc}	2.37 ± 0.23 ^{abc}	2.20 ± 0.13 ^{abc}	1.48 ± 0.19 ^{bc}	1.13 ± 0.05 ^{bc}	0.888 ± 0.166 ^{bc}
Heptanal	2.21 ± 0.97 ^{ab}	1.81 ± 0.11 ^{abc}	2.29 ± 0.57 ^{a}	1.42 ± 0.18 ^{abcd}	1.63 ± 0.80 ^{abcd}	1.16 ± 0.16 ^{abcd}	1.10 ± 0.36 ^{abcd}	0.909 ± 0.041 ^{bcd}	0.582 ± 0.107 ^{cd}	0.358 ± 0.112 ^{d}
Methional	0.33 ± 0.19 ^{a}	0.267 ± 0.064 ^{ab}	0.277 ± 0.056 ^{ab}	0.215 ± 0.042 ^{abc}	0.084 ± 0.055 ^{bcd}	0.060 ± 0.015 ^{cd}	0.046 ± 0.027 ^{cd}	0.053 ± 0.005 ^{cd}	0.013 ± 0.004 ^{cd}	0.00 ± 0.00 ^{d}
Benzaldehyde	5.34 ± 2.48 ^{a}	5.67 ± 1.93 ^{a}	4.80 ± 1.74 ^{a}	6.46 ± 0.56 ^{a}	5.19 ± 1.25 ^{a}	5.10 ± 0.81 ^{a}	4.12 ± 0.96 ^{a}	2.72 ± 0.28 ^{a}	3.49 ± 0.73 ^{a}	3.36 ± 0.31 ^{a}
(Z)-6-octen-2-one	1.07 ± 0.44 ^{a}	0.414 ± 0.273 ^{bc}	0.720 ± 0.150 ^{ab}	0.465 ± 0.115 ^{abc}	0.684 ± 0.277 ^{ab}	0.591 ± 0.299 ^{abc}	0.683 ± 0.093 ^{ab}	0.00 ± 0.00 ^{c}	0.891 ± 0.093 ^{ab}	0.275 ± 0.080 ^{bc}
(E,E)-2,4-heptadienal	9.84 ± 1.88 ^{a}	6.11 ± 0.79 ^{b}	7.15 ± 2.01 ^{ab}	9.05 ± 0.62 ^{a}	0.879 ± 0.326 ^{c}	0.770 ± 0.225 ^{c}	0.596 ± 0.247 ^{c}	1.31 ± 0.12 ^{c}	0.306 ± 0.097 ^{c}	0.225 ± 0.079 ^{c}
Octanal	2.73 ± 0.93 ^{a}	1.67 ± 0.17 ^{abc}	1.64 ± 0.67 ^{abc}	1.67 ± 0.08 ^{abc}	1.88 ± 0.71 ^{ab}	1.14 ± 0.16 ^{bcd}	1.25 ± 0.41 ^{bcd}	1.04 ± 0.09 ^{bcd}	0.428 ± 0.016 ^{cd}	0.294 ± 0.033 ^{d}
2-Nonanone	1.38 ± 0.47 ^{a}	1.04 ± 0.29 ^{ab}	1.14 ± 0.30 ^{ab}	0.578 ± 0.015 ^{bc}	1.41 ± 0.56 ^{a}	0.694 ± 0.140 ^{abc}	0.757 ± 0.158 ^{abc}	0.236 ± 0.041 ^{c}	0.569 ± 0.062 ^{bc}	0.149 ± 0.037 ^{c}
(E,Z)-2,6-nonadienal	3.72 ± 0.97 ^{a}	2.24 ± 0.13 ^{b}	2.54 ± 0.78 ^{ab}	1.36 ± 0.17 ^{bc}	0.699 ± 0.294 ^{cd}	0.580 ± 0.073 ^{cd}	0.523 ± 0.044 ^{cd}	0.461 ± 0.044 ^{cd}	0.030 ± 0.006 ^{d}	0.030 ± 0.009 ^{d}

Values within the same row with different letters (a–e) are significantly different $p < 0.05$ (Tukey test).

Then, regarding the ED treatments, no significant difference was globally observed between the hydrolysate at a given pH at the initial time and the hydrolysate treated with or without current at the final time. More precisely, if attention is paid to the ED treatments conducted without current, the fact that no decrease in the content of the targeted volatile compounds was observed would suggest that a simple circulation of the hydrolysate solution for 240 min, independently of its pH, was not sufficient enough to allow a loss of these molecules due to their volatile state. Concerning ED conducted with current, no change in the content of volatile compounds occurred except for (Z)-6-octen-2-one, (E,E)-2,4-heptadienal and (E,Z)-2,6-nonadienal whose abundance was inferior at final time for ED on the hydrolysate at pH 4 ($p < 0.05$). These results could be representative of the non-migration of volatile compounds. Since ED is a process based on the migration of charged compounds, and since the targeted compounds were not supposed to be charged under the conditions tested, it was not surprising, at the first glance, to obtain such results. However, in their studies, Cros et al. (2005) and Chindapan et al. (2011) observed that an ED treatment could lead to a drop of volatile compounds even if they are not charged [25,26]. Several points could explain the discrepancy between these two studies and the present one. The first could be that the compounds whose abundance dropped during ED treatments in the studies of Cros et al. (2005) and Chindapan et al. (2011) were not the same compounds targeted in the present study. Indeed, Cros et al. (2005) observed a significant decrease in the non ionizable (Z)-4-heptenal, 2,3-butanedione, 3-octen-2-one and limonene compounds. The only compound that this study and the present one had in common was (Z)-4-heptenal. Nevertheless, it is noteworthy to mention that Cros et al. (2005) noticed the important decrease of the compounds listed before only while the LCD was reached. Different hypotheses were formulated to explain such a decrease under this specific condition in this study. The first was that the formation of protons H^+ and hydroxyls OH^- resulting from water dissociation under this critical condition could have altered volatile compounds, explaining their decrease. Another was that the LCD could have also brought about a local membrane heating, potentially leading to a thermal degradation of volatile compounds. Or, simply, the volatile compounds could have been adsorbed on the membranes through hydrophobic and ionic interactions [26]. The migration of these molecules was not considered to be a potential explanation, as none of them were found in the recovery solution. Regarding (Z)-4-heptenal only, Cros et al. (2005) hypothesized its hydrogenation in heptanal as a possible explanation for its decrease [26]. Since not all the ED treatments conducted in the present study seemed to have evidenced reaching LCD, it was not possible to totally verify all the hypotheses formulated by Cros et al. (2005). However, as suggested by the analyses of the parameters of the ED treatment conducted on the hydrolysate at pH 7, it seems that this condition experienced the reaching of LCD. As no change regarding the volatile compound content was observed in that case, it may indicate that the LCD was not a sufficient condition to lead to a decrease in the abundance of these compounds. In addition, the fact that both ED treatments conducted with current presented water dissociation in the present study could show that the generation of H^+ and OH^- species could not be effectively responsible for the alteration of the volatile compounds. On the contrary to Cros et al. (2005), Chindapan et al. (2011) did not reach the LCD condition in their study. Despite this fact, they observed a significant decrease in 2,6-dimethylpyrazine, phenol and carboxylic acids (acetic acid, butanoic acid, 2-methylbutanoic acid, pentanoic acid, 4-methylpentanoic acid) while the ED treatment was performed, to reach a salt concentration of 2% in the treated fish sauce. Chindapan et al. (2011) gave two main reasons for the loss of these compounds: either their adsorption on the membranes or their transport through the membranes occurring at the same time as electroosmosis. Nevertheless, as no mention concerning the composition of the recovery solution was made, it was not possible to know if the latter reason was plausible in that case [25]. Concerning the decreased in (Z)-6-octen-2-one, (E,E)-2,4-heptadienal and (E,Z)-2,6-nonadienal observed in the hydrolysate at pH 4 treated with current at final time, three hypotheses could be made, based on those previously mentioned. The first one would be that a slight loss of these molecules due to their volatile state happened during the ED process. However, as no decrease of these compounds was observed for other conditions, this hypothesis does not seem highly plausible. The second hypothesis would be that preferential interactions occurred at pH 4 between these three volatile compounds and other constituents of the HMH. Finally, the last hypothesis would be that these compounds preferentially adsorb on the membranes due to hydrophobic interactions. This last hypothesis appears to be the most probable, based on the membrane conductivity evolution discussed previously.

The content in volatile compounds of the KCl recovery solutions was analyzed for each condition, and is listed in Table 7. The results show the unchanged presence of 3-methylbutanal, hexanal, heptanal, benzaldehyde, octanal and 2-nonanone in KCl solution at final time independently of the ED treatment. It is worthwhile to mention that none of these compounds were detected in the KCl solution at the initial time. Except for benzaldehyde and hexanal, in all cases, the presence of volatile compounds in the recovery solution can be considered to be trace. This should probably be due to a punctual contamination of these compounds due to their volatility from the hydrolysate to the KCl solution. The fact that this phenomenon could be considered to be punctual was accredited by the generally high values of standard deviations proportionally to those of means, and even sometimes the higher values of standard deviations compared to the corresponding means. However, another explanation could be involved for hexanal and benzaldehyde. Regarding hexanal, its presence in the KCl solution could be due to its diffusion or migration. However, based on the membrane conductivity analysis discussed previously, it seemed that some interactions with membrane components also occurred during the different ED treatments. Therefore, another explanation could be that hexanal may have interacted with the sulfonic groups present in the CEMs, resulting in its release into the KCl compartment thereafter. Interestingly, the same trend was not found for compounds similar to hexanal, such as pentanal and heptanal. In that case, the differences observed should probably be due to the presence of hexanal in higher quantity in HMH, compared to pentanal and heptanal. Regarding benzaldehyde, the same explanations as those mentioned for hexanal could be involved. Nevertheless, on the contrary to hexanal, the fact that benzaldehyde was found in higher abundance only in KCl solution of the hydrolysate at pH 4 treated with current may indicate that a special mechanism was involved in that case. Initially, as benzaldehyde is not charged, it was not supposed to migrate. However, its recovery in the KCl solution might suggest that benzaldehyde could have either established interactions with another positively charged constituent that migrated into the CEMs, or that benzaldehyde established an interaction with the sulfonic groups of the CEMs, resulting in its release into the KCl compartment thereafter. Nonetheless, assuming that an interaction with another constituent could explain the presence of benzaldehyde in that case, this interaction could have been broken once this compound was finally in the KCl solution, as its detection was still allowed. Indeed, as mentioned previously in this study, while volatile compounds interact with other constituents, it hinders their detection [69]. Moreover, the results could show that this potential interaction occurred only at pH 4, as a similar trend was not found at pH 7. The charged compounds in that case could be histidine, present mainly in its free form in the HMH [1] as, at pH 4, its side-chain was totally protonated (pKa ~6.0), allowing its migration to the cathode through CEMs, while at pH 7 this latter was in its non-charged form. Interestingly, the presence of (Z)-6-octen-2-one, (E,E)-2,4-heptadienal and (E,Z)-2,6-nonadienal, whose abundance was lower in the hydrolysate at pH 4 treated with current at the final time, was not found in the corresponding KCl solution. Therefore, supposing that the hypothesis formulated before aiming that these compounds could have established interactions with membranes, this could indicate that none of these three compounds were released into the KCl compartment thereafter.

Finally, the performance of ED to decrease the abundance of the most potent odor-active compounds of the HMH was compared to that of a deaerator (Table 4). In this case, in addition to pH 4 and 7, the treatment was also conducted on the hydrolysate at pH 10. Compared to pH 4-hydrolysate at the initial time, the deaerator allowed the decrease in seven compounds ($p < 0.05$), namely 3-methylbutanal, 2-methylbutanal, 2,3-pentanedione, pentanal, hexanal, 2-nonanone and (E,Z)-2,6-nonadienal. A similar trend was observed for pH 7-hydrolysate, for which the deaerator allowed a drop in the abundance of the seven following compounds ($p < 0.05$): 3-methylbutanal,

2-methylbutanal, 1-methyl-1H-tetrazole, pentanal, hexanal, (Z)-6-octen-2-one and 2-nonanone. It is of interest to note that, in this case, the deaerator conducted the total loss of 1-methyl-1H-tetrazole, pentanal and (Z)-6-octen-2-one. That could suggest that the ability of such device to remove volatile compounds was better at pH 7 than at pH 4. As it was mentioned previously, the hypothesis that looked more plausible to explain the difference in volatile compound content between the initial hydrolysate at different pH values was the following: while pH increased, higher interactions between the volatiles and other constituents occurred. This could be in line with the results of the deaerator for the hydrolysate at pH 4 and pH 7. Indeed, it seemed that this device could break weak interactions occurring between volatile compounds and other compounds present in the hydrolysate, resulting in a better decrease rate at pH 7 than at pH 4. Nevertheless, the deaerator did not lead to a decrease in the volatile content while the hydrolysate was treated at pH 10. This could indicate that, at pH 10, the chosen hypothesis was not enough to totally explain the mechanisms involved. It could be supposed that, at pH 10, a certain proportion of volatile compounds could take part in strong interactions, such as covalent bonds, but at the same time, some of them could have been lost due to their promoted passage in the headspace of the hydrolysate solution as well, or simply altered, hindering their detection.

Table 7. Abundance of the most-potent odor active compounds of HMH recovered in the KCl solution ($\times 10^7$ A.U) after the different ED treatments (mean ± standard deviation).

		pH 4		pH 7	
	Initial Time	**Final Time (ED + Current)**	**Final Time (ED No Current)**	**Final Time (ED + Current)**	**Final Time (ED No Current)**
3-Methylbutanal	0.00 ± 0.00 ^{a}	0.027 ± 0.143 ^{a}	0.133 ± 0.030 ^{a}	0.960 ± 1.318 ^{a}	0.097 ± 0.151 ^{a}
Hexanal	0.00 ± 0.00 ^{b}	2.19 ± 0.47 ^{a}	1.71 ± 0.70 ^{a}	2.33 ± 0.17 ^{a}	1.86 ± 0.12 ^{a}
Heptanal	0.00 ± 0.00 ^{a}	0.156 ± 0.188 ^{a}	0.025 ± 0.082 ^{a}	0.026 ± 0.091 ^{a}	0.095 ± 0.138 ^{a}
Benzaldehyde	0.00 ± 0.00 ^{c}	4.54 ± 0.46 ^{a}	0.731 ± 0.019 ^{b}	0.595 ± 0.090 ^{b}	0.684 ± 0.042 ^{b}
Octanal	0.00 ± 0.00 ^{b}	0.202 ± 0.089 ^{ab}	0.058 ± 0.055 ^{ab}	0.214 ± 0.143 ^{ab}	0.333 ± 0.182 ^{a}
2-Nonanone	0.00 ± 0.00 ^{c}	0.026 ± 0.00 ^{a}	0.017 ± 0.003 ^{ab}	0.003 ± 0.012 ^{bc}	0.003 ± 0.005 ^{bc}

Values within the same row with different letters (a–c) are significantly different, $p < 0.05$ (Tukey test).

TMAO, TMA and DMA

The TMAO, TMA and DMA contents of HMH are shown in Table 8. Firstly, concerning the hydrolysate at the initial time for the three tested pH values, their concentration in TMAO, TMA and DMA was similar ($p > 0.05$). The only difference observed was related to the content of TMAO of the hydrolysate at pH 10, which was 20 times lower ($p < 0.05$) than those of the hydrolysate at pH 4 and 7 at the initial time. In this context, it is worth to mention that the procedures used for the analysis of TMAO, TMA and DMA recommend to alkalize samples of interest, to allow a better detection of these molecules based on their higher release into the sample headspace [41]. Therefore, the huge decrease in TMAO content observed in the hydrolysate at pH 10 could be related to a loss following their release into the headspace of the hydrolysate solution due to its high volatility.

Then, concerning the content in TMAO, TMA and DMA after the four ED treatments, no difference was observed between the hydrolysate at initial and final times. ED treatments were especially designed to assess whether TMA and DMA, two positively charged compounds at pH 4 and 7, were able to migrate. However, the results indicated that no migration happened while experiments were conducted with current. As suggested by the ED parameter analyses, some water dissociation took place during the treatments conducted at pH 4 and 7 with current. Therefore, it could be hypothesized that TMA and DMA had been in competition with the generated H^+ to migrate into the CEMs, and that H^+ could have prevailed over TMA and DMA. Another explanation could be that fouling occurring on CEMs, as suggested by the membrane conductivity analysis, hindering the migration of TMA and DMA, thus explaining such results. Chindapan et al. (2011) experienced, in their study, a decrease in TMA, and explained this result by its loss occurring during ED due to its high volatility [25]. However, the results obtained in the present study may indicate that TMA could not be lost as easily, since ED treatments carried out without current did not evidence any change in the content of this compound. Moreover, the fact that no change in the concentration of TMAO was observed between the hydrolysate at the initial and final times treated with current was more expected. Indeed, this molecule is a zwitterion, and the absence of global charge makes it less likely to migrate during an ED process.

The contents in TMAO, TMA and DMA of KCl recovery solution were analyzed (Table 9). The results show that the initial solution was free of these compounds, while the KCl solution at the final time of all the tested conditions only evidenced the presence of TMAO. The presence of TMAO in the recovery solution of treatments conducted with current was not expected, as the global charge of this compound was neutral. However, the fact that TMAO was present in the recovery solution of treatments carried out without current as well could suggest that another mechanism than electromigration could be involved. In addition, the concentration of TMAO in the different KCl recovery solutions was surprisingly as important as those of the corresponding hydrolysate at the initial time, and since the concentration of this compound in the hydrolysate did not evolve during the different ED treatments, this could suggest that new TMAO was generated over the time. The most logical explanation at the first glance could have been that some TMA evidenced oxidation, resulting in the formation of much more TMAO. However, this was not possible in the case of this study, as the initial hydrolysate, independently of its pH, had too low a content of TMA. This means that more complex mechanisms occurred. TMAO is traditionally produced from nitrogenous compounds, such as choline, betaine or carnitine, through metabolism pathways involving enzymes and gut microbiota [75]. Interestingly, HMH contains phospholipids whose choline can be a constituent and carnitine as well (Table 1). Even if metabolism pathways could not be involved in that case, it could be supposed that some TMAO was generated from the choline of phospholipids and carnitine through other reactions, such as oxidation. Nevertheless, this could only explain the occurrence of much more TMAO compared to the initial time, and not its recovery in the different KCl solutions. Another hypothesis could be that reactions between constituents of HMH, such as the choline of phospholipids or carnitine, as mentioned before, and those of CEMs could have taken place. This latter hypothesis seems to be even more plausible, as the analysis of ED parameters, and more specifically those regarding membrane conductivity, revealed that some interactions happened between hydrolysate constituents and membranes. However, at this stage, it is not possible to effectively favor one hypothesis rather than another one. A last point that is worth mentioning is that the absence of TMA and DMA in the KCl recovery solutions of treatments conducted with current was effectively representative of their non-migration.

Finally, the comparison of the performance of ED with those of a deaerator was assessed. The results are presented in Table 8. They indicate that the deaerator was only effective in decreasing the concentration of TMAO ($p < 0.05$) of the hydrolysate at pH 4 and 7. Regarding the hydrolysate at pH 10, this device had no effect on its composition. These results were consistent with those obtained for the most potent odor-active compounds, as discussed before. However, in that case, the fact that no impact regarding the TMAO content of the hydrolysate was observed gave credit to its loss following its release into the headspace of the hydrolysate sample, promoted by alkaline conditions and occurring before the deaerator treatment, as mentioned previously.

Table 8. TMA, DMA and TMAO content of HMH at the initial time and after the different treatments (mean ± standard deviation).

	pH 4				pH 7				pH 10	
	Initial Time	Final Time (ED + Current)	Final Time (ED No Current)	Final Time Deaerator	Initial Time	Final Time (ED + Current)	Final Time (ED No Current)	Final Time Deaerator	Initial Time	Final Time Deaerator
TMA (ppm)	3.55 ± 0.60 ^{abc}	3.08 ± 0.99 ^{abcd}	2.39 ± 2.05 ^{cd}	3.67 ± 0.37 ^{abc}	4.88 ± 0.27 ^{abc}	5.54 ± 0.59 ^{a}	4.95 ± 1.01 ^{ab}	2.50 ± 0.61 ^{bcd}	2.63 ± 0.20 ^{bcd}	0.72 ± 0.04 ^{d}
DMA (ppm)	5.31 ± 0.33 ^{a}	4.58 ± 2.04 ^{a}	4.12 ± 3.57 ^{a}	5.63 ± 1.05 ^{a}	2.54 ± 0.25 ^{a}	2.76 ± 0.50 ^{a}	2.51 ± 0.19 ^{a}	3.02 ± 0.31 ^{a}	4.26 ± 0.11 ^{a}	3.78 ± 0.09 ^{a}
TMAO (ppm)	942.36 ± 135.9 ^{a}	957.00 ± 153.79 ^{a}	1019.29 ± 155.12 ^{a}	155.63 ± 46.51 ^{b}	1023.67 ± 63.10 ^{a}	1004.00 ±31.24 ^{a}	948.83 ± 75.03 ^{a}	121.91 ± 44.55 ^{b}	52.87 ± 15.03 ^{b}	101.37 ± 26.05 ^{b}

Values within the same row with different letters (a–d) are significantly different $p < 0.05$ (Tukey test).

Table 9. TMA, DMA and TMAO content of KCl recovery solution at initial time and after the different ED treatments (mean ± standard deviation).

	pH 4			pH 7	
	Initial Time	Final Time (ED + Current)	Final Time (ED No Current)	Final Time (ED + Current)	Final Time (ED No Current)
TMA (ppm)	<0.02 ^{a}	<0.02 ^{a}	<0.02 ^{a}	<0.02 ^{a}	<0.02 ^{a}
DMA (ppm)	<1.00 ^{a}	<1.00 ^{a}	<1.00 ^{a}	<1.00 ^{a}	<1.00 ^{a}
TMAO (ppm)	<2.50 ^{b}	1083.64 ± 158.89 ^{a}	967.67 ± 127.79 ^{a}	1026.73 ± 143.80 ^{a}	1013.88 ± 142.96 ^{a}

Values within the same row with different letters (a–b) are significantly different $p < 0.05$ (Tukey test).

4. Conclusions

GC-MS analysis allowed the identification of a total of 86 volatile compounds in the HMH. Among these 86, the following 15 were determined to be the most potent odor-active compounds of this hydrolysate by GC-O, combined with the detection frequency method: 3-methylbutanal, 2-methylbutanal, 1-methyl-1H-tetrazole, 2,3-pentanedione, pentanal, hexanal, (Z)-4-heptenal, heptanal, methional, benzaldehyde, (Z)-6-octen-2-one, (E,E)-2,4-heptadienal, octanal, 2-nonanone and (E,Z)-2,6-nonadienal. In addition, the HS-GC-NPD analysis revealed the presence of TMAO, TMA and DMA in the HMH. Furthermore, the performance of ED as a deodorization method was compared to that of a deaerator device. The results showed that pH had a huge impact on the volatile compound contents of the hydrolysate at the initial time. In fact, the abundance of the targeted molecules was lower at pH 10 than at pH 4, and intermediate at pH 7. While the pH increased from pH 4 to 7, volatile compounds were more involved in interactions with amino-acid-containing materials, explaining their lower availability and thus their lower abundance. However, at pH 10, more than one mechanism could be involved. Indeed, part of the targeted odor-active compounds should have been lost due to their volatility, while another part participated in irreversible bonds or was altered, hampering their detection. Regarding TMAO specifically, its lower content in the hydrolysate at pH 10 could be related to its loss resulting from a greater release into the headspace solution. On the other hand, ED did not affect the volatile compound contents of HMH. Concerning ED treatments conducted with current, no migration of volatile compounds, and more precisely no migration of TMA and DMA, occurred. Two phenomena were considered to be the main possible limiting process conditions regarding the removal of the targeted compounds. The first was the occurrence of fouling on IEM due to both electrostatic and hydrophobic interactions between IEM and HMH constituents, including volatile compounds. The second was the occurrence of water dissociation on the IEM interfaces due to the reaching of LCD, as well as the presence of water dissociation catalyzers involved in both IEM and HMH constituents. Moreover, the fact that ED treatments without current did not impact the volatile compound contents implied that no loss of these molecules due to their volatile nature happened during the circulation of the hydrolysate solution in an ED system. Interestingly, independently of the pH and current conditions of the ED treatments, it appeared that new TMAO was generated over the time. Two hypotheses were considered. The first would be that TMAO was generated directly in the hydrolysate solution from its precursors, while the second could involve its precursors, as well as the constituents of CEMs. On the contrary to ED, treatments conducted by deaerator significantly decreased the abundance of the targeted compounds at pH 4 and 7, but had no effect at pH 10. Therefore, the conditions leading to the best removal levels of the targeted volatile compounds were the deaerator treatment performed on the hydrolysate at pH 7, and the alkalization of this latter until pH 10. Despite the fact that the relevance of ED to be used as a deodorization method of HMH was not proven at that stage, it appeared that the establishment of strategies to avoid both fouling and water dissociation phenomena could lead to a better process efficiency. However, this supposes to deepen the knowledge regarding, especially, the fouling resulting from interactions between IEM and HMH constituents, with particular interest in those involving volatile compounds, by performing Attenuated Total Reflection–Fourier Transform Infrared (ATR-FTIR), as well as identifying the mechanisms leading to TMAO formation during ED. This is currently under investigation. Finally, the use of electromembrane processes, other than conventional electrodialysis, could be another promising solution that is worth further investigation. Therefore, electromembrane processes have a chance to become an effective deodorization method in the future.

Author Contributions: Conceptualization, S.T. and L.B.; methodology, S.T. and L.B.; software, S.T.; validation, S.T., V.P., M.B., P.D., C.G., Y.B. and L.B.; formal analysis, S.T.; investigation, S.T.; resources, S.T., V.P., M.B., P.D., C.G. and L.B.; data curation, S.T.; writing—original draft preparation, S.T.; writing—review and editing, S.T., V.P., M.B., P.D., C.G., Y.B. and L.B.; visualization, S.T. and L.B.; supervision, L.B.; project administration, L.B.; funding acquisition, L.B. All authors have read and agreed to the published version of the manuscript.

Funding: This work was funded by the Natural Sciences and Engineering Research Council of Canada (NSERC) Industrial Research Chair on ElectroMembrane processes aimed at the ecoefficiency improvement of biofood production lines.

Acknowledgments: The NSERC Industrial Research Chair on ElectroMembrane processes aimed at the ecoefficiency improvement of biofood production lines is acknowledged. The authors are grateful to Ocean NutraSciences for providing the herring milt hydrolysate powder. The authors thank Diane GAGNON and Jacinthe THIBODEAU, professional researchers at Université Laval, for their involvement in this project.

Conflicts of Interest: The authors declare no conflict of interest.

References

1. Durand, R.; Fraboulet, E.; Marette, A.; Bazinet, L. Simultaneous double cationic and anionic molecule separation from herring milt hydrolysate and impact on resulting fraction bioactivities. *Sep. Purif. Technol.* **2019**, *210*, 431–441. [CrossRef]
2. Nasri, R.; Younes, I.; Jridi, M.; Trigui, M.; Bougatef, A.; Nedjar-Arroume, N.; Dhulster, P.; Nasri, M.; Karra-Châabouni, M. ACE inhibitory and antioxidative activities of Goby (Zosterissessor ophiocephalus) fish protein hydrolysates: Effect on meat lipid oxidation. *Food Res. Int.* **2013**, *54*, 552–561. [CrossRef]
3. Picot, L.; Bordenave, S.; Didelot, S.; Fruitier-Arnaudin, I.; Sannier, F.; Thorkelsson, G.; Bergé, J.P.; Guérard, F.; Chabeaud, A.; Piot, J.M. Antiproliferative activity of fish protein hydrolysates on human breast cancer cell lines. *Process Biochem.* **2006**, *41*, 1217–1222. [CrossRef]
4. Roblet, C.; Akhtar, M.J.; Mikhaylin, S.; Pilon, G.; Gill, T.; Marette, A.; Bazinet, L. Enhancement of glucose uptake in muscular cell by peptide fractions separated by electrodialysis with filtration membrane from salmon frame protein hydrolysate. *J. Funct. Foods* **2016**, *22*, 337–346. [CrossRef]
5. Sampath Kumar, N.S.; Nazeer, R.A.; Jaiganesh, R. Purification and biochemical characterization of antioxidant peptide from horse mackerel (Magalaspis cordyla) viscera protein. *Peptides* **2011**, *32*, 1496–1501. [CrossRef]
6. Sila, A.; Hedhili, K.; Przybylski, R.; Ellouz-Chaabouni, S.; Dhulster, P.; Bougatef, A.; Nedjar-Arroume, N. Antibacterial activity of new peptides from barbel protein hydrolysates and mode of action via a membrane damage mechanism against Listeria monocytogenes. *J. Funct. Foods* **2014**, *11*, 322–329. [CrossRef]
7. Villamil, O.; Váquiro, H.; Solanilla, J.F. Fish viscera protein hydrolysates: Production, potential applications and functional and bioactive properties. *Food Chem.* **2017**, *224*, 160–171. [CrossRef]
8. Durand, R.; Pellerin, G.; Thibodeau, J.; Fraboulet, E.; Marette, A.; Bazinet, L. Screening for metabolic syndrome application of a herring by-product hydrolysate after its separation by electrodialysis with ultrafiltration membrane and identification of novel anti-inflammatory peptides. *Sep. Purif. Technol.* **2020**, *235*, 116205. [CrossRef]
9. Chen, D.; Chen, X.; Chen, H.; Cai, B.; Wan, P.; Zhu, X.; Sun, H.; Sun, H.; Pan, J. Identification of odor volatile compounds and deodorization of Paphia undulata enzymatic hydrolysate. *J. Ocean Univ. China* **2016**, *15*, 1101–1110. [CrossRef]
10. Yarnpakdee, S.; Benjakul, S.; Nalinanon, S.; Kristinsson, H.G. Lipid oxidation and fishy odour development in protein hydrolysate from Nile tilapia (Oreochromis niloticus) muscle as affected by freshness and antioxidants. *Food Chem.* **2012**, *132*, 1781–1788. [CrossRef]
11. Sghaier, L.; Vial, J.; Sassiat, P.; Thiebaut, D.; Watiez, M.; Breton, S.; Rutledge, D.N.; Cordella, C.B.Y. Analysis of target volatile compounds related to fishy off-flavor in heated rapeseed oil: A comparative study of different headspace techniques. *Eur. J. Lipid Sci. Technol.* **2016**, *118*, 906–918. [CrossRef]
12. Cros, S.; Lignot, B.; Razafintsalama, C.; Jaouen, P.; Bourseau, P. Electrodialysis Desalination and Reverse Osmosis Concentration of an Industrial Mussel Cooking Juice: Process Impact on Pollution Reduction and on Aroma Quality. *J. Food Sci.* **2004**, *69*, C435–C442. [CrossRef]
13. Chung, K.-H.; Lee, K.-Y. Removal of trimethylamine by adsorption over zeolite catalysts and deodorization of fish oil. *J. Hazard. Mater.* **2009**, *172*, 922–927. [CrossRef]
14. Baliño-Zuazo, L.; Barranco, A. A novel liquid chromatography–mass spectrometric method for the simultaneous determination of trimethylamine, dimethylamine and methylamine in fishery products. *Food Chem.* **2016**, *196*, 1207–1214. [CrossRef] [PubMed]
15. Wu, T.H.; Bechtel, P.J. Ammonia, Dimethylamine, Trimethylamine, and Trimethylamine Oxide from Raw and Processed Fish By-Products. *J. Aquat. Food Prod. Technol.* **2008**, *17*, 27–38. [CrossRef]

16. Marsili, R.T.; Laskonis, C.R. Odorant Synergy Effects as the Cause of Fishy Malodors in Algal Marine Oils. *J. Agric. Food Chem.* **2014**, *62*, 9676–9682. [CrossRef]
17. Fukami, K.; Funatsu, Y.; Kawasaki, K.; Watabe, S. Improvement of Fish-sauce Odor by Treatment with Bacteria Isolated from the Fish-sauce Mush (Moromi) Made from Frigate Mackerel. *J. Food Sci.* **2004**, *69*, fms45–fms49. [CrossRef]
18. Hirano, T.; Kurosawa, H.; Nakamura, K.; Amano, Y. Simultaneous removal of hydrogen sulfide and trimethylamine by a bacterial deodorant. *J. Ferment. Bioeng.* **1996**, *81*, 337–342. [CrossRef]
19. Schrader, K.K.; Davidson, J.W.; Rimando, A.M.; Summerfelt, S.T. Evaluation of ozonation on levels of the off-flavor compounds geosmin and 2-methylisoborneol in water and rainbow trout Oncorhynchus mykiss from recirculating aquaculture systems. *Aquac. Eng.* **2010**, *43*, 46–50. [CrossRef]
20. Fan, W.; Chi, Y.; Zhang, S. The use of a tea polyphenol dip to extend the shelf life of silver carp (Hypophthalmicthys molitrix) during storage in ice. *Food Chem.* **2008**, *108*, 148–153. [CrossRef]
21. Fang, Y.; Gu, S.; Zhang, J.; Liu, S.; Ding, Y.; Liu, J. Deodorisation of fish oil by nanofiltration membrane process: focus on volatile flavour compounds and fatty acids composition. *Int. J. Food Sci. Technol.* **2018**, *53*, 692–699. [CrossRef]
22. Jarrault, C.; Dornier, M.; Labatut, M.-L.; Giampaoli, P.; Lameloise, M.-L. Coupling nanofiltration and osmotic evaporation for the recovery of a natural flavouring concentrate from shrimp cooking juice. *Innov. Food Sci. Emerg. Technol.* **2017**, *43*, 182–190. [CrossRef]
23. Vandanjon, L.; Cros, S.; Jaouen, P.; Quéméneur, F.; Bourseau, P. Recovery by nanofiltration and reverse osmosis of marine flavours from seafood cooking waters. *Desalination* **2002**, *144*, 379–385. [CrossRef]
24. Walha, K.; Ben Amar, R.; Massé, A.; Bourseau, P.; Cardinal, M.; Cornet, J.; Prost, C.; Jaouen, P. Aromas potentiality of tuna cooking juice concentrated by nanofiltration. *LWT Food Sci. Technol.* **2011**, *44*, 153–157. [CrossRef]
25. Chindapan, N.; Devahastin, S.; Chiewchan, N.; Sablani, S.S. Desalination of Fish Sauce by Electrodialysis: Effect on Selected Aroma Compounds and Amino Acid Compositions. *J. Food Sci.* **2011**, *76*, S451–S457. [CrossRef]
26. Cros, S.; Lignot, B.; Bourseau, P.; Jaouen, P.; Prost, C. Desalination of mussel cooking juices by electrodialysis: effect on the aroma profile. *J. Food Eng.* **2005**, *69*, 425–436. [CrossRef]
27. Bazinet, L. *Concepts de Génie Alimentaire : Procédés Associés, Application à la Conservation et Transformation des Aliments*; Presses polytechniques: Montreal, QC, Canada, 2019; Chapter 17, pp. 579–628.
28. Chang, G.W.; Chang, W.L.; Lew, K.B.K. Trimethylamine-Specific Electrode for Fish Quality Control. *J. Food Sci.* **1976**, *41*, 723–724. [CrossRef]
29. Tremblay, A.; Corcuff, R.; Goulet, C.; Godefroy, S.B.; Doyen, A.; Beaulieu, L. Valorization of snow crab (Chionoecetes opilio) cooking effluents for food applications. *J. Sci. Food Agric.* **2020**, *100*, 384–393. [CrossRef]
30. Dufton, G.; Mikhaylin, S.; Gaaloul, S.; Bazinet, L. Systematic Study of the Impact of Pulsed Electric Field Parameters (Pulse/Pause Duration and Frequency) on ED Performances during Acid Whey Treatment. *Membranes* **2020**, *10*, 14. [CrossRef]
31. Horwitz, W.; Chichilo, P.; Reynolds, H. *Official Methods of Analysis of the Association of Official Analytical Chemists*; Association of Official Analytical Chemists: Washington, DC, USA, 1970.
32. Lemay, N.; Mikhaylin, S.; Bazinet, L. Voltage spike and electroconvective vortices generation during electrodialysis under pulsed electric field: Impact on demineralization process efficiency and energy consumption. *Innov. Food Sci. Emerg. Technol.* **2019**, *52*, 221–231. [CrossRef]
33. Zhou, X.; Chong, Y.; Ding, Y.; Gu, S.; Liu, L. Determination of the effects of different washing processes on aroma characteristics in silver carp mince by MMSE–GC–MS, e-nose and sensory evaluation. *Food Chem.* **2016**, *207*, 205–213. [CrossRef] [PubMed]
34. Giri, A.; Osako, K.; Okamoto, A.; Ohshima, T. Olfactometric characterization of aroma active compounds in fermented fish paste in comparison with fish sauce, fermented soy paste and sauce products. *Food Res. Int.* **2010**, *43*, 1027–1040. [CrossRef]
35. Varlet, V.; Prost, C.; Serot, T. Volatile aldehydes in smoked fish: Analysis methods, occurence and mechanisms of formation. *Food Chem.* **2007**, *105*, 1536–1556. [CrossRef]
36. Le Guen, S.; Prost, C.; Demaimay, M. Critical Comparison of Three Olfactometric Methods for the Identification of the Most Potent Odorants in Cooked Mussels (*Mytilus edulis*). *J. Agric. Food Chem.* **2000**, *48*, 1307–1314. [CrossRef]

37. Peinado, I.; Miles, W.; Koutsidis, G. Odour characteristics of seafood flavour formulations produced with fish by-products incorporating EPA, DHA and fish oil. *Food Chem.* **2016**, *212*, 612–619. [CrossRef] [PubMed]
38. Varlet, V.; Knockaert, C.; Prost, C.; Serot, T. Comparison of Odor-Active Volatile Compounds of Fresh and Smoked Salmon. *J. Agric. Food Chem.* **2006**, *54*, 3391–3401. [CrossRef]
39. Cha, Y.J.; Cadwallader, K.R. Volatile Components in Salt-Fermented Fish and Shrimp Pastes. *J. Food Sci.* **1995**, *60*, 19–24. [CrossRef]
40. daCosta, K.-A.; Vrbanac, J.J.; Zeisel, S.H. The measurement of dimethylamine, trimethylamine, and trimethylamine N-oxide using capillary gas chromatography-mass spectrometry. *Anal. Biochem.* **1990**, *187*, 234–239. [CrossRef]
41. Chung, S.W.C.; Chan, B.T.P. Trimethylamine oxide, dimethylamine, trimethylamine and formaldehyde levels in main traded fish species in Hong Kong. *Food Addit. Contam. Part B Surveill. Commun.* **2009**, *2*, 44–51. [CrossRef]
42. Aro, T.; Brede, C.; Manninen, P.; Kallio, H. Determination of Semivolatile Compounds in Baltic Herring (*Clupea harengus membras*) by Supercritical Fluid Extraction−Supercritical Fluid Chromatography−Gas Chromatography−Mass Spectrometry. *J. Agric. Food Chem.* **2002**, *50*, 1970–1975. [CrossRef]
43. Varlet, V.; Serot, T.; Cardinal, M.; Knockaert, C.; Prost, C. Olfactometric Determination of the Most Potent Odor-Active Compounds in Salmon Muscle (*Salmo salar*) Smoked by Using Four Smoke Generation Techniques. *J. Agric. Food Chem.* **2007**, *55*, 4518–4525. [CrossRef] [PubMed]
44. Fukami, K.; Ishiyama, S.; Yaguramaki, H.; Masuzawa, T.; Nabeta, Y.; Endo, K.; Shimoda, M. Identification of Distinctive Volatile Compounds in Fish Sauce. *J. Agric. Food Chem.* **2002**, *50*, 5412–5416. [CrossRef] [PubMed]
45. Jiang, J.-J.; Zeng, Q.-X.; Zhu, Z.-W. Analysis of Volatile Compounds in Traditional Chinese Fish Sauce (Yu Lu). *Food Bioprocess Technol* **2011**, *4*, 266–271. [CrossRef]
46. Lapsongphon, N.; Cadwallader, K.R.; Rodtong, S.; Yongsawatdigul, J. Characterization of Protein Hydrolysis and Odor-Active Compounds of Fish Sauce Inoculated with Virgibacillus sp. SK37 under Reduced Salt Content. *J. Agric. Food Chem.* **2013**, *61*, 6604–6613. [CrossRef]
47. Pham, A.J.; Schilling, M.W.; Yoon, Y.; Kamadia, V.V.; Marshall, D.L. Characterization of Fish Sauce Aroma-Impact Compounds Using GC-MS, SPME-Osme-GCO, and Stevens' Power Law Exponents. *J. Food Sci.* **2008**, *73*, C268–C274. [CrossRef]
48. Cha, Y.-J. Volatile Compounds in Oyster Hydrolysate Produced by Commercial Protease. *J. Korean Soc. Food Sci. Nutr.* **1995**, *24*, 420–426.
49. Chung, H.Y.; Yeung, C.W.; Kim, J.-S.; Chen, F. Static headspace analysis-olfactometry (SHA-O) of odor impact components in salted-dried white herring (Ilisha elongata). *Food Chem.* **2007**, *104*, 842–851. [CrossRef]
50. Liu, S.; Liao, T.; McCrummen, S.T.; Hanson, T.R.; Wang, Y. Exploration of volatile compounds causing off-flavor in farm-raised channel catfish (Ictalurus punctatus) fillet. *Aquac. Int.* **2017**, *25*, 413–422. [CrossRef]
51. Iglesias, J.; Medina, I.; Bianchi, F.; Careri, M.; Mangia, A.; Musci, M. Study of the volatile compounds useful for the characterisation of fresh and frozen-thawed cultured gilthead sea bream fish by solid-phase microextraction gas chromatography–mass spectrometry. *Food Chem.* **2009**, *115*, 1473–1478. [CrossRef]
52. Aidos, I.; Jacobsen, C.; Jensen, B.; Luten, J.B.; van der Padt, A.; Boom, R.M. Volatile oxidation products formed in crude herring oil under accelerated oxidative conditions. *Eur. J. Lipid Sci. Technol.* **2002**, *104*, 808–818. [CrossRef]
53. Fernandez, X.; Breme, K.; Varlet, V. Couplage chromatographie en phase gazeuse/olfactométrie. *Tech. De L'ingénieur* **2009**, *P1488*, 1–18.
54. Chung, H.Y.; Yung, I.K.S.; Ma, W.C.J.; Kim, J.-S. Analysis of volatile components in frozen and dried scallops (Patinopecten yessoensis) by gas chromatography/mass spectrometry. *Food Res. Int.* **2002**, *35*, 43–53. [CrossRef]
55. Hallier, A.; Courcoux, P.; Sérot, T.; Prost, C. New gas chromatography–olfactometric investigative method, and its application to cooked Silurus glanis (European catfish) odor characterization. *J. Chromatogr. A* **2004**, *1056*, 201–208. [CrossRef]
56. Persico, M.; Mikhaylin, S.; Doyen, A.; Firdaous, L.; Nikonenko, V.; Pismenskaya, N.; Bazinet, L. Prevention of peptide fouling on ion-exchange membranes during electrodialysis in overlimiting conditions. *J. Membr. Sci.* **2017**, *543*, 212–221. [CrossRef]
57. Tanaka, Y. Water dissociation reaction generated in an ion exchange membrane. *J. Membr. Sci.* **2010**, *350*, 347–360. [CrossRef]

58. Dufton, G.; Mikhaylin, S.; Gaaloul, S.; Bazinet, L. How electrodialysis configuration influences acid whey deacidification and membrane scaling. *J. Dairy Sci.* **2018**, *101*, 7833–7850. [CrossRef]
59. Lemay, N.; Mikhaylin, S.; Mareev, S.; Pismenskaya, N.; Nikonenko, V.; Bazinet, L. How demineralization duration by electrodialysis under high frequency pulsed electric field can be the same as in continuous current condition and that for better performances? *J. Membr. Sci.* **2020**, *603*, 117878. [CrossRef]
60. Zhang, W.; Miao, M.; Pan, J.; Sotto, A.; Shen, J.; Gao, C.; der Bruggen, B.V. Separation of divalent ions from seawater concentrate to enhance the purity of coarse salt by electrodialysis with monovalent-selective membranes. *Desalination* **2017**, *411*, 28–37. [CrossRef]
61. Dufton, G.; Mikhaylin, S.; Gaaloul, S.; Bazinet, L. Positive Impact of Pulsed Electric Field on Lactic Acid Removal, Demineralization and Membrane Scaling during Acid Whey Electrodialysis. *Int. J. Mol. Sci.* **2019**, *20*, 797. [CrossRef]
62. Persico, M.; Mikhaylin, S.; Doyen, A.; Firdaous, L.; Hammami, R.; Chevalier, M.; Flahaut, C.; Dhulster, P.; Bazinet, L. Formation of peptide layers and adsorption mechanisms on a negatively charged cation-exchange membrane. *J. Colloid Interface Sci.* **2017**, *508*, 488–499. [CrossRef]
63. Kattan Readi, O.M.; Gironès, M.; Nijmeijer, K. Separation of complex mixtures of amino acids for biorefinery applications using electrodialysis. *J. Membr. Sci.* **2013**, *429*, 338–348. [CrossRef]
64. Persico, M.; Mikhaylin, S.; Doyen, A.; Firdaous, L.; Hammami, R.; Bazinet, L. How peptide physicochemical and structural characteristics affect anion-exchange membranes fouling by a tryptic whey protein hydrolysate. *J. Membr. Sci.* **2016**, *520*, 914–923. [CrossRef]
65. Pismenskaya, N.; Nikonenko, V.; Volodina, E.; Pourcelly, G. Electrotransport of weak-acid anions through anion-exchange membranes. *Desalination* **2002**, *147*, 345–350. [CrossRef]
66. Eliseeva, T.V.; Shaposhnik, V.A. Effects of circulation and facilitated electromigration of amino acids in electrodialysis with ion-exchange membranes. *Russ. J. Electrochem.* **2000**, *36*, 64–67. [CrossRef]
67. Langevin, M.-E.; Bazinet, L. Ion-exchange membrane fouling by peptides: A phenomenon governed by electrostatic interactions. *J. Membr. Sci.* **2011**, *369*, 359–366. [CrossRef]
68. Chopin, C.; Kone, M.; Serot, T. Study of the interaction of fish myosin with the products of lipid oxidation: The case of aldehydes. *Food Chem.* **2007**, *105*, 126–132. [CrossRef]
69. Gianelli, M.P.; Flores, M.; Toldrá, F. Interactions of Soluble Peptides and Proteins from Skeletal Muscle on the Release of Volatile Compounds. *J. Agric. Food Chem.* **2003**, *51*, 6828–6834. [CrossRef]
70. Guichard, E. Interactions between flavor compounds and food ingredients and their influence on flavor perception. *Food Rev. Int.* **2002**, *18*, 49–70. [CrossRef]
71. Guichard, E.; Langourieux, S. Interactions between β-lactoglobulin and flavour compounds. *Food Chem.* **2000**, *71*, 301–308. [CrossRef]
72. Jouenne, E.; Crouzet, J. Effect of pH on Retention of Aroma Compounds by β-Lactoglobulin. *J. Agric. Food Chem.* **2000**, *48*, 1273–1277. [CrossRef]
73. Pérez-Juan, M.; Flores, M.; Toldrá, F. Effect of pork meat proteins on the binding of volatile compounds. *Food Chem.* **2008**, *108*, 1226–1233. [CrossRef]
74. Meynier, A.; Rampon, V.; Dalgalarrondo, M.; Genot, C. Hexanal and t-2-hexenal form covalent bonds with whey proteins and sodium caseinate in aqueous solution. *Int. Dairy J.* **2004**, *14*, 681–690. [CrossRef]
75. Velasquez, M.T.; Ramezani, A.; Manal, A.; Raj, D.S. Trimethylamine N-Oxide: The Good, the Bad and the Unknown. *Toxins (Basel)* **2016**, *8*, 326. [CrossRef] [PubMed]

Article

A 2D Convection-Diffusion Model of Anodic Oxidation of Organic Compounds Mediated by Hydroxyl Radicals Using Porous Reactive Electrochemical Membrane

Ekaterina Skolotneva [1], Clement Trellu [2], Marc Cretin [3] and Semyon Mareev [1,*]

1 Physical Chemistry Department, Kuban State University, 149 Stavropolskaya str., 350040 Krasnodar, Russia; ek.skolotneva@gmail.com

2 Laboratoire Géomatériaux et Environnement (EA 4508), Université Gustave Eiffel, 77454 Marne la Vallée, France; clement.trellu@univ-eiffel.fr

3 Institut Europeen des Membranes, IEM-UMR 5635, ENSCM, CNRS, Univ Montpellier, 34095 Montpellier, France; marc.cretin@umontpellier.fr

* Correspondence: mareev-semyon@bk.com; Tel.: +7-861-519-9573

Received: 27 April 2020; Accepted: 13 May 2020; Published: 16 May 2020

Abstract: In recent years, electrochemical methods utilizing reactive electrochemical membranes (REM) have been considered as a promising technology for efficient degradation and mineralization of organic compounds in natural, industrial and municipal wastewaters. In this paper, we propose a two-dimensional (2D) convection-diffusion-reaction model concerning the transport and reaction of organic species with hydroxyl radicals generated at a TiO_x REM operated in flow-through mode. It allows the determination of unknown parameters of the system by treatment of experimental data and predicts the behavior of the electrolysis setup. There is a good agreement in the calculated and experimental degradation rate of a model pollutant at different permeate fluxes and current densities. The model also provides an understanding of the current density distribution over an electrically heterogeneous surface and its effect on the distribution profile of hydroxyl radicals and diluted species. It was shown that the percentage of the removal of paracetamol increases with decreasing the pore radius and/or increasing the porosity. The effect becomes more pronounced as the current density increases. The model highlights how convection, diffusion and reaction limitations have to be taken into consideration for understanding the effectiveness of the process.

Keywords: reactive electrochemical membrane; porous electrode; anodic oxidation; hydroxyl radicals

---

1. Introduction

The removal of biorefractory emerging organic pollutants requires the implementation of a novel advanced treatment system for drinking water production and wastewater treatment. The use of membrane processes also requires an appropriate pretreatment in order to reduce membrane fouling [1–3]. In the last decades, anodic oxidation has been considered as a highly efficient electrochemical advanced oxidation process used in the treatment of natural and wastewaters containing organic pollutants [4,5]. The process involves the generation of hydroxyl radicals during the electrochemical decomposition of water and the consequent decomposition of organic compounds [4]. The hydroxyl radical is a highly reactive substance that allows the non-selective oxidation of most bio-degradable and bio-refractory organic pollutants [6]. Comparative studies show that among various methods, anodic oxidation allows achieving the highest mineralization of such compounds [5,7–10].

Despite all the advantages, the use of anodic oxidation in water purification processes is hampered by a number of drawbacks. First of all, hydroxyl radicals are formed only on the surface of the anode and have a short lifetime. Therefore, they are only present in a thin boundary layer (<1 μm) [11–14]. As a result, the oxidation of organic pollutants occurs only near the surface of the electrode. The process is limited by convective-diffusion delivery of contaminants from the solution to the reaction area. The most effective way to reduce mass transport limitations is to increase the electrochemically active surface area of the anodes by using porous electrodes, so-called reactive electrochemical membranes (REM), in flow-through configurations [15,16]. The effluent to treat is flowing through the porous anode. Several recent studies have focused on electrodes made from Magneli phases [17,18]. The results obtained over the past five years have shown that it is a very promising technology for electro-oxidation of organic pollutants in water treatment systems [19]. Such a material has a high oxygen evolution potential, which makes possible the generation of hydroxyl radicals, and thus, the oxidation of persistent organic compounds [20]. It has also been shown that porous anodes based on sub-stoichiometric titanium oxide exhibit suitable properties for efficient oxidation of organic pollutants [17,21].

Operating conditions drastically affect the efficiency of the electro-oxidation process. Electro-oxidation processes can be either operated under mass transport or current (reaction) limitation. In the case of porous electrodes, the flux density of the permeate and its residence time in the reaction zone are crucial parameters influencing mass transfer and reaction phenomena. An important role is also played by the distribution of the electric field strength in the pores of the electrode, the value of which is rather large, mainly on the side of the anode facing the cathode [16]. Nevertheless, it has been reported that the current density can be similar within the whole membrane depth when the process is operated under current limitation and without significant ohmic drop [18]. Another specificity of porous electrodes operated under current limitation has been reported: pollutants that are not degraded on the electrode surface can go deep into the anode, where the voltage is lower and not enough to generate hydroxyl radicals, but optimal for electropolymerization and further deposition of polymerized compounds [16,21,22], which might cause fouling and reduce process efficiency [21,23]. In [16], it was shown that a high current density significantly increased energy consumption, but allowed avoiding electro-polymerization due to an excessive amount of hydroxyl radicals and operation of the process under mass transport limitation. In recent studies, the influence of specific system parameters (flow rate, solution concentration and current intensity) has been experimentally studied as regards to both degree of purification achieved and energy consumption [19,24].

Mathematical modeling in the field of anodic oxidation for water treatment is also developing and has already made possible the estimation of the thickness of the reaction zone of anodic oxidation for dense electrodes [11–13], the development of transmission line models for estimation of the reactive region and REM fouling [16], the estimation of one-dimensional (1D) distribution of potential and current density within REM [18], the evolution of the concentration of organic compounds [18,25] or the prediction of permeate flux in tubular porous electrodes [26]. Convective-diffusion models that take into account hydroxyl radical mediated chemical reactions at the REM are absent, but their need has already been indicated in previous studies [21,23].

In this study, we present a theoretical analysis based on two-dimensional (2D) convection-diffusion-reaction model representing an anodic oxidation device with a porous electrode operated under galvanostatic condition. The model was calibrated based on experimental data previously obtained with a sub-stoichiometric titanium oxide REM [21]. Pore size, porosity and reaction rate between hydroxyl radicals and the model pollutant were taken from previous experimental studies. The model provides a better understanding of the distribution of current density over the heterogeneous porous surface. The model was then used for prediction of process effectiveness as a function of pore radius and porosity of the electrode.

2. Mathematical Model

2.1. Geometry of the System Under Study

The system under study is a cross-flow electrolyzer (Figure 1), which utilizes REM as an anode in inside-outside cross-flow filtration mode operated under galvanostatic conditions. The experimental setup was described in detail in a previous study [21]. The REM is a porous electrode (in our case, it is a tubular electrode with a defined inner, r_1, and outer, r_2, radii and length, L). A rod of radius r_3 is used as a cathode and placed at the center of the REM. Anode and cathode form a tubular electrolyzer channel with a thickness $(r_1 - r_3)$. A feed solution flows at a constant cross-flow rate, U. The permeate flux, expressed as the linear velocity V, is controlled by transmembrane pressure and depends on operating conditions of the experiment. The feed solution contains an organic compound with initial concentration c_0, which is the target pollutant, and supporting electrolyte, which is assumed to do not participate in any chemical reaction and is used only to decrease the total resistance of the system.

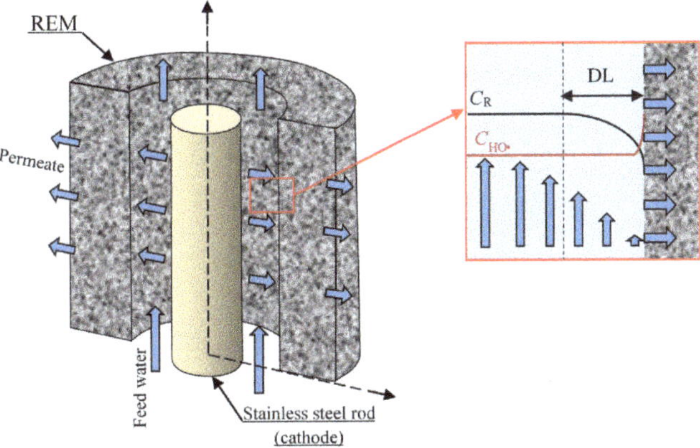

Figure 1. Schematic representation of the system under study. C_R represents the concentration of an organic compound. The red square shows the area taken into consideration for the model: the electrode and the diffusion layer (DL).

The real REM consists of a huge number of more or less heterogeneous pores, for which a three-dimensional (3D) model would be too complex to solve mathematically. The REM bulk might be considered as a homogeneous porous media characterized by only one structural parameter, for example, the inertial resistance factor [26]. In this case, the velocity and concentration distributions of the solution cannot be calculated in the individual pores, and it makes it more difficult to correctly define the effect of pore radius. Thus, in this work, we have applied other assumptions. The system under study consisted of a REM with the adjacent diffusion layer (DL) (Figure 1). The transition from a random to a hexagonal distribution of pores and conductive areas over the REM surface is shown in Figure 2a,b. The unit cell on the surface can be considered as consisting of a pore of radius R_1 within a ring of anode material of external radius R_2. The value of the curvature radius of the electrode surface was taken equal to the pore radius (Figure 2c) according to the image of the REM surface (Figure 2a). The DL thickness is δ. Then, the 3D unit cell was considered as a cylinder of radius R_2, involving a pore and the surrounding part of the REM consisting of the solution with the DL of thickness δ (Figure 2c) calculated using the Leveque equation (Equation A15 in Appendix A). After the application of the cylindrical symmetry assumption, the system under study may be presented in two coordinates r and z (Figure 2d). The porosity of the modeled REM is calculated as $\varepsilon = (R_1/R_2)^2$. The approach is applicable for different systems and allows us to describe the transport phenomena in systems with a complex structure [27,28].

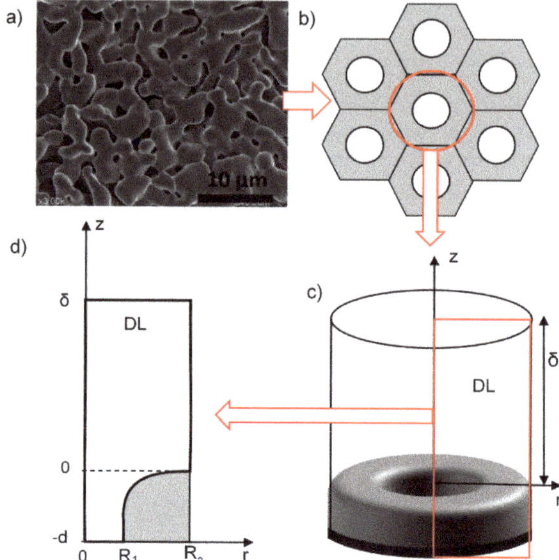

Figure 2. The schematic representation of the transition from a complex surface structure of the reactive electrochemical membranes (REM) (**a**) to the simplified uniformly distributed pattern (**b**) and three-dimensional (3D) model representation with axial symmetry (**c**) and two-dimensional (2D) unit cell (**d**).

2.2. Problem Formulation

According to previous theoretical investigations [11–13], the following simplifying assumptions are made:

- The oxidation of organic compounds proceeds only via the assistance of hydroxyl radicals;
- The transport number of the organic compound is negligible comparing to the transport number of supporting electrolyte. Thus, only convection and diffusion fluxes are taken into account.
- Only the faradaic current is considered; the charging current is not taken into account;
- The temperature, activity coefficient and density gradients are ignored.
- The reaction of organic compounds by direct electron transfer on the REM surface is neglected.

Mathematical modeling of the mass transfer of chemical species in the vicinity of the REM includes the Fick's law with the convective term (Equation (1)), the matter conservation (the continuity) law (Equation (2)), the Ohm's law (Equation (3)), the charge conservation law (Equation (4)), the mass conservation law (Equation (6)) and the Navier-Stokes equation (Equation (5)), as follows:

$$\vec{J}_i = -D_i \nabla c_i + c_i \vec{V} \quad (1)$$

$$\frac{\partial c_i}{\partial t} = -\mathrm{div}\,\vec{J}_i + v_i \quad (2)$$

$$\vec{j} = -\kappa \nabla \varphi \quad (3)$$

$$\mathrm{div}\,\vec{j} = 0 \quad (4)$$

$$\frac{\partial \vec{V}}{\partial t} + \left(\vec{V}\,\nabla\right)\vec{V} = -\frac{1}{\rho_0}\nabla P + \nu \Delta \vec{V} \quad (5)$$

$$\nabla \cdot \vec{V} = 0 \tag{6}$$

where D_i, c_i, and $\vec{J}_i$ are the diffusion coefficient, molar concentration and flux density of the i-th species, respectively; v_i is the chemical reaction, in which i-th specie is included; φ is the electric potential; t is the time; $\vec{V}$ is the linear velocity of the solution; ρ_0 is the solution density (assumed to be constant); P is the pressure; ν is the kinematic viscosity; κ is the conductivity of the matter.

$P, \vec{V}, \varphi, \vec{j}, \vec{J}_i$ and c_i are functions of t, z and r. Equations (1)–(4) describe the concentration and potential fields, and Equations (5) and (6) describe the velocity field. These two groups of equations are coupled by the velocity in the Equation (1) and boundary conditions (Appendix A).

With regards to the modeling of (electro)-chemical reactions, the following reactions were considered. Hydroxyl radicals are generated from water discharge at the anode surface, according to the reaction [12]:

$$H_2O \rightarrow HO_{(aq)} + H^+_{(aq)} + e^- \tag{7}$$

Free hydroxyl radicals can react with each other to form hydrogen peroxide:

$$HO + HO \overset{k_{HO}}{\rightarrow} H_2O_2 \tag{8}$$

where k_{HO} is the rate constant of reaction (8).

In parallel to the reaction (8), hydroxyl radicals can also react with an organic compound (R) and its degradation by-products:

$$R \underset{HO}{\overset{k_R}{\rightarrow}} \text{by-products} \underset{HO}{\overset{k_i}{\rightarrow}} \text{end-products} \tag{9}$$

Reaction kinetics are formulated assuming a second-order rate expression depending on the concentration of the hydroxyl radicals, the organic compound (i = 1) and its possible by-products ($1 < i \leq n$):

$$v_{HO} = k_{HO} c_{HO}^2 + \sum_{i=1}^{n} k_i c_{HO} c_i \tag{10}$$

where $c_{HO\bullet}$ is the hydroxyl radical concentration and c_i and k_i are the concentration and the oxidation rate constant of the intermediate products in reaction (9).

Each by-product may have a distinct reaction coefficient k_i according to its reactivity with hydroxyl radicals [14]. For simplification, we used an average k_Σ and integral c_Σ (which is considered equal to the concentration of the organic compound c_R) in our calculations. Thus, the second term of Equation (10) can be rewritten:

$$\sum_{i=1}^{n} k_i c_{HO} c_i = \alpha k_\Sigma c_{HO} c_R \tag{11}$$

where α represents the number of hydroxyl radicals involved in the degradation of the initial organic compounds into end-products; k_Σ is the average rate constant of reaction (9); c_R is the concentration of the organic compound.

Thus, the rate of degradation of the initial organic compound oxidation reads:

$$v_R = k_R c_R c_{HO} \tag{12}$$

In the stationary state condition, Equation (2) with Equations (10)–(12) becomes as follows:

$$\operatorname{div} \vec{J}_{HO\cdot} = k_{HO} c_{HO}^2 + \alpha k_\Sigma c_R c_{HO} \tag{13}$$

$$\operatorname{div} \vec{J}_R = k_R c_R c_{HO} \tag{14}$$

where subindexes $HO^{\bullet}$ and R are related to the hydroxyl radical and organic compound, respectively.

The boundary conditions are presented in Appendix A.

3. Results and Discussions

Equation system (1)–(6) under the boundary conditions (Appendix A) is solved numerically using Comsol Multiphysics 5.5 software package.

3.1. *Degrdation of Paracetamol and its by-Products*

In the calculations, the oxidation of paracetamol (PCT) was considered. Brillas et al. [29,30] proposed several possible ways for PCT oxidation by hydroxyl radicals. Its well-known degradation by-product 1,4-benzoquinone (Figure 3) is strongly refractory to direct electron transfer. But some end-products such as carboxylic acids have a lower oxidation rate by $HO^{\bullet}$ and faster mineralization can be achieved by DET [31]. Therefore, it was calculated that 28 hydroxyl radicals participate in the oxidation of PCT into oxalic and oxamic acids. Thereby, we used $\alpha = 28$ in all the calculations presented in this study.

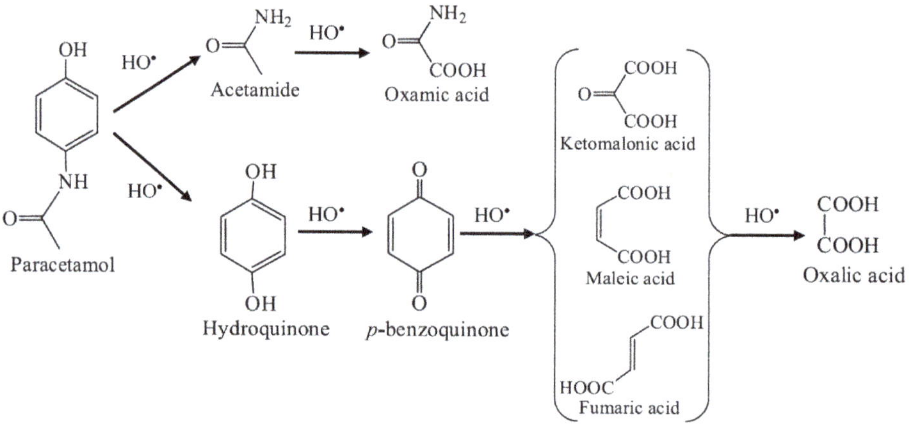

Figure 3. Proposed reaction sequence for paracetamol degradation in acidic aqueous medium by hydroxyl radicals oxidation processes [30]. In our study, we have considered the hydroxyl radical mediated oxidation of paracetamol into oxalic and oxamic acids.

3.2. *Treatment of Experimental Data*

The experimental data are taken from [21]. The parameters of the experimental setup were as follows: the REM was Magnelli phase anode with $r_1 = 3$ mm, $r_2 = 5$ mm, $L = 9$ cm, the stainless steel rod was used as cathode with $r_3 = 1.5$ mm, the cross-flow rate was $U = 0.88$ m·s^{-1}, the permeate flux rate (V) varied in the range 0.44–5.4 × 10^{-4} m·s^{-1}, the concentration of PCT was $c_R{}^0 = 0.19$ mM, concentration of the supporting electrolyte (sodium sulfate) was 50 mM, current density (j) was in the range 60–300 A·m^{-2}. The REM has a monomodal pore size distribution ranging between 0.8 and 1.9 μm. The median pore size $R_1 = 0.7$ μm was considered for the model. The relative porous fraction considered at the REM/liquid interface was $\varepsilon = 0.2$ (value obtained from the analysis of SEM images, Supplementary Materials).

The value of $\delta = 30$ μm is obtained from the Lévêque approximate solution (A15) according to hydrodynamic conditions and geometric parameters of the device used in the experiments. The rate constant of PCT oxidation reaction k_R was evaluated in numerous articles [21,32–34], and it differs in the range between 2×10^6 and 1.4×10^7 m^{3}mol^{-1}·s^{-1}. We used an average value of $k_R = 1 \times 10^7$ m^{3}mol^{-1}·s^{-1}.

The value of k_Σ was considered as a fitting parameter and was evaluated by comparing experimental and theoretical results (Figure 4).

Table 1. Parameters of the system used in the simulations. PCT: paracetamol.

R_1, μm	ε	k_R, $m^3 mol^{-1} \cdot s^{-1}$	k_Σ, $m^3 mol^{-1} \cdot s^{-1}$	$k_{HO\bullet}$, $m^3 mol^{-1} \cdot s^{-1}$	$c_R{}^0$, mM	D_{PCT}, m^2/s	$D_{HO\bullet}$, m^2/s	α	δ, μm
0.7 [21]	0.2 Suppl. M.	1×10^7 [21,32–34]	6.5×10^6 fitting parameter	5.5×10^6 [35]	0.19 [21]	0.65×10^{-9} [36]	2.2×10^{-9} [12]	28	30 Equation (A15)

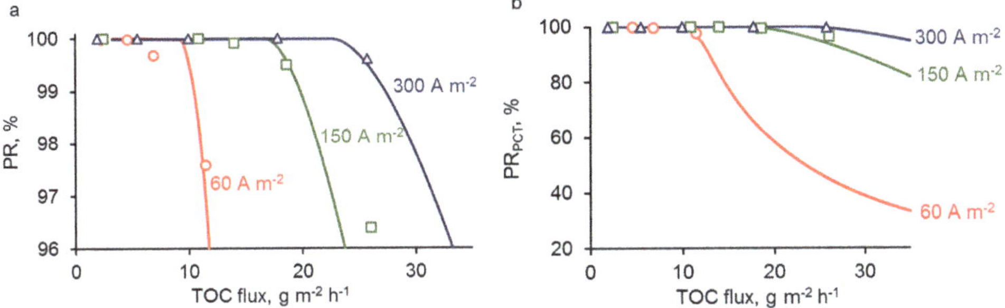

Figure 4. Experimental (dots) and theoretical (lines) dependencies of percentage removal of PCT (PR_{PCT}) on total organic carbon (TOC) flux at different current densities (mentioned in the figure). All other parameters are presented in Table 1. (**a**) is a zoomed version of (**b**).

Experimental and theoretical curves of (PR) of PCT as a function of total organic carbon (TOC) flux through the REM are shown in Figure 4. During the calculations, we varied the permeate flux (velocity) of the solution while the concentration remained constant. The graphs show that both in the experiment and theory, the PR strongly depends on permeate flux, which both increases the amount of organic compounds to oxidize and decreases the residence time of organic compounds in the reaction zone. As expected, PR decreases when increasing the permeate flux. Interestingly, the PR starts to decrease below 99% only from a threshold value that strongly depends on the current density. At low permeate flux, the process is operated under mass transport limitation, meaning that the amount of organic compounds to oxidize is limited by the transport of these organic compounds to the reaction zone. At high permeate flux, the process can switch from mass transport to reaction limitation, meaning that the oxidation rate is limited by the amount of hydroxyl radicals generated in the reaction zone. This latter limitation results in the decrease of the PR below 99%. As the formation of hydroxyl radicals is promoted by the increase in current density, it is consistent to observe that the reaction limitation occurs at higher permeate flux for high current density. A complete explanation of the results should also take into consideration both diffusive and convective mass transport of organic compounds. If the velocity of the solution is high, the convective flux exceeds the diffusion, and some organic molecules may not have the time to reach the reactive zone at the REM surface. This phenomenon will be further discussed below.

3.3. Concentration and Current Density Distribution

The current lines in the system are distributed non-uniformly over the electrode surface (Figure 5a). They condense at the upper side and become sparser in the bulk of the pore. The heterogeneous distribution of the current density leads to strong increases of local current densities on the REM surface. As can be seen, the model developed in this study predicts that the active area of the REM surface is located at the entrance of the pore (Figure 5a) [16]. Thereby, the local flux of hydroxyl radicals, which depends on the current density (Equation (15)), also enhances the reactive area at the upper side of the electrode/solution interface.

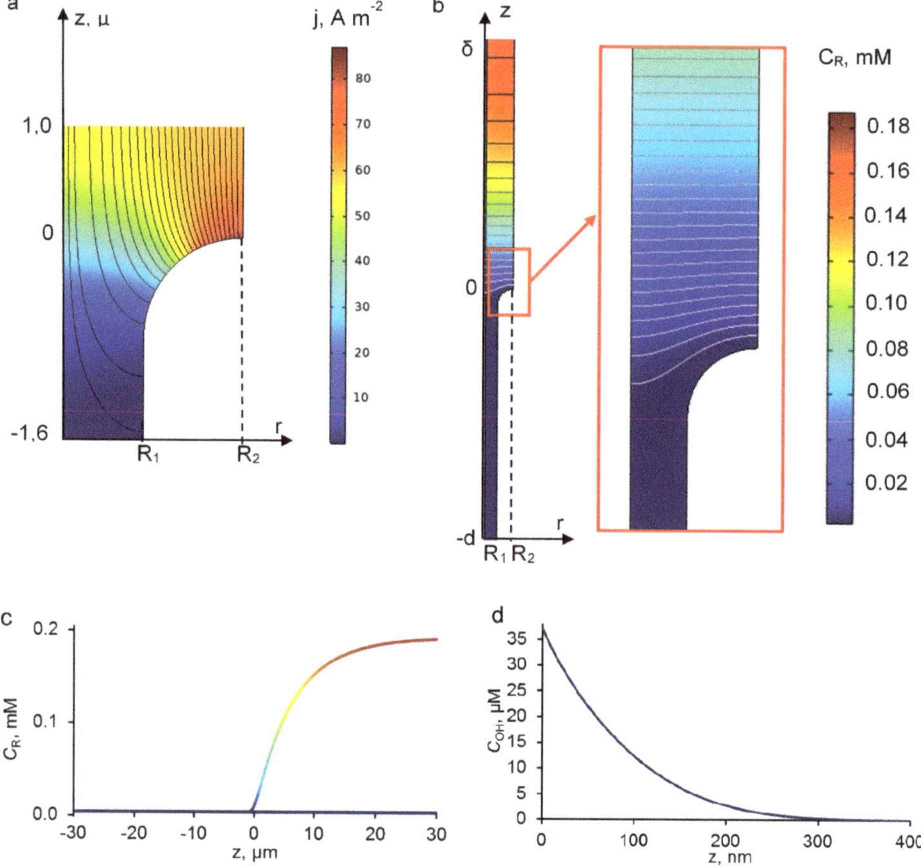

Figure 5. Calculated current density distribution and current lines near the entrance of the pore (**a**). Theoretical concentration distribution of PCT in the system (**b**), at the symmetry axis at $r = 0$ (**c**). Theoretical concentration distribution of $HO^{\bullet}$ radicals at $r = R_2$ (**d**). The calculations were performed at $j = 60$ A·m^{-2}; $V = 0.0001$ m/s. All other parameters are presented in Table 1.

Previous studies have highlighted the competition between reactions (8) (hydroxyl radical dimerization) and (9) (hydroxyl radical reaction with organic compounds) and the influence of organic compounds on the distribution profile of hydroxyl radicals: the higher the concentration of organic compounds, the lower the reactive zone thickness is [10–12]. The analytical Equation (15), which takes into account both reactions, was considered for taking into account the influence of the concentration of the organic compound (c_R) on the hydroxyl radical concentration distribution (Appendix B):

$$c_{HO} = \frac{4\alpha k_{\Sigma} c_R c^s_{HO}}{\left[\sqrt{\frac{2}{3}k_{HO}c^s_{HO} + zk_{\Sigma}c_R(1-A)} + \sqrt{zk_{\Sigma}c_R}(1+A)\right]^2} A;\ A = \exp\left(-x\sqrt{\frac{\alpha k_{\Sigma} c_R}{D_{HO}}}\right) \quad (15)$$

where c^s_{HO} is the concentration of $HO^{\bullet}$ on the anode surface.

The concentration profiles of hydroxyl radicals and organic compounds are shown in Figure 5. The concentration of hydroxyl radicals exponentially depends on the distance from the electrode surface (Figure 5d). The distribution is typical for the anodic oxidation systems [11–13]. Previous theoretical studies demonstrated that the thickness of the reaction zone is a function of current density, organic compound concentration, diffusion coefficient, reaction rate constant and number of hydroxyl radicals involved in the oxidation of one molecule. A good agreement was obtained between our results and previous studies: the reaction zone thickness varied between 1 nm (in the presence of 1 M of organic compound) and 1 µm in the absence of organics [12]. In the experimental conditions detailed above, results from the model indicates that hydroxyl radicals are only present in a reaction zone thickness lower than 300 nm from the electrode surface. As this thickness is lower than pore radius or DL thickness, it means that diffusion of the organic compounds might be a crucial phenomenon. Therefore, parameters such as convection rate, pore size and porosity of the electrode have to be considered in order to ensure that organic pollutants have sufficient time for diffusing to the reaction zone.

3.4. *Effect of REM Pore Radius*

For different current densities (j = 60, 150 and 300 A·m^2), we have predicted from the model the evolution of the PR of PCT at fixed porosity (ε = 0.2) and various length of pore radius R_1. As can be seen in Figure 6, the PR of PCT decreases with increasing R_1 at fixed TOC flux. The effect is due to the diffusion limitation of organic compound delivery to the surface of the electrode, where the reactive zone is located. The characteristic time for PCT diffusion to the electrode surface decreases with low pore size because the distance from the center of the pore to the electrode surface is shorter. Thus, at a fixed permeate flux (convection), the probability of a molecule to reach the reaction zone increases with decreasing R_1.

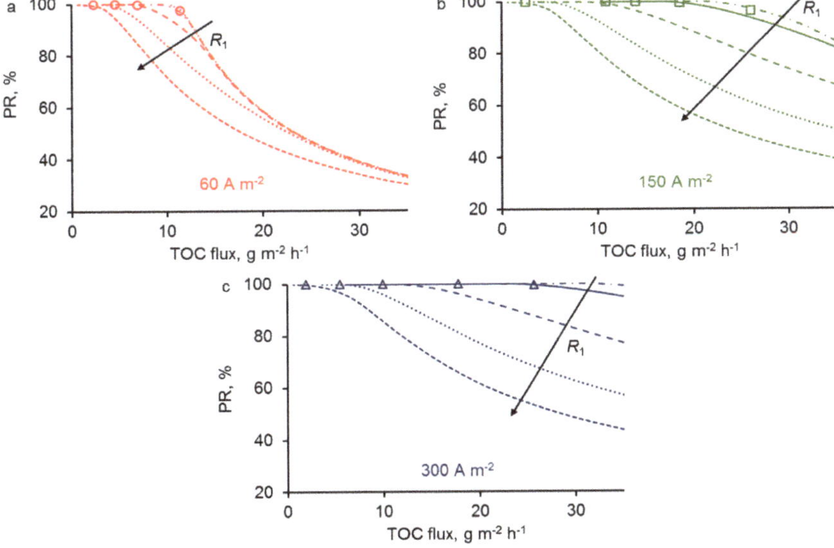

Figure 6. The influence of pore radii on the PR of the PCT at fixed porous fraction ε = 0.2 and j = 60 (**a**), 150 (**b**) and 300 (**c**) A·m^{-2}. The current density is mentioned in each figure. The arrow represents an increment of pore radius: R_1 = 0.5, 0.7, 1.5, 3 and 5 µm. All other parameters are presented in Table 1.

This limitation from the diffusion also depends on the permeate flux. The limitation from the diffusion actually starts when the convection becomes too high compared to the times required for diffusion. At low permeate flux, organic compounds will have sufficient time to diffuse to the electrode surface, whatever the length of pore radius (in the range studied). Thus, at TOC flux below 5 g·m^{-2}·h^{-1}, similar PR of PCT was predicted for all values of pore radius. Interestingly, at low current density (60 A·m^{-2}), the model did not predict any significant effect of pore radius in the range 0.5–1.5 µm.

It means that the diffusion was not the limiting phenomenon in this case. Instead, limitation from the reaction was the predominant phenomenon controlling the PR of PCT. At higher current density, limitation from the reaction is reached at higher permeate flux, thus making possible the limitation from the diffusion to occur and to affect the PR of PCT. For example, with a pore radius of 5 μm, similar trends were obtained whatever the current density, meaning that the limitation from the diffusion was the predominant phenomenon in this different case.

These results predicted from the model highlight that convection, diffusion and reaction phenomena have to be taken into consideration in order to optimize crucial parameters of the process such as pore size, permeate flux and current density.

3.5. Effect of REM Porosity

Similar to the previous case, we fixed one of the geometric parameters, namely the pore radius: $R_1 = 0.7$ μm. We analyzed the effect of ε on the PR of PCT at different current density (Figure 7). Results predicted from the models indicate that the effectiveness of the process should be improved when increasing the porosity. For a constant pore radius, increasing the porosity means that the number of pores over a same surface is increasing. For the same permeate flux, it involves that the water velocity (convection) in each pore is lower, and thus, pollutants have more time to diffuse to the electrode surface. Therefore, increasing the porosity allows the limitation from the diffusion to occur only at higher permeate flux. As observed for pore radius, the effect of porosity is not significant for the lowest current density (60 A·m^{-2}) because in this case the PR of PCT is almost only controlled by the limitation from the reaction that occurs at low permeate flux.

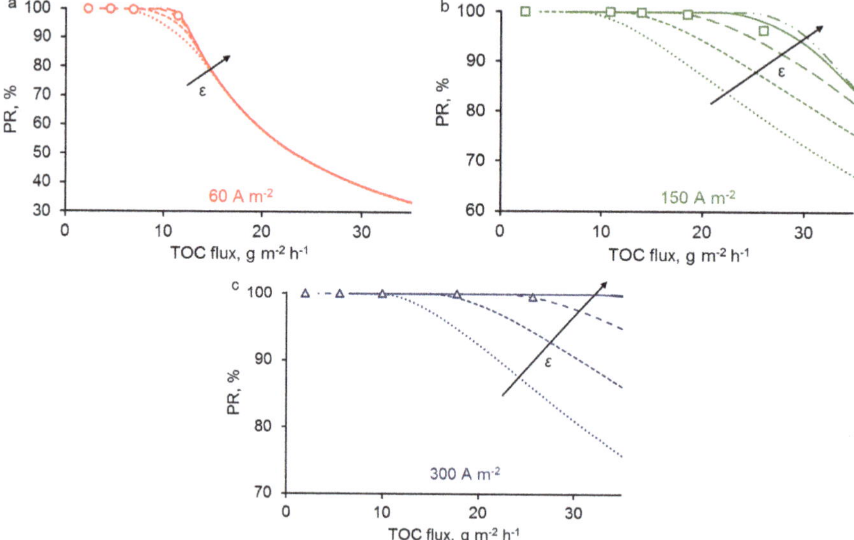

Figure 7. The influence of porosity on the PR at fixed pore size $R_1 = 0.7$ μm and $j = 60$ (**a**), 150 (**b**) and 300 (**c**) A·m^{-2}. The arrow represents an increment of porosity: $\varepsilon = 0.05, 0.1, 0.2, 0.4, 0.6$. All other parameters are presented in Table 1.

Overall, these theoretical results highlight how current density, pore radius and porosity influence the process effectiveness of the REM by taking into consideration convection, diffusion and reaction phenomena. Such models might help to the design of novel efficient REM minimizing mass transport limitations.

Nevertheless, it might be also important to take into consideration an additional effect, which is the influence of the porosity on the distribution of the current density (Figure 8). When the porosity increases, the surface of the electrode material at z = 0 decreases. As the active area of the REM is mainly located at the entrance of the pore, it leads to an increase of local current densities (Figure 8). This effect might have adverse effects on the lifetime of the electrode since high local current density might promote passivation and corrosion of the electrode.

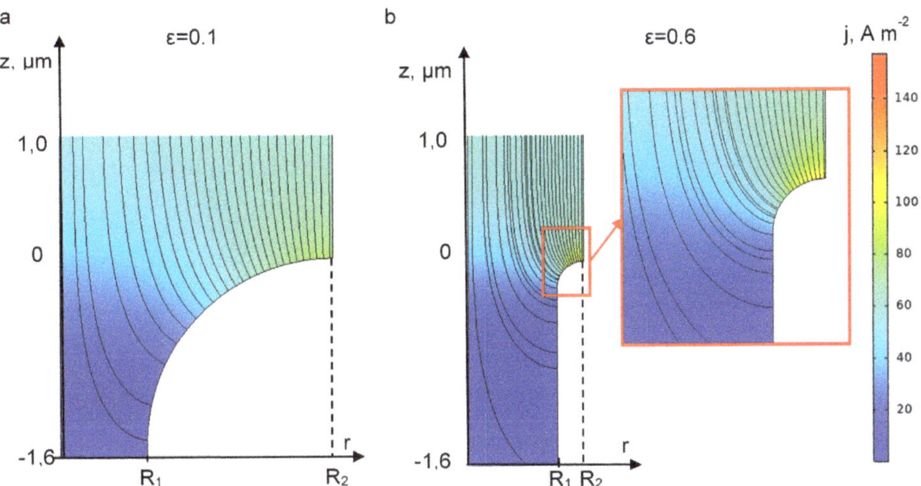

Figure 8. The influence of the REM porosity (mentioned in the pictures) on the local current density distributions at j = 60 A·m^{-2} and ε = 0.1 (**a**) and 0.6 (**b**). All other parameters are presented in Table 1.

4. Conclusions

In the paper, we proposed a 2D stationary model of transport of diluted species in the electrolysis system with a REM. The model is based on the Fick-Navier-Stokes equations and takes into account the local geometrical and hydrodynamic properties of the system as well as chemical reactions related to the oxidation of organic compounds by hydroxyl radicals. Porosity and pore radius of the REM were considered as two crucial parameters.

We highlighted theoretically the competitive mechanisms relative to the delivery of an organic compound to the reactive zone, where the oxidation reaction occurs. Convection, diffusion and reaction limitations have to be taken into consideration in order to explain the results. The calculated dependencies of the degradation rate of a model organic compound on TOC flux are in good agreement with experimental data.

Porosity and pore radius of the REM significantly affect the removal of the target organic compound: the degradation rate decreases with increasing pore radius or decreasing porosity. These adverse effects are due to the diffusion time of organic compounds to the pore surface that becomes limiting compared to the characteristic time of convection. These theoretical predictions might be useful for the conception of novel REM minimizing mass transport limitations.

Supplementary Materials: The following are available online at http://www.mdpi.com/2077-0375/10/5/102/s1, Figure S1: (a) The microphotograph of REM was exposed to a procedure the example of which is presented in the Figure. The pore entrances were painted in red color (in form of oval (b) or in random form (c)) with the help of Corel PHOTO-PAINT software, and the areas corresponding to electrode material in white color (d,e). The presence of only a two-color image allowed us to determine the ratio of pores on the REM surface. In both cases (ovals or random form) the relative porous fraction was close to 0.2 (0.195 and 0.21, respectively).

Author Contributions: Conceptualization, E.S., C.T., M.C. and S.M.; methodology, M.C., C.T. and S.M.; software, E.S. and S.M.; validation, E.S. and S.M.; formal analysis, S.M. and C.T.; investigation, E.S., S.A. and C.T.; resources, S.M.; writing—original draft preparation, S.M.; writing—review and editing, C.T., S.M. and E.S.; visualization, E.S. and S.M.; supervision, S.M.; project administration, M.C. and S.M.; funding acquisition, S.M. All authors have read and agreed to the published version of the manuscript.

Funding: This research was funded by the Russian Science Foundation, grant number 19-79-00268.

Acknowledgments: The authors are grateful to V.V. Nikonenko (Kuban State University, Krasnodar, Russia) for help at various stages of the preparation of this article.

Conflicts of Interest: The authors declare no conflict of interest

Appendix A

Boundary Conditions

1. At the interface between the DL and electrode surface, the following boundary conditions are employed:

- The nonslip condition:

$$\vec{V} = 0 \tag{A1}$$

- The flux continuity condition for the electric potential:

$$-\kappa_{solution}\nabla\varphi\big|_{solution} = -\kappa_{electrode}\nabla\varphi\big|_{electrode} \tag{A2}$$

- The flux of hydroxyl radicals can be expressed in terms of current density:

$$-\vec{n}J_{HO} = \frac{-\vec{j}}{F} \tag{A3}$$

- The direct electron transfer is not taken into account; hence the organic compound flux is equal to zero:

$$-\vec{n}J_R = 0 \tag{A4}$$

2. At the boundary between the DL and bulk solution (at $z = \delta$):

- Potential value equals to zero:

$$\varphi = 0 \tag{A5}$$

- The concentration of the organic compound is constant and equals to the bulk concentration:

$$c_R = c_R^0 \tag{A6}$$

- Hydroxyl radicals are absent in the bulk solution:

$$c_{HO} = 0 \tag{A7}$$

- The velocity of the solution in the z-direction is equal to the transmembrane value:

$$V_z = V_0 \tag{A8}$$

3. At the outer boundaries between the unit cells of the system, the zero-flux condition is assumed:

$$-\vec{n}\cdot J_{HO^{\cdot}} = 0 \tag{A9}$$

$$-\vec{n}\cdot J_R = 0 \tag{A10}$$

$$-\vec{n} \cdot V = 0 \tag{A11}$$

$$-\vec{n} \cdot j = 0 \tag{A12}$$

4. In the bulk of the pore (at $z = -d$, $r \in (0, R_1)$

   - The flux continuity equation gives:

$$-\vec{n} \cdot J_{HO^{\cdot}} = c_{HO^{\cdot}} V_z \tag{A13}$$

$$-\vec{n} \cdot J_R = c_R V_z \tag{A14}$$

The Diffusion Layer Thickness in the Electrolyzer

The DL thickness must be known to solve the problem presented above. If the cross-flow velocity is several times higher than the transmembrane velocity of the solution, then in the case of steady laminar flow between two parallel plane homogeneous permselective walls, the value of δ may be calculated, when a limiting current density occurs. In this case, the system of Equations (1)–(6) may be solved with the boundary condition setting nearly zero electrolyte concentration at the walls. The approximate formula for the average effective thickness of the diffusion layer, δ, reads:

$$\delta = 0.71\, h\left(\frac{LD_{PCT}}{h^2 U}\right)^{1/3} \tag{A15}$$

where U is the average cross-flow fluid velocity, h is the distance between the walls (in our case, it is equal to $(r_1 - r_3)$). Equation (A15) holds for relatively "short" channels ($L \le 0.02h^2U/D_{PCT}$). It was deduced by Lévêque in 1926 [37] for heat transfer and adapted by Newman [38] for electrode systems.

Appendix B

Derivation of the equation for the distribution of the HO$^\bullet$ concentration as a function of the distance from the electrode surface, x.

Let us consider the processes occurring near the surface of the anode. We suggest additional approximations to obtain an analytical solution of the problem:

- The coordinate perpendicular to the membrane surface is considered, the tangential fluxes are neglected;
- The concentration of the organic compound c_R is constant;
- The convective flux is negligible; only the diffusion of HO$^\bullet$ is taken into account.

The boundary conditions for the differential Equation (13) are obtained from the assumption that at a sufficiently large distance from the electrode ($x = \infty$), the concentration of hydroxyl radicals is equal to zero (A16), and on the electrode surface ($x = 0$) the concentration of hydroxyl radicals is constant, c^s_{HO}(A17):

$$c_{HO} = 0 \; at \; x = \infty \tag{A16}$$

$$c_{HO} = c^s_{HO} at x = 0 \tag{A17}$$

Using the boundary conditions (A16) and (A17), Equation (13) has the following analytical solution:

$$c_{HO} = \frac{4\alpha k_\Sigma c_R c^s_{HO}}{\left[\sqrt{\frac{2}{3}k_{HO}c^s_{HO} + \alpha k_\Sigma c_R}(1-A) + \sqrt{\alpha k_\Sigma c_R}(1+A)\right]^2} A;\; A = \exp\left(-x\sqrt{\frac{\alpha k_\Sigma c_R}{D_{HO}}}\right)$$

where the concentration of hydroxyl radicals is an explicit function of the distance from the electrode surface (Figure A1a).

To find the concentration of hydroxyl radicals at the electrode surface, we must calculate the concentration gradient:

$$\frac{dc_{HO}}{dx} = -4\frac{c_{HO}^{s}\sqrt{(\alpha k_{\Sigma}c_{R})^{3}}}{\sqrt{D_{HO}}}\frac{\sqrt{\frac{2}{3}k_{HO}c_{HO}^{s}+\alpha k_{\Sigma}c_{R}}(1+A)+\sqrt{\alpha k_{\Sigma}c_{R}}(1-A)}{\left[\sqrt{\frac{2}{3}k_{HO}c_{HO}^{s}+\alpha k_{\Sigma}c_{R}}(1-A)+\sqrt{\alpha k_{\Sigma}c_{R}}(1+A)\right]^{3}}A \tag{A18}$$

Then, the flux of hydroxyl radicals on the electrode surface (at $x = 0$) reads:

$$[J_{HO}]_{x=0} = -D_{HO}\left[\frac{dc_{HO}}{dx}\right]_{x=0} = \sqrt{D_{HO}}c_{HO}^{s}\sqrt{\frac{2}{3}k_{HO}c_{HO}^{s}+\alpha k_{\Sigma}c_{R}} \tag{A19}$$

Comparing (A19) with (A3), we obtain the expression for the concentration of hydroxyl radicals on the electrode surface:

$${c_{HO}^{s}}^{3}+\frac{3}{2}\frac{\alpha k_{\Sigma}c_{R}}{k_{HO}}{c_{HO}^{s}}^{2}-\frac{3}{2}\frac{j^{2}}{D_{HO}k_{HO}F^{2}}=0 \tag{A20}$$

This cubic equation may be solved using the Vieta-Cardano formulas (Figure A1b).

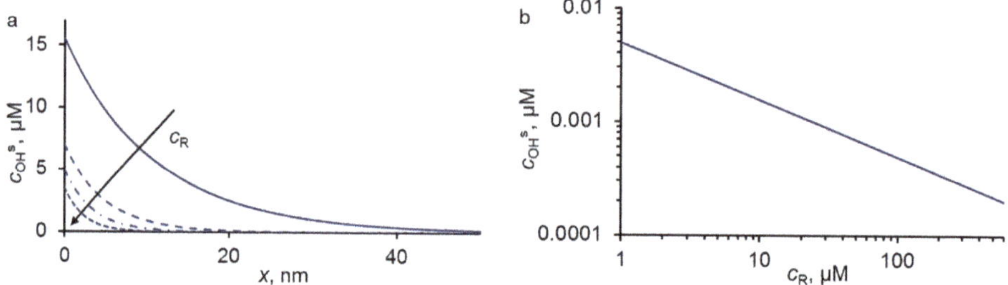

Figure A1. Calculated concentration profiles (**a**) and concentration dependence of hydroxyl radicals on the REM surface (c_{HO}^{S}) on the concentration of organic compounds c_R (**b**). Parameters used in the calculations: j = 300 A·m^{-2}; D_{HO} = 2.2 × 10^{-9} m^2·s^{-1}, k_{HO} = 5.5 × 10^6 m^{3}mol^{-1}·s^{-1}, k_{Σ} = 6.5 × 10^6 m^{3}mol^{-1}·s^{-1}, α = 28. The arrow represents an increment of the concentration of organic compounds: c_R = 0.1, 0.5, 1, 2 mM.

References

1. Chun, Y.; Mulcahy, D.; Zou, L.; Kim, I.S. A short review of membrane fouling in forward osmosis processes. *Membranes* **2017**, *7*, 30. [CrossRef] [PubMed]
2. Bdiri, M.; Perreault, V.; Mikhaylin, S.; Larchet, C.; Hellal, F.; Bazinet, L.; Dammak, L. Identification of phenolic compounds and their fouling mechanisms in ion-exchange membranes used at an industrial scale for wine tartaric stabilization by electrodialysis. *Sep. Purif. Technol.* **2020**, *233*, 115995. [CrossRef]
3. Bdiri, M.; Dammak, L.; Larchet, C.; Hellal, F.; Porozhnyy, M.; Nevakshenova, E.; Pismenskaya, N.; Nikonenko, V. Characterization and cleaning of anion-exchange membranes used in electrodialysis of polyphenol-containing food industry solutions; comparison with cation-exchange membranes. *Sep. Purif. Technol.* **2019**, *210*, 636–650. [CrossRef]
4. Panizza, M.; Cerisola, G. Direct and Mediated Anodic Oxidation of Organic Pollutants. *Chem. Rev.* **2009**, *109*, 6541–6569. [CrossRef] [PubMed]
5. Chaplin, B.P. Critical review of electrochemical advanced oxidation processes for water treatment applications. *Environ. Sci. Process. Impacts* **2014**, *16*, 1182–1203. [CrossRef] [PubMed]
6. Ghrib, F.; Saied, T.; Bellakhal, N. Experimental Design Methodology Applied to the Degradation of a Cytostatic Agent the Imatinib Mesylate Using Fenton Process. *Chem. Africa* **2019**, *2*, 103–111. [CrossRef]
7. Fernandes, A.; Pacheco, M.J.; Ciríaco, L.; Lopes, A. Anodic oxidation of a biologically treated leachate on a boron-doped diamond anode. *J. Hazard. Mater.* **2012**, *199–200*, 82–87. [CrossRef]

8. Oturan, N.; Brillas, E.; Oturan, M.A. Unprecedented total mineralization of atrazine and cyanuric acid by anodic oxidation and electro-Fenton with a boron-doped diamond anode. *Environ. Chem. Lett.* **2012**, *10*, 165–170. [CrossRef]
9. Trellu, C.; Péchaud, Y.; Oturan, N.; Mousset, E.; Huguenot, D.; van Hullebusch, E.D.; Esposito, G.; Oturan, M.A. Comparative study on the removal of humic acids from drinking water by anodic oxidation and electro-Fenton processes: Mineralization efficiency and modelling. *Appl. Catal. B Environ.* **2016**, *194*, 32–41. [CrossRef]
10. Trellu, C.; Ganzenko, O.; Papirio, S.; Pechaud, Y.; Oturan, N.; Huguenot, D.; van Hullebusch, E.D.; Esposito, G.; Oturan, M.A. Combination of anodic oxidation and biological treatment for the removal of phenanthrene and Tween 80 from soil washing solution. *Chem. Eng. J.* **2016**, *306*, 588–596. [CrossRef]
11. Mascia, M.; Vacca, A.; Palmas, S.; Polcaro, A.M. Kinetics of the electrochemical oxidation of organic compounds at BDD anodes: Modelling of surface reactions. *J. Appl. Electrochem.* **2007**, *37*, 71–76. [CrossRef]
12. Kapałka, A.; Fóti, G.; Comninellis, C. The importance of electrode material in environmental electrochemistry. *Electrochim. Acta* **2009**, *54*, 2018–2023. [CrossRef]
13. Donaghue, A.; Chaplin, B.P. Effect of Select Organic Compounds on Perchlorate Formation at Boron-doped Diamond Film Anodes. *Environ. Sci. Technol.* **2013**, *47*, 12391–12399. [CrossRef] [PubMed]
14. Groenen-Serrano, K.; Weiss-Hortala, E.; Savall, A.; Spiteri, P. Role of Hydroxyl Radicals During the Competitive Electrooxidation of Organic Compounds on a Boron-Doped Diamond Anode. *Electrocatalysis* **2013**, *4*, 346–352. [CrossRef]
15. Zaky, A.M.; Chaplin, B.P. Porous substoichiometric TiO_2 anodes as reactive electrochemical membranes for water treatment. *Environ. Sci. Technol.* **2013**, *47*, 6554–6563. [CrossRef] [PubMed]
16. Jing, Y.; Chaplin, B.P. Electrochemical impedance spectroscopy study of membrane fouling and electrochemical regeneration at a sub-stoichiometric TiO_2 reactive electrochemical membrane. *J. Membr. Sci.* **2016**, *510*, 238–249. [CrossRef]
17. Nayak, S.; Chaplin, B.P. Fabrication and characterization of porous, conductive, monolithic Ti_4O_7 electrodes. *Electrochim. Acta* **2018**, *263*, 299–310. [CrossRef]
18. Misal, S.N.; Lin, M.H.; Mehraeen, S.; Chaplin, B.P. Modeling electrochemical oxidation and reduction of sulfamethoxazole using electrocatalytic reactive electrochemical membranes. *J. Hazard. Mater.* **2020**, *384*, 121420. [CrossRef]
19. Trellu, C.; Chaplin, B.P.; Coetsier, C.; Esmilaire, R.; Cerneaux, S.; Causserand, C.; Cretin, M. Electro-oxidation of organic pollutants by reactive electrochemical membranes. *Chemosphere* **2018**, *208*, 159–175. [CrossRef]
20. Bejan, D.; Malcolm, J.D.; Morrison, L.; Bunce, N.J. Mechanistic investigation of the conductive ceramic Ebonex® as an anode material. *Electrochim. Acta* **2009**, *54*, 5548–5556. [CrossRef]
21. Trellu, C.; Coetsier, C.; Rouch, J.-C.C.; Esmilaire, R.; Rivallin, M.; Cretin, M.; Causserand, C. Mineralization of organic pollutants by anodic oxidation using reactive electrochemical membrane synthesized from carbothermal reduction of TiO_2. *Water Res.* **2018**, *131*, 310–319. [CrossRef]
22. Ahmed, F.; Lalia, B.S.; Kochkodan, V.; Hilal, N.; Hashaikeh, R. Electrically conductive polymeric membranes for fouling prevention and detection: A review. *Desalination* **2016**, *391*, 1–15. [CrossRef]
23. Zaky, A.M.; Chaplin, B.P. Mechanism of p-substituted phenol oxidation at a Ti_4O_7 reactive electrochemical membrane. *Environ. Sci. Technol.* **2014**, *48*, 5857–5867. [CrossRef] [PubMed]
24. Fu, W.; Wang, X.; Zheng, J.; Liu, M.; Wang, Z. Antifouling performance and mechanisms in an electrochemical ceramic membrane reactor for wastewater treatment. *J. Membr. Sci.* **2019**, *570–571*, 355–361. [CrossRef]
25. Cruz-Díaz, M.R.; Rivero, E.P.; Rodríguez, F.A.; Domínguez-Bautista, R. Experimental study and mathematical modeling of the electrochemical degradation of dyeing wastewaters in presence of chloride ion with dimensional stable anodes (DSA) of expanded meshes in a FM01-LC reactor. *Electrochim. Acta* **2018**, *260*, 726–737. [CrossRef]
26. Wei, X.; Wang, H.; Yin, Z.; Qaseem, S.; Li, J. Tubular electrocatalytic membrane reactor for alcohol oxidation: CFD simulation and experiment. *Chin. J. Chem. Eng.* **2017**, *25*, 18–25. [CrossRef]
27. Mareev, S.A.; Nichka, V.S.; Butylskii, D.Y.; Urtenov, M.K.; Pismenskaya, N.D.; Apel, P.Y.; Nikonenko, V.V. Chronopotentiometric Response of an Electrically Heterogeneous Permselective Surface: 3D Modeling of Transition Time and Experiment. *J. Phys. Chem. C* **2016**, *120*, 13113–13119. [CrossRef]
28. Mareev, S.A.; Butylskii, D.Y.; Pismenskaya, N.D.; Larchet, C.; Dammak, L.; Nikonenko, V.V. Geometric heterogeneity of homogeneous ion-exchange Neosepta membranes. *J. Membr. Sci.* **2018**, *563*, 768–776. [CrossRef]

29. Brillas, E.; Sirés, I.; Arias, C.; Cabot, P.L.; Centellas, F.; Rodríguez, R.M.; Garrido, J.A. Mineralization of paracetamol in aqueous medium by anodic oxidation with a boron-doped diamond electrode. *Chemosphere* **2005**, *58*, 399–406. [CrossRef]
30. Sirés, I.; Garrido, J.A.; Rodríguez, R.M.; Cabot, P.L.; Centellas, F.; Arias, C.; Brillas, E. Electrochemical Degradation of Paracetamol from Water by Catalytic Action of Fe^{2+}, Cu^{2+}, and UVA Light on Electrogenerated Hydrogen Peroxide. *J. Electrochem. Soc.* **2006**, *153*, D1. [CrossRef]
31. Weiss, E.; Groenen-Serrano, K.; Savall, A.; Comninellis, C. A kinetic study of the electrochemical oxidation of maleic acid on boron doped diamond. *J. Appl. Electrochem.* **2007**, *37*, 41–47. [CrossRef]
32. Land, E.J.; Ebert, M. Pulse radiolysis studies of aqueous phenol. Water elimination from dihydroxycyclohexadienyl radicals to form phenoxyl. *Trans. Faraday Soc.* **1967**, *63*, 1181. [CrossRef]
33. Hamdi El Najjar, N.; Touffet, A.; Deborde, M.; Journel, R.; Karpel Vel Leitner, N. Kinetics of paracetamol oxidation by ozone and hydroxyl radicals, formation of transformation products and toxicity. *Sep. Purif. Technol.* **2014**, *136*, 137–143. [CrossRef]
34. Andreozzi, R.; Caprio, V.; Marotta, R.; Vogna, D. Paracetamol oxidation from aqueous solutions by means of ozonation and H_2O_2/UV system. *Water Res.* **2003**, *37*, 993–1004. [CrossRef]
35. Buxton, G.V.; Greenstock, C.L.; Helman, W.P.; Ross, A.B. Critical Review of rate constants for reactions of hydrated electrons, hydrogen atoms and hydroxyl radicals (OH/O—In Aqueous Solution. *J. Phys. Chem. Ref. Data* **1988**, *17*, 513–886. [CrossRef]
36. Ribeiro, A.C.F.; Barros, M.C.F.; Veríssimo, L.M.P.; Santos, C.I.A.V.; Cabral, A.M.T.D.P.V.; Gaspar, G.D.; Esteso, M.A. Diffusion coefficients of paracetamol in aqueous solutions. *J. Chem. Thermodyn.* **2012**, *54*, 97–99. [CrossRef]
37. Lévêque, A. Les lois de la transmission de chaleur par convection. *Ann. Mines* **1928**, *13*, 201–299.
38. Newman, J. *Electrochemical Systems*; Prentice Englewood Cliffs: New York, NY, USA, 1973.

Article

Influence of Surface Modification of MK-40 Membrane with Polyaniline on Scale Formation under Electrodialysis

Marina A. Andreeva [1,*], Natalia V. Loza [1], Natalia D. Pis'menskaya [1], Lasaad Dammak [2] and Christian Larchet [2]

1 Physical Chemistry Department, Faculty of Chemistry and High Technologies, Kuban State University, 149 Stavropolskaya st., 350040 Krasnodar, Russia; nata_loza@mail.ru (N.V.L.); n_pismen@mail.ru (N.D.P.)

2 Institut de Chimie et des Matériaux Paris-Est (ICMPE) UMR 7182 CNRS, Université Paris-Est, 2 Rue Henri Dunant, 94320 Thiais, France; dammak@u-pec.fr (L.D.); larchet@u-pec.fr (C.L.)

* Correspondence: andreeva_marina_90@bk.ru

Received: 12 June 2020; Accepted: 3 July 2020; Published: 7 July 2020

Abstract: A comprehensive study of the polyaniline influence on mineral scaling on the surface of the heterogeneous MK-40 sulfocationite membrane under electrodialysis has been conducted. Current-voltage curves and chronopotentiograms have been obtained and analyzed for the pristine MK-40 membrane and the MK-40 membrane which is surface-modified by polyaniline. The study of the electrochemical behavior of membranes has been accompanied by the simultaneous control of the pH of the solution outcoming from the desalination compartment. The mixture of Na_2CO_3, KCl, $CaCl_2$, and $MgCl_2$ is used as a model salt solution. Two limiting states are observed on the current-voltage curve of the surface-modified membrane. There is the first pseudo-limiting state in the range of small values of the potential drop. The second limiting current is comparable with that of the limiting current for the pristine membrane. It is shown that chronopotentiometry cannot be used as a self-sufficient method for membrane scaling identification on the surface-modified membrane at high currents. A mineral scale on the surfaces of the studied membranes has been found by scanning electron microscopy. The amount of precipitate is higher in the case of the surface-modified membrane compared with the pristine one.

Keywords: ion-exchange membrane; polyaniline; surface modification; mineral scaling; voltammetry; chronopotentiometry

---

1. Introduction

To date, methods of electromembrane technology are effective, environmentally cleanest, and cost-effective [1,2]. Electrodialysis (ED) processes are widely implemented for brine concentrations in sea-salt production [2], the food industry [3,4], the extraction of precious or toxic substances, such as organic acids [5,6], wastewater treatment, especially for the removal of heavy metals [7,8], and production of acids and bases [9]. ED is based on the selective migration of ions through ion-exchange membranes (IEMs) under the action of applied electric field as the driving force. Low-energy consumption and high-current efficiency are the advantages of the ED process. However, despite the efforts of many other researchers to optimize ED performance for various applications, there are several weaknesses that restrict ED usefulness, involving selectivity, membrane fouling, and mineral scaling.

Mineral scaling on IEMs happens when salts from the solution precipitate and settle on the membrane surface and inside the ion-conductive pathways (pores) of the membrane [10]. Scaling is closely related to the development of one of the coupled effects of the concentration polarization in the electromembrane system, namely, water splitting [11,12]. pH variations of the solution close to the interface boundary result in the creation of conditions where precipitation of sparingly soluble salts occurs. The precipitate occurs when the solute concentrations exceed the solubility of the sparingly soluble solids [13]. The multivalent ions such as barium, calcium, magnesium, ferric, bicarbonate, and sulfate are the major scaling ions. The presence of mineral scales on the membrane surface and inside the membrane pores reduces the membrane working area and operation lifetime, causes additional resistance to the solution flux, and mass transfer.

Another coupled effect of concentration polarization, namely, electroconvection, hinders the process of water splitting [14–16]. Thus, it is possible to reduce the negative effect of scaling by enhancing electroconvection [17–19]. It was shown that the coating of a homogeneous Nafion® film on the surface of a heterogeneous MK-40 membrane leads to the electroconvection intensification and water splitting decrease due to a relatively more hydrophobic and less electrically inhomogeneous surface compared to the original MK-40 membrane, and as a result, to scaling mitigation [19].

The most used methods for reducing the membrane scaling are changing regimes of electrodialysis treatment, mechanical action, pretreatment using pressure-driven membrane processes, cleaning agents, and the modification of IEMs [10]. Despite the fact that researches show the successful tendency for IEM clogging mitigation by means of electrodialysis reversal and pulsed electrical field [20], IEMs scaling is still a limiting factor for the wide construction of ED.

Surface topology of the IEMs also influences the scale formation. The rougher the membrane surface, the more susceptible it is to scaling due to the presence of more active sites for surface nucleation [13,18,19]. The smoother membrane surface results in weaker scale-membrane adhesion. Thereby the development of membranes with improved scaling resistance is a promising alternative for scaling mitigation.

Another way of membrane surface modification leading to the monovalent membrane selectivity is used for membrane clogging mitigation [21]. It is known from the literature that the formation of a thin oppositely charged layer on the surface of IEMs could improve the membrane selectivity towards the transfer of monovalent ions [22]. The monovalent ion selectivity stems from the kinetic effect of electrostatic repulsion of the charged surface layer towards divalent cations in the solution [22]. Thus, the transfer of multicharged ions from the desalination compartment to the concentration compartment reduces, mitigates, or prevents scaling in the concentration compartment. Some studies have also found that the above modification helps drastically decrease the peptide fouling during electrodialysis as a result of changes in the nature of the electrostatic interaction between proteins and the membrane surface [21]. It has been shown that polymerization of polyaniline (PANI) on the surface of the cation-exchange membrane results in the appearance of selectivity for the transfer of singly charged ions [23,24] and the decrease of water transport number [25]. At the same time, as a result of modifying the membrane with PANI, there was no significant decrease in the permselectivity of the cation exchange membrane [26]. Nevertheless, the formation of an internal bipolar interface between the cation-and anion-exchange layers led to catalytic water splitting that could stimulate the rate of scale formation on the IEM [27]. Potentially, these membranes may be used as monovalent-cation-selective ones. However, their behavior has not been studied from the point of view of possible scaling. In this context, the main goal of the article is to study how the PANI modification of heterogeneous membranes affects the scaling process.

2. Materials and Methods

2.1. Membranes

Heterogeneous ion-exchange MK-40 and MK-40/PANI membranes are tested in the study. The heterogeneous cation-exchange MK-40 membrane, manufactured by UCC "Shchekinoazot" (urban locality Pervomaysky, Tula oblast, Russia), is a mixture of the polyethylene and cation exchange resin KU-2-8 based on sulfonated polystyrene crosslinked by divinylbenzene. To ensure the mechanical strength, the membrane is reinforced with a nylon mesh. MK-40 membranes are used in the electrodialysis concentration and demineralization of electrolyte solutions [28–31]. The closest analog of this membrane is the Ralex-CMH membrane produced by MEGA a.s. (Straz pod Ralskem, Czech Republic). The properties of the MK-40 membrane are described elsewhere, for example in [32–36]. The MK-40/PANI membrane is obtained by synthesizing PANI on the surface of the MK-40 membrane under conditions of electrodiffusion of the monomer and the oxidizer in an external electric field [37]. Thus, the MK-40/PANI composite has an asymmetric structure—the MK-40 membrane is on the one side, and the PANI layer is on the other side. The surfaces of the tested membranes are illustrated in Figure 1.

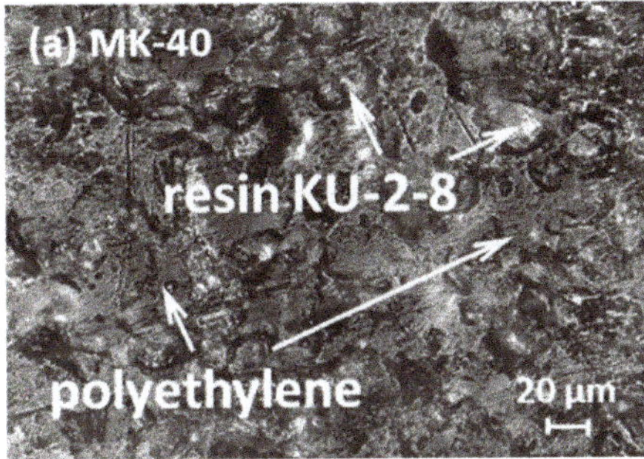

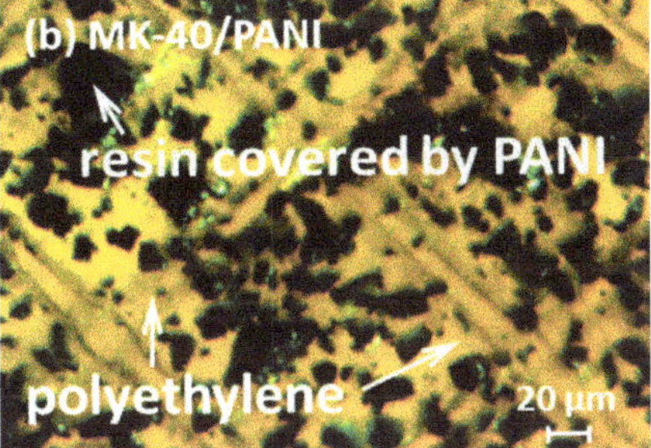

Figure 1. Image of the surface of swollen (**a**) MK-40 and (**b**) MK-40/PANI membranes obtained with a Zeiss AxioCam MRc5 light microscope.

The heterogeneous anion-exchange commercial MA-41 membrane, manufactured by Shchekinoazot (Russia) and the MK-40 membrane are used as the auxiliary membranes in the chronopotentiometry and voltammetry measurements. The functional groups of the MA-41 membrane are quaternary ammonium groups.

2.2. Analysis Methods

2.2.1. Membrane Thickness

The membrane thickness, L, is measured with a Micrometer MK 0-25 (Model 102, Chelyabinsk, Russia). The membrane thickness values are averaged from ten measurements at different locations on the effective surface of each membrane.

2.2.2. Membrane Electrical Conductivity

The membrane conductivity is determined by the difference method. The membrane electrical conductivity, κ, could be calculated as

$$\kappa = \frac{L}{R_m S}, \quad (1)$$

where S and L are the membrane area and thickness, respectively, R_m—the electric resistance of the membrane.

The electrical resistance is measured with a specially designed cell coupled to the Immitance meter (RLC) E7-21 (MNIPI, Minsk, Republic of Belarus), and 0.1 M NaCl reference solution is used. According to the procedure developed by Lteif et al. [38], the electric resistance of the membrane is calculated according to:

$$R_m = R_{m+s} - R_s, \quad (2)$$

where R_{m+s} and R_s are the electrical resistance of the solution with membrane and without membrane, respectively.

2.2.3. Contact Angle Measurements

Contact angles, θ, of the wet membranes are measured by the sessile drop method in the sodium form [39]. Data are treated using the ImageJ software.

2.2.4. Scanning Electron Microscopy

Visualization of the membrane surface covered by the precipitate is made by a scanning electron microscope (Merlin, Carl Zeiss Microscopy GmbH, Oberkochen, Germany) equipped with an energy dispersive spectrometer. The energy dispersive spectrometer conditions are 6 kV accelerating voltage with a 9.9-mm working distance. The dried membrane samples are coated with a thin layer of platinum in order to make them electrically conductive and to improve the quality of the microscopy photographs.

2.3. *Electrodialysis Cell and Experimental Setup*

The investigations are carried out in the same flow-through four-compartment electrodialysis cell, which is used for the membrane modification [19]. The cell comprises the tested MK-40 or MK-40/PANI membrane and two auxiliary membranes. The intermembrane distance in the cell compartments is 0.64 cm; the membrane area exposed to the current flow is 4 cm^2. The anode and the cathode are platinum-polarizing electrodes. The experimental setup involves two closed loops containing a model salt solution and a 0.04 M NaCl electrolyte solution. The model solution circulates across the central desalination compartment—the 0.04 M NaCl solution—through an auxiliary and two electrode compartments in parallel, where the solution flow rate through each compartment is 30 mL min^{-1}. The model salt solution is composed of Na_2CO_3 (1000 mg L^{-1}), KCl (800 mg L^{-1}), $CaCl_2$*$2H_2O$ (4116 mg L^{-1}), and $MgCl_2$*$6H_2O$ (2440 mg L^{-1}). In this solution, the Mg^{2+}/Ca^{2+} molar concentrations ratio is 2/5, and the total concentration of $MgCl_2$ and $CaCl_2$ is 0.04 M to ensure membrane scaling formation [40]. The pH of the solution is adjusted to 6.5 by adding HCl. The mineral composition of this solution corresponds approximately to that of trice the concentrated milk. Each closed loop is connected to a separated external plastic reservoir, allowing continuous recirculation.

2.4. *Protocol*

The procedures for measuring I-V curves and chronopotentiograms (ChPs) of MK-40 and MK-40/PANI membranes are described in [39]. A KEITHLEY 220 current source is used to supply the current between the polarizing electrodes. The potential drop (PD) across the membrane under study, $\Delta\varphi$, is measured using Ag/AgCl electrodes. These electrodes are placed in the Luggin capillaries. The Luggin tips are installed at both sides of the membrane under study in its geometric center at a distance of about 0.5 mm from the surface. PD is registered by a HEWLETT PACKARD 34401A multimeter. The I-V curves are recorded when the current is swept from 0 to 10.5 mA cm^{-2} at a scan rate of 0.8 μA s^{-1}; the PD remains < 3 V. The chronopotentiometric measurements are made in the range of current densities from 0.2 i_{lim} to 1.5 i_{lim}, where i_{lim} is the experimentally determined limiting current density. The experiment is carried out at 20 °C. All I-V curves and ChPs for the MK-40/PANI membrane are measured in such a way that the PANI layer of the composite is oriented toward the desalination compartment.

3. Results and Discussion

3.1. *Physico-Chemical Characteristics of Ion-Exchange Membranes*

As shown in Figure 1, the green PANI chains are observed only on the particles of ion-exchange resin at the membrane surface, excluding polyethylene. The PANI layer within the MK-40/PANI membrane has weak anion-exchange properties. Thus, its formation leads to a bipolar interface, in this case, the main layer is the non-modified part of the pristine cation-exchange membrane. The thickness of MK-40/PANI membrane rises by 0.02 mm compared to the the pristine MK-40 membrane (Table 1).

The conductivity of MK-40/PANI is about 16% lower than that of the MK-40 membrane. These data correlate with earlier results [41]. Shkirskaya et al. had studied the influence of the PANI layer both on the homogeneous and on the heterogeneous surface of sulfocationite membranes on their electric conductivity in an NaCl solution. A decrease in the electric conductivity in 10% of the MK-40/PANI composite was observed in the diluted solutions (less than 0.3 M). A further increase in the solution concentration leveled out these differences. However, for homogeneous MF-4SK membranes (Russian analog of Nafion®), the electric conductivity decreased significantly after modification by PANI. The authors suggested that membrane heterogeneity is a more significant factor than the polyaniline synthesis conditions. Along with this, the contact angle of MK-40 and MK-40/PANI is nearly the same, close to 55°, indicating the invariability of the hydrophilic properties of the membrane surface. The contact angle does not change because polyethylene covers about 80% of both MK-40 and MK-40/PANI membranes.

Table 1. The main properties of the MK-40 and MK-40/PANI membranes.

Membrane	L, mm	κ, S m^{-1}	θ, °
MK-40	0.470 ± 0.005	0.36 ± 0.03	55 ± 3
MK-40/PANI	0.490 ± 0.005	0.30 ± 0.02	54 ± 3

3.2. Voltammetry

There are three regions on the I-V curve of monopolar IEMs, which are generally distinguished in the literature [12,16]. An initial linear region is followed by a plateau of limiting current density, and then by a region of higher growth of current density. When approaching i_{lim}, the electrolyte interfacial concentration is nearly zero that initiates coupled effects of concentration polarization—current-induced convection (electroconvection and gravitational convection) and water splitting [12]. The shape of the I-V curve is influenced by many factors. The plateau of limiting current density becomes less appreciable, the slope appears, and the length of the plateau decreases when the concentration of the solution and the stirring rate increase [42]. Similar effects are observed in solutions of complex compositions, for example, containing surfactants [43] or multicharged ions [44]. In this case, the method of numerical differentiation is used to evaluate the limiting current, and the value of the limiting current is determined as an extremum on the curves expressed in $d(\Delta\varphi)/di$-i_{av} coordinates.

Figure 2 shows the I-V curves obtained for MK-40 and MK-40/PANI membranes in the model salt solution and the corresponding dependence of the pH of the desalinated solution. In case of the

MK-40 membrane, the shape of the I-V curve is the typical one. Water splitting at the depleted membrane interface, begins after reaching i_{lim}. The I-V curve obtained for the MK-40/PANI membrane has two inflection points (Figure 2).

The first inflection point is in the region of small PD about 300 mV. Apparently, it is related to the depletion of the bipolar interface between the cation- and anion-exchange layers within the cation exchange particles on the membrane surface. Under the action of external electric field, the cations migrate from the bipolar interface into the bulk of the cation-exchange layer; the anions migrate from this interface into the bulk of the anion-exchange layer. At a certain current density, the ion concentration at this interface becomes sufficiently low leading to an essential increase in the membrane resistance. The latter is seen by an appearing plateau on the I-V curve in the range of current densities close to 4 mA cm^{-2}. This critical current density, which is caused by the depletion of the bipolar interface, may be called the pseudo-limiting current density, $i_{\text{pseudo-lim}}$. When the concentration of the salt ions at the depleted bipolar interface becomes sufficiently low, water splitting occurs similar to that which takes place in the bipolar membranes [11,45]. This is confirmed by a change in the pH of the solution coming out of the desalination compartment at $i \approx 4$ mA cm^{-2} (Figure 2, green dashed line). This process produces H^+ and OH^- ions, which are additional current carriers and whose emergence result in decreasing the membrane resistance. This is confirmed by the fact that the derived I-V curve of the modified membrane goes down after reaching $i_{\text{pseudo-lim}}$, which indicates a decrease in the resistance of the electromembrane system (Figure 3). Thus, the bipolar interface between the cation and anion exchange layers causes a significant increase in the water splitting rate compared to the MK-40 membrane (Figure 2). Similar phenomena were observed for the asymmetric bipolar membranes, in which the thickness of one of the layers is much larger [46].

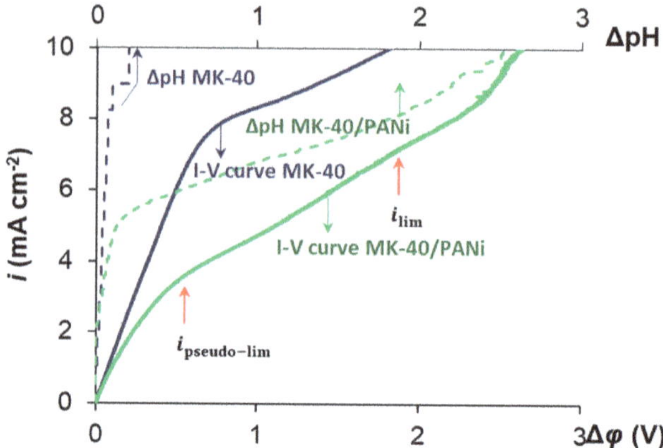

Figure 2. I-V curves for MK-40 and MK-40/PANI membranes in the model solution and corresponding dependence of the pH of desalinated solution.

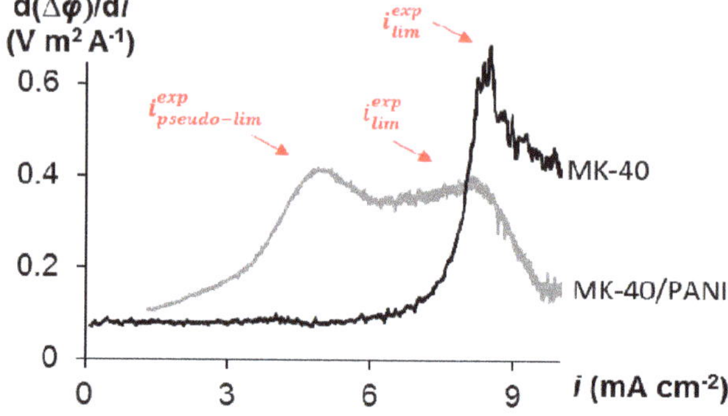

Figure 3. Derived I-V curves, $d(\Delta\varphi)/di$, for MK-40 and MK-40/PANI membranes in the model solution at the corresponding current density.

As the current continues to increase, the ion concentration at the external membrane surface decreases to a low value that causes a new increase in the system resistance in the case of the MK-40/PANI membrane. Since the anion exchange layer has a very small thickness compared to the cation exchange layer, this leads to the fact that through this layer, cations are transported by the diffusion mechanism. In this case, we see the development of concentration polarization according to the classical type for monopolar membranes. This state relates to the "classical" limiting current density, i_{lim}, observed in case of conventional monopolar IEMs [12]. The presence of two limiting current densities can be seen more clearly on the derived I-V curves for the studied membranes in the model solution shown in Figure 3.

The values of both limiting current densities may be determined by the point of intersection of the tangents drawn to the linear parts of the I-V curve on the left and on the right of the region, where a sharp change in the rate of the current growth with potential drop occurs (Figure 2). These values for both limiting current densities are present in Table 2.

Table 2. The values of limiting current densities for MK-40 and MK-40/PANI membranes in the model solution at 20 °C.

Membrane	$i_{pseudo\text{-}lim}$, mA cm^{-2}	i_{lim}, mA cm^{-2}
MK-40	-	7.7
MK-40/PANI	3.2	7.0

At the same time, there are no two extrema on the derived I-V curves for asymmetric bipolar membranes compared to the modified membrane studied in this work. This may be because PANI synthesis has been carried out directly on the surface of the IEM, and the localization of the modifier is the particles of cation-exchange resin. As a result, a structure is formed where the anion and cation layers are much closer to each other compared to the classic and asymmetric bipolar membranes. Classic bipolar membranes are mechanically composed of anion and cation exchange layers. A thin anion exchange layer is deposited on the cation substrate in the asymmetric bipolar membranes. However, a more significant change in the pH of the desalination solution in the case of the MK-40/PANI membrane compared to the MK-40 membrane indicates continued generation of H^+ and OH^- ions at the internal bipolar interface. In examining the polarization behavior of the MK-40/PANI membrane, a change in the shape of the I-V curve is noted during the experiments. The amount of electricity flowing through the system required to change the shape of the I-V curve is equal to 2000 C. There is no $i_{pseudo\text{-}lim}$ on the derived I-V curve and the value of i_{lim} becomes higher in the case of the used MK-40/PANI membrane compared to the fresh MK-40/PANI membrane (Figure 4). Probably, the clearly expressed bipolar interface of the composite is vanishing; therefore, the pseudo-limiting current disappears on the I-V curve. The resistance of the ohmic region decreases, and the plateau region becomes longer and more pronounced.

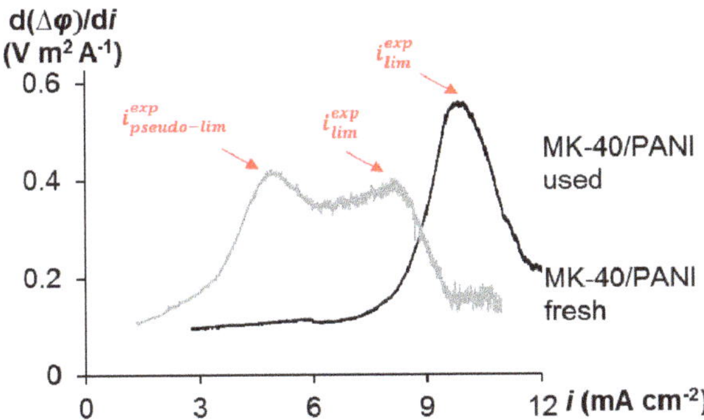

Figure 4. Derived I-V curves, d(Δφ)/d*i*, for fresh and used MK-40/PANI membranes.

3.3. Chronopotentiometry

The ChP of the pristine MK-40 membrane and its derivative measured in the 0.04 M $MgCl_2$ solution at an overlimiting current density (1.9 i_{lim}) is presented in Figure 5. The curve behavior has a classical form for monopolar IEMs at $i \geq i_{lim}$, discussed in the literature [47,48]. When the current starts to flow, there is a speedy growth of PD related to the ohmic potential drop, $\Delta\varphi_{ohm}$, over the membrane and two adjacent solutions, and there is no influence of concentration polarization. The value of PD does not increase significantly during one experimental run at a current density less the limiting value. However, the shape of ChP changes differently at $i \geq i_{lim}$. The electrolyte concentration at the depleted membrane/solution interface decreases by degrees with time, the process is governed by electrodiffusion. As a result, there is a drastic rise in PD on the ChP. The electrolyte concentration at the depleted membrane/solution interface gets low enough by a certain time called the transition time, τ, [47,48]. This time corresponds to the appearance of an additional mechanism of ion transport, namely, the current-induced convection. Therefore, the growth rate of PD falls, and PD tends to a quasi-steady state value, $\Delta\varphi_{st}$.

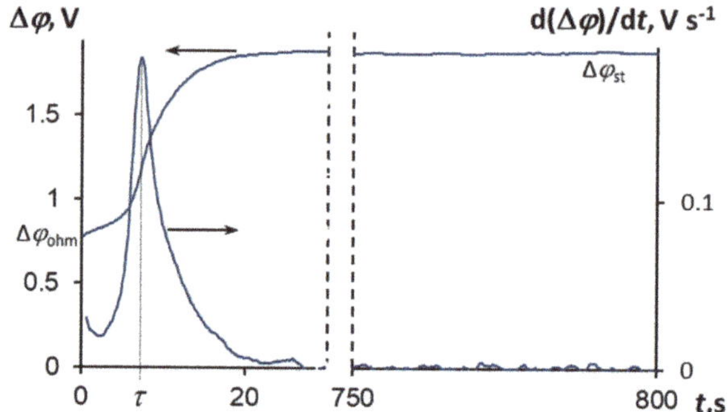

Figure 5. Chronopotentiogram of the MK-40 membrane in the 0.04 M $MgCl_2$ solution, $\Delta\varphi$ vs. t, and its derivative, $d(\Delta\varphi)/dt$ vs. t, in the model solution at $i = 1.9\ i_{lim}$. $\Delta\varphi_{ohm}$ is the ohmic potential drop just after the current is switched on; $\Delta\varphi_{st}$ is the potential drop at the quasi-steady state; τ is the transition time.

Figure 6 shows the ChPs for the MK-40 and MK-40/PANI membranes measured in the model salt solution at different current densities and the corresponding time-dependence of the pH of the desalinated solution. To enhance the presentation of the obtained ChPs, $\Delta\varphi_{ohm}$ is deducted from the measured value of $\Delta\varphi$ at a corresponding time and ChPs are represented at $\Delta\varphi'$ vs. t coordinates. $\Delta\varphi'$ is the reduced potential drop and can be calculated as follows:

$$\Delta\varphi'(t) = \Delta\varphi(t) - \Delta\varphi_{ohm}. \tag{3}$$

In case of the MK-40 membrane, the chronopotentiogram has a classical form. There is no transition time in the current density below i_{lim}, and the pH of the desalinated solution does not change within the error (Figure 6a). In case of the MK-40/PANI membrane, the transition time is observed on the ChP at a current density higher that $i_{pseudo\text{-}lim}$ but lower than i_{lim} (Figure 6a). Also, there is a noticeable alkalization of the solution in the desalination compartment in the case of the MK-40/PANI membrane (Figure 6a). This point corresponds to the appearance of an additional mechanism of ion transport—the emergence of H^+ and OH^- ions due to water splitting at the internal bipolar interface occurring when the ion concentration at this boundary becomes sufficiently low. This is consistent with the results of voltammetry.

Also, the transition time is observed on the ChP for MK-40/PANI membrane at $i = i_{lim}$, but there is a more significant change in the pH of the desalinated solution compared to the lower current densities (Figure 6b). It is also noted that PD slowly increases and does not reach a quasi-steady state value in time for the MK-40/PANI membrane (Figure 6b). A similar shape of the ChP was observed during the sedimentation of mineral salts on the surface and inside the pores of IEMs [49]. It could be assumed that the similar behavior of the ChPs in our study is associated with sedimentation on the membrane surface. A small amount of scale on the surface of MK-40/PANI membrane has been found after the experiment by visual inspection.

The PD slowly increases and does not reach a quasi-steady state value in time at currents higher than the limiting current for the MK-40 membrane (Figure 6c), which indicates the scale formation. However, the scale has not been detected in the measuring cell during the experimental run and on the membrane surface after the experiment by visual inspection. It should be noted that the pH of the desalinated solution does not change within the error in this case. The slow growth of PD in time on the ChP for the MK-40/PANI membrane is observed only at $i = i_{lim}$, when the solution in the desalination compartment is significantly alkalized (Figure 6b).

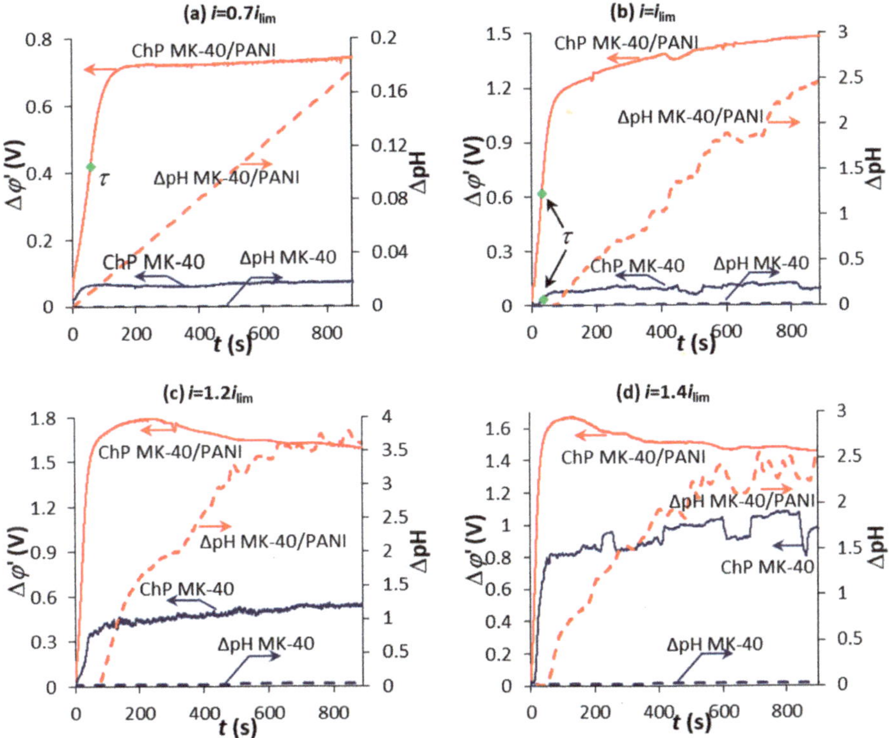

Figure 6. Chronopotentiograms of MK-40 and MK-40/PANI membranes in the model solution and the corresponding time-dependence of the pH of the desalinated solution at $i = 0.7\ i_{lim}$ (**a**); $i = i_{lim}$ (**b**); $i = 1.2\ i_{lim}$ (**c**); $i = 1.4\ i_{lim}$ (**d**).

At further growth in current density, the PD on the ChP for the MK-40/PANI membrane increases during the first 100–200 sec, and then gradually decreases reaching a quasi-steady state (Figure 6c,d). Also, the rate of change in the pH of the desalinated solution increases compared to that observed at $i = i_{lim}$. H^+ and OH^- ions appear in the system due to water splitting at the internal MK-40/PANI boundary. There is an apparent contradiction between the behavior of the membrane presented in Figure 6b,c. Even though the shape of the ChP in Figure 6c does not indicate the appearance of mineral scale, the muddy solution in the desalination compartment and precipitate on the surface-modified membrane was visually observed at $i > i_{lim}$. Water splitting at the internal MK-40/PANI boundary at high currents has a dominant effect on the shape of the ChP compared to the mineral scaling on the membrane. The value of the PD decreases due to the appearance of new charge carriers (H^+ and OH^- ions) instead of its increase due to scale formation.

The surfaces of the studied membranes have been checked by scanning electron microscopy after chronopotentiometry to validate the present assumptions. The analysis of the obtained results of scanning electron microscopy confirms the precipitate on both the pristine and surface-modified membranes (Figure 7). The precipitate, having a lamellar structure typical of calcite, as well as more complex polycrystalline forms of calcium carbonate, were found [50,51]. Moreover, the mineral scale covers the part of the working surface area of the MK-40 membrane (Figure 7a–c) and its entire surface of the MK-40/PANI membrane (Figure 7d–f).

Figure 7. SEM images of the MK-40 (**a–c**) and MK-40/PANI (**d–f**) membrane surfaces facing the desalination compartment after 5 h of electrodialysis of model solution at $i = 1.4\ i_{lim}$.

According to energy-dispersive analysis, the precipitate on the membrane surfaces is a mixture of calcium carbonate and magnesium hydroxide with a predominance of calcium carbonate (Figure 8). There were similar findings obtained by C. Casademont et al. [52–55].

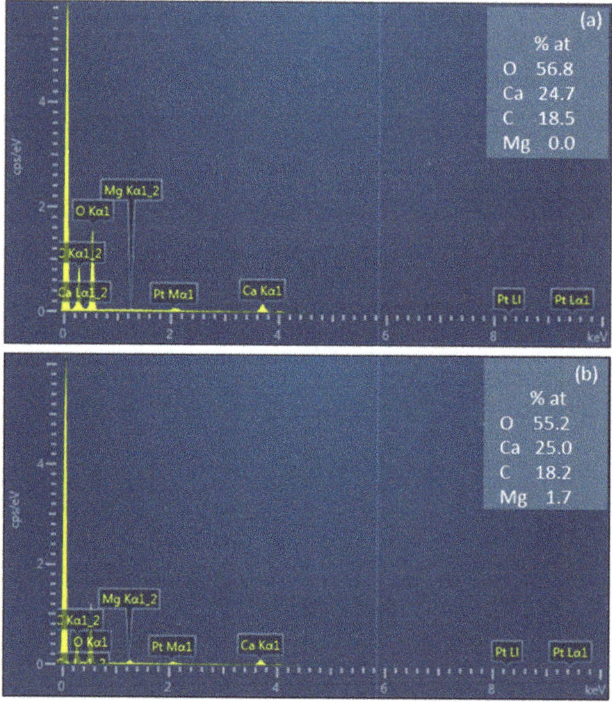

Figure 8. Energy dispersive X-ray spectroscopy of precipitate on the MK-40 membrane surface.

4. Conclusions

The scale formation on the surface of two ion-exchange membranes, a heterogeneous MK-40 and its modification MK-40/PANI, during electrodialysis of a solution containing the scale-forming cations, Ca^{2+} and Mg^{2+}, was investigated by the method of voltammetry and chronopotentiometry.

The I-V curve had two inflection points in the case of the MK-40/PANI membrane in comparison with the MK-40 membrane. The first pseudo-limiting current density is in the region of small displacements of the potential from equilibrium. It was related to the depletion of the bipolar interface within the membrane. As the current continued to increase, the ion concentration at the external membrane surface decreased to a low value that caused a new increase in the system resistance, and the "classical" limiting current density was observed. The disappearance of the pseudo-limiting current with the simultaneous increase in the value of the limiting current density was observed for the MK-40/PANI membrane at the long-term usage.

The precipitate was detected at a current density of more than 0.8 i_{lim} for the MK-40/PANI, and more than i_{lim} for the MK-40 membrane. However, at current densities less than 0.8 i_{lim}, no scaling was observed for the MK-40/PANI membrane. This was evidenced by the growth of the potential drop across the membrane at $i = i_{lim}$ for MK-40/PANI and at $i = 1.2\ i_{lim}$ for the MK-40 membranes instead of reaching the quasi-steady state. However, the shape of the chronopotentiogram for the MK-40/PANI membrane changed at currents above the limit. At the same time, the water splitting rate at the MK-40/PANI bipolar boundary increased. All these facts indicated that the water splitting reaction was a determining factor in the electrochemical behavior of the membrane system than mineral scaling at currents above the limit. Despite the fact that the amount of precipitate is higher on the surface of the MK-40/PANI membrane, the values of the limiting current and electrical conductivity of the MK-40/PANI membrane were comparable with the values of the MK-40 membrane. Thus, it is possible to recommend the MK-40/PANI composite membrane for monovalent-ion-selective electrodialysis at the underlimiting current densities.

Author Contributions: Conceptualization, N.D.P. and C.L.; formal analysis, N.D.P. and N.V.L.; investigation, M.A.A.; writing—original draft preparation, M.A.A.; writing—review and editing, M.A.A., N.V.L., N.D.P.; supervision, L.D. All authors have read and agreed to the published version of the manuscript.

Funding: This research was supported by the Russian Foundation for Basic Research and Krasnodar Region Administration (projects No 19-48-230040 r_a).

Acknowledgments: The authors gratefully acknowledge Victor Nikonenko (Kuban State University, Russia) for the support with experiments and discussions.

Conflicts of Interest: The authors declare no conflict of interest.

References

1. Al-Amshawee, S.; Yunus, M.Y.B.M.; Azoddein, A.A.M.; Hassell, D.G.; Dakhil, I.H.; Hasan, H.A. Electrodialysis desalination for water and wastewater: A review. *Chem. Eng. J.* **2020**, *380*, 122231. [CrossRef]
2. Campione, A.; Gurreri, L.; Ciofalo, M.; Micale, G.; Tamburini, A.; Cipollina, A. Electrodialysis for water desalination: A critical assessment of recent developments on process fundamentals, models and applications. *Desalination* **2018**, *434*, 121–160. [CrossRef]
3. Kravtsov, V.; Kulikova, I.; Mikhaylin, S.; Bazinet, L. Alkalinization of acid whey by means of electrodialysis with bipolar membranes and analysis of induced membrane fouling. *J. Food Eng.* **2020**, *277*, 109891. [CrossRef]
4. Merkel, A.; Ashrafi, A.M.; Ečer, J. Bipolar membrane electrodialysis assisted pH correction of milk whey. *J. Memb. Sci.* **2018**, *555*, 185–196. [CrossRef]
5. Fehér, J.; Červeňanský, I.; Václavík, L.; Markoš, J. Electrodialysis applied for phenylacetic acid separation from organic impurities: Increasing the recovery. *Sep. Purif. Technol.* **2020**. [CrossRef]
6. Kattan Readi, O.M.; Gironès, M.; Nijmeijer, K. Separation of complex mixtures of amino acids for biorefinery applications using electrodialysis. *J. Memb. Sci.* **2013**, *429*, 338–348. [CrossRef]

7. Al-Saydeh, S.A.; El-Naas, M.H.; Zaidi, S.J. Copper removal from industrial wastewater: A comprehensive review. *J. Ind. Eng. Chem.* **2017**, *56*, 35–44. [CrossRef]
8. Fu, F.; Wang, Q. Removal of heavy metal ions from wastewaters: A review. *J. Environ. Manag.* **2011**, *92*, 407–418. [CrossRef]
9. Sheldeshov, N.V.; Zabolotsky, V.I.; Kovalev, N.V.; Karpenko, T.V. Electrochemical characteristics of heterogeneous bipolar membranes and electromembrane process of recovery of nitric acid and sodium hydroxide from sodium nitrate solution. *Sep. Purif. Technol.* **2020**, *241*, 116648. [CrossRef]
10. Mikhaylin, S.; Bazinet, L. Fouling on ion-exchange membranes: Classification, characterization and strategies of prevention and control. *Adv. Colloid Interface Sci.* **2016**, *229*, 34–56. [CrossRef]
11. Tanaka, Y. Chapter 8 Water Dissociation. In *Membrane Science and Technology*; Elsevier: Amsterdam, The Netherlands, 2007; Volume 12, pp. 139–186. ISBN 9780444519825.
12. Krol, J.J.; Wessling, M.; Strathmann, H. Concentration polarization with monopolar ion exchange membranes: Current-voltage curves and water dissociation. *J. Memb. Sci.* **1999**. [CrossRef]
13. Tong, T.; Wallace, A.F.; Zhao, S.; Wang, Z. Mineral scaling in membrane desalination: Mechanisms, mitigation strategies, and feasibility of scaling-resistant membranes. *J. Memb. Sci.* **2019**, *579*, 52–69. [CrossRef]
14. Nikonenko, V.V.; Mareev, S.A.; Pis'menskaya, N.D.; Uzdenova, A.M.; Kovalenko, A.V.; Urtenov, M.K.; Pourcelly, G. Effect of electroconvection and its use in intensifying the mass transfer in electrodialysis (Review). *Russ. J. Electrochem.* **2017**, *53*, 1122–1144. [CrossRef]
15. Vasil'eva, V.I.; Zhil'tsova, A.V.; Malykhin, M.D.; Zabolotskii, V.I.; Lebedev, K.A.; Chermit, R.K.; Sharafan, M.V. Effect of the chemical nature of the ionogenic groups of ion-exchange membranes on the size of the electroconvective instability region in high-current modes. *Russ. J. Electrochem.* **2014**, *50*, 120–128. [CrossRef]
16. Balster, J.; Yildirim, M.H.; Stamatialis, D.F.; Ibanez, R.; Lammertink, R.G.H.; Jordan, V.; Wessling, M. Morphology and microtopology of cation-exchange polymers and the origin of the overlimiting current. *J. Phys. Chem. B* **2007**, *111*, 2152–2165. [CrossRef]
17. Belashova, E.; Mikhaylin, S.; Pismenskaya, N.; Nikonenko, V.; Bazinet, L. Impact of cation-exchange membrane scaling nature on the electrochemical characteristics of membrane system. *Sep. Purif. Technol.* **2017**, *189*, 441–448. [CrossRef]
18. Mikhaylin, S.; Nikonenko, V.; Pismenskaya, N.; Pourcelly, G.; Choi, S.; Kwon, H.J.; Han, J.; Bazinet, L. How physico-chemical and surface properties of cation-exchange membrane affect membrane scaling and electroconvective vortices: Influence on performance of electrodialysis with pulsed electric field. *Desalination* **2016**, *393*, 102–114. [CrossRef]
19. Andreeva, M.A.; Gil, V.V.; Pismenskaya, N.D.; Nikonenko, V.V.; Dammak, L.; Larchet, C.; Grande, D.; Kononenko, N.A. Effect of homogenization and hydrophobization of a cation-exchange membrane surface on its scaling in the presence of calcium and magnesium chlorides during electrodialysis. *J. Memb. Sci.* **2017**, *540*, 183–191. [CrossRef]
20. Suwal, S.; Amiot, J.; Beaulieu, L.; Bazinet, L. Effect of pulsed electric field and polarity reversal on peptide/amino acid migration, selectivity and fouling mitigation. *J. Memb. Sci.* **2016**, *510*, 405–416. [CrossRef]
21. Persico, M.; Bazinet, L. Fouling prevention of peptides from a tryptic whey hydrolysate during electromembrane processes by use of monovalent ion permselective membranes. *J. Memb. Sci.* **2018**, *549*, 486–494. [CrossRef]
22. Luo, T.; Abdu, S.; Wessling, M. Selectivity of ion exchange membranes: A review. *J. Memb. Sci.* **2018**, *555*, 429–454. [CrossRef]
23. Nagarale, R.K.; Gohil, G.S.; Shahi, V.K.; Trivedi, G.S.; Rangarajan, R. Preparation and electrochemical characterization of cation- and anion-exchange/polyaniline composite membranes. *J. Colloid Interface Sci.* **2004**. [CrossRef] [PubMed]
24. Sata, T. Composite Membranes Prepared from Cation Exchange Membranes and Polyaniline and Their Transport Properties in Electrodialysis. *J. Electrochem. Soc.* **1999**, *146*, 585. [CrossRef]
25. Berezina, N.P.; Shkirskaya, S.A.; Kolechko, M.V.; Popova, O.V.; Senchikhin, I.N.; Roldugin, V.I. Barrier effects of polyaniline layer in surface modified MF-4SK/Polyaniline membranes. *Russ. J. Electrochem.* **2011**, *47*, 995–1005. [CrossRef]
26. Demina, O.A.; Shkirskaya, S.A.; Kononenko, N.A.; Nazyrova, E.V. Assessing the selectivity of composite ion-exchange membranes within the framework of the extended three-wire model of conduction. *Russ. J. Electrochem.* **2016**. [CrossRef]

27. Kononenko, N.A.; Dolgopolov, S.V.; Loza, N.V.; Shel'deshov, N.V. Effects of ph variation in solutions under the polarization conditions of the MF-4SK membrane with surface modified by polyaniline. *Russ. J. Electrochem.* **2015**, *51*, 19–24. [CrossRef]
28. Shestakov, K.V.; Lazarev, S.I.; Polyanskiy, K.K. Study of kinetic and structural characteristics of membranes in purification process of copper-containing solutions by electrodialysis. *Izv. Vyss. Uchebnykh Zaved. Seriya Khimiya i Khimicheskaya Tekhnologiya* **2019**. [CrossRef]
29. Asraf-Snir, M.; Gilron, J.; Oren, Y. Scaling of cation exchange membranes by gypsum during Donnan exchange and electrodialysis. *J. Memb. Sci.* **2018**, *567*, 28–38. [CrossRef]
30. Melnikov, S.; Sheldeshov, N.; Zabolotsky, V.; Loza, S.; Achoh, A. Pilot scale complex electrodialysis technology for processing a solution of lithium chloride containing organic solvents. *Sep. Purif. Technol.* **2017**, *189*, 74–81. [CrossRef]
31. Melnikov, S.; Loza, S.; Sharafan, M.; Zabolotskiy, V. Electrodialysis treatment of secondary steam condensate obtained during production of ammonium nitrate. Technical and economic analysis. *Sep. Purif. Technol.* **2016**. [CrossRef]
32. Sarapulova, V.; Shkorkina, I.; Mareev, S.; Pismenskaya, N.; Kononenko, N.; Larchet, C.; Dammak, L.; Nikonenko, V. Transport characteristics of fujifilm ion-exchange membranes as compared to homogeneous membranes AMXand CMXand to heterogeneous membranes MK-40 and MA-41. *Membranes (Basel)* **2019**, *9*, 84. [CrossRef] [PubMed]
33. Akberova, E.M.; Vasil'eva, V.I.; Zabolotsky, V.I.; Novak, L. Effect of the sulfocation-exchanger dispersity on the surface morphology, microrelief of heterogeneous membranes and development of electroconvection in intense current modes. *J. Memb. Sci.* **2018**, *566*, 317–328. [CrossRef]
34. Falina, I.V.; Demina, O.A.; Zabolotskii, V.I. Verification of a capillary model for the electroosmotic transport of a free solvent in ion-exchange membranes of different natures. *Colloid J.* **2017**. [CrossRef]
35. Berezina, N.P.; Kononenko, N.A.; Dyomina, O.A.; Gnusin, N.P. Characterization of ion-exchange membrane materials: Properties vs structure. *Adv. Colloid Interface Sci.* **2008**. [CrossRef]
36. Gnusin, N.P.; Berezina, N.P.; Kononenko, N.A.; Dyomina, O.A. Transport structural parameters to characterize ion exchange membranes. *J. Memb. Sci.* **2004**. [CrossRef]
37. Loza, N.V.; Dolgopolov, S.V.; Kononenko, N.A.; Andreeva, M.A.; Korshikova, Y.S. Effect of surface modification of perfluorinated membranes with polyaniline on their polarization behavior. *Russ. J. Electrochem.* **2015**, *51*, 538–545. [CrossRef]
38. Lteif, R.; Dammak, L.; Larchet, C.; Auclair, B. Conductivitéélectrique membranaire: Étude de l'effet de la concentration, de la nature de l'électrolyte et de la structure membranaire. *Eur. Polym. J.* **1999**, *35*, 1187–1195. [CrossRef]
39. Belashova, E.D.; Melnik, N.A.; Pismenskaya, N.D.; Shevtsova, K.A.; Nebavsky, A.V.; Lebedev, K.A.; Nikonenko, V.V. Overlimiting mass transfer through cation-exchange membranes modified by Nafion film and carbon nanotubes. *Electrochim. Acta* **2012**, *59*, 412–423. [CrossRef]
40. Mikhaylin, S.; Nikonenko, V.; Pourcelly, G.; Bazinet, L. Intensification of demineralization process and decrease in scaling by application of pulsed electric field with short pulse/pause conditions. *J. Memb. Sci.* **2014**, *468*, 389–399. [CrossRef]
41. Shkirskaya, S.A.; Senchikhin, I.N.; Kononenko, N.A.; Roldugin, V.I. Effect of polyaniline on the stability of electrotransport characteristics and thermochemical properties of sulfocationite membranes with different polymer matrices. *Russ. J. Electrochem.* **2017**, *53*, 78–85. [CrossRef]
42. Barragán, V.M.; Ruíz-Bauzá, C. Current–Voltage Curves for Ion-Exchange Membranes: A Method for Determining the Limiting Current Density. *J. Colloid Interface Sci.* **1998**, *205*, 365–373. [CrossRef] [PubMed]
43. Kononenko, N.; Berezina, N.; Loza, N. Interaction of surfactants with ion-exchange membranes. *Colloids Surfaces A Physicochem. Eng. Asp.* **2004**, *239*, 59–64. [CrossRef]
44. Zerdoumi, R.; Oulmi, K.; Benslimane, S. Electrochemical characterization of the CMX cation exchange membrane in buffered solutions: Effect on concentration polarization and counterions transport properties. *Desalination* **2014**, *340*, 42–48. [CrossRef]
45. Strathmann, H. Preparation and Characterization of Ion-Exchange Membranes. In *Membrane Science and Technology*; Elsevier: Amsterdam, The Netherlands, 2004; Volume 9, pp. 89–146.
46. Melnikov, S.S.; Sheldeshov, N.V.; Zabolotskii, V.I. Theoretical and experimental study of current-voltage characteristics of asymmetric bipolar membranes. *Desalin. Water Treat.* **2018**, *123*, 1–13. [CrossRef]

47. Krol, J. Chronopotentiometry and overlimiting ion transport through monopolar ion exchange membranes. *J. Memb. Sci.* **1999**, *162*, 155–164. [CrossRef]
48. Pismenskaia, N.; Sistat, P.; Huguet, P.; Nikonenko, V.; Pourcelly, G. Chronopotentiometry applied to the study of ion transfer through anion exchange membranes. *J. Memb. Sci.* **2004**, *228*, 65–76. [CrossRef]
49. Martí-Calatayud, M.C.; García-Gabaldón, M.; Pérez-Herranz, V. Effect of the equilibria of multivalent metal sulfates on the transport through cation-exchange membranes at different current regimes. *J. Memb. Sci.* **2013**, *443*, 181–192. [CrossRef]
50. Paul, D.; Halder, S.; Das, G. Whey protein directed in vitro vaterite biomineralization: Influence of external parameters on phase transformation. *Colloid Interface Sci. Commun.* **2020**, *36*, 100255. [CrossRef]
51. Huang, F.; Liang, Y.; He, Y. On the Pickering emulsions stabilized by calcium carbonate particles with various morphologies. *Colloids Surfaces A Physicochem. Eng. Asp.* **2019**, *580*, 123722. [CrossRef]
52. Casademont, C.; Pourcelly, G.; Bazinet, L. Bilayered self-oriented membrane fouling and impact of magnesium on CaCO3 formation during consecutive electrodialysis treatments. *Langmuir* **2010**. [CrossRef] [PubMed]
53. Casademont, C.; Sistat, P.; Ruiz, B.; Pourcelly, G.; Bazinet, L. Electrodialysis of model salt solution containing whey proteins: Enhancement by pulsed electric field and modified cell configuration. *J. Memb. Sci.* **2009**, *328*, 238–245. [CrossRef]
54. Casademont, C.; Farias, M.; Pourcelly, G.; Bazinet, L. Impact of electrodialytic parameters on cation migration kinetics and fouling nature of ion-exchange membranes during treatment of solutions with different magnesium/calcium ratios. *J. Memb. Sci.* **2008**, *325*, 570–579. [CrossRef]
55. Casademont, C.; Pourcelly, G.; Bazinet, L. Effect of magnesium/calcium ratio in solutions subjected to electrodialysis: Characterization of cation-exchange membrane fouling. *J. Colloid Interface Sci.* **2007**, *315*, 544–554. [CrossRef] [PubMed]

Article

Impacts of Flow Rate and Pulsed Electric Field Current Mode on Protein Fouling Formation during Bipolar Membrane Electroacidification of Skim Milk

Vladlen S. Nichka [1,2], Thibaud R. Geoffroy [1], Victor Nikonenko [2] and Laurent Bazinet [1,*]

1 Department of Food Sciences, Laboratoire de Transformation Alimentaire et Procédés ÉlectroMembranaires (LTAPEM, Laboratory of Food Processing and Electromembrane Processes), Institute of Nutrition and Functional Foods (INAF), Dairy Research Center (STELA), Université Laval, Québec, QC G1V 0A6, Canada; vladlen.nichka.1@ulaval.ca (V.S.N.); thibaud.geoffroy.1@ulaval.ca (T.R.G.)
2 Physical Chemistry Department, Kuban State University, 149 Stavropolskaya str., 350040 Krasnodar, Russia; v_nikonenko@mail.ru
* Correspondence: laurent.bazinet@fsaa.ulaval.ca; Tel.: +1-418-656-2131 (ext. 407445); Fax: +1-418-656-3353

Received: 7 August 2020; Accepted: 24 August 2020; Published: 26 August 2020

Abstract: Fouling is one of the major problems in electrodialysis. The aim of the present work was to investigate the effect of five different solution flow rates (corresponding to Reynolds numbers of 162, 242, 323, 404 and 485) combined with the use of pulsed electric field (PEF) current mode on protein fouling of bipolar membrane (BPM) during electrodialysis with bipolar membranes (EDBM) of skim milk. The application of PEF prevented the fouling formation by proteins on the cationic interface of the BPM almost completely, regardless of the flow rate or Reynolds number. Indeed, under PEF mode of current the weight of protein fouling was negligible in comparison with CC current mode (0.07 ± 0.08 mg/cm^2 versus 5.56 ± 2.40 mg/cm^2). When a continuous current (CC) mode was applied, Reynolds number equals or higher than 323 corresponded to a minimal value of protein fouling of BPM. This positive effect of both increasing the flow rate and using PEF is due to the facts that during pauses, the solution flow flushes the accumulated protein from the membrane while in the same time there is a decrease in concentration polarization (CP) and consequently decrease in H^+ generation at the cationic interface of the BPM, minimizing fouling formation and accumulation.

Keywords: electrochemical acidification; electrodialysis; casein; concentration polarization; ion-exchange membrane; fouling; Reynolds number; mode of current; flow flush

1. Introduction

Bovine milk is one of the most important raw materials in the food industry, which is composed of water, lactose, lipids, proteins, and minerals [1]. Caseins are the main proteins of milk; their contents are around 80% of total proteins [2]. Caseins are of great interest due to their nutritional value and functional properties. These proteins are used in processed cheese, coffee whiteners, infant formulas, and in pharmaceutical products [3]. They are also used in the manufacture of paper coatings, textile fabrics, adhesives, concrete, paints, and cosmetics [4]. One of the ways of casein production from milk is using chemical acidification. However, this method has drawbacks such as producing large amounts of salts, which have to be separated from the acid whey resulting in undesired waste streams [5,6].

Bazinet et al. [7] developed a different method of acid casein production using electrodialysis with bipolar membranes (EDBM) of milk. EDBM combines conventional electrodialysis with the special properties of bipolar membranes (BPM) to split water with the protons leading to protein precipitation while the selectivity of the monopolar membranes allows demineralization of the final acid whey. The major problems during EDBM are protein fouling and scaling of membranes. Scaling and colloidal fouling lead to an increase in electrical resistance, a decrease in permselectivity and membrane alteration [8,9]. The presence of fouling significantly increases the cost of electrodialysis in the food industry. According to Mikhaylin and Bazinet [10] the costs of membrane regeneration and replacement amount from 20–30% (for pressure driven processes) to 40–50% (for electrically driven processes) of the total costs for the membrane processes in food industry.

There are different ways to prevent or minimize fouling formation during ED treatment. It could be membrane modification, cleaning procedures, pretreatment (for example with pressure-driven processes), mechanical actions, ED with reverse polarity, overlimiting current regimes and others [10]. All of these methods have different disadvantages and operations limitations. Using these methods can lead to additional expenses, membrane deterioration, generation of additional effluents, or they are not suitable with ED systems where BPMs are stacked. A recent effective way to prevent or minimize fouling formation is the use of pulsed electric fields (PEF) [11–13]. In this non-stationary current mode, continuous current (CC) pulses alternate with pauses during which the current is equal to zero. The positive effect of PEF mode of current is associated with a decrease in CP phenomenon and hence a decrease in water splitting and an increase in ED power efficiency [14]. Indeed, during the pause lapse, when there is no water splitting, the ion concentration at the membrane interfaces can be partially restored, reducing the CP phenomena and membrane fouling during the subsequent pulse lapse [15].

It was also demonstrated that an optimization of hydrodynamic conditions during electrodialysis can also help to minimize the membrane fouling and scaling formation. The influence of flow rate on the electroacidification parameters (duration of the process, anode/cathode voltage difference and conductivity of milk) was previously investigated by Bazinet et al. [7] during EDBM of skim milk for bovine casein production, but the authors did not study the effect of current mode on the process. In the paper of Mikhaylin et al. [16] authors observed that the use of higher flow rates during EDBM of skim milk coupled with an ultrafiltration module leads to more than 38% decrease in CEM scaling formation in comparison to the conventional treatment due to the creation of unfavorable hydrodynamic conditions for the scaling attachment and growth, but the authors did not study protein fouling due to the use of UF module prior EDBM. Increasing of flow rate has similar effect in the ED cell as the use of PEF. It has two main advantages: decrease of water splitting rate and prevention of fouling formation due to the higher mixing of solution during the process. The decrease in the rate of water splitting is due to the decrease in the diffusion layer thickness (decrease of CP phenomenon) caused by the increase of the flow rate [17]. However, the coupled effect of mode of current supply and hydrodynamic conditions on the protein fouling formation has never been studied before.

In this context, the aim of the present study was to evaluate the influence of solution flow rate coupled with the mode of current (CC and PEF) on the protein fouling formation on BPM during EDBM of skim milk. The specific objectives of the work were: (1) to study the impact of different flow rates on protein fouling at the interface of the BPM; (2) to test the effect of PEF on protein fouling formation and to compare the results with those in CC mode; (3) to characterize the membrane properties before and after EDBM; (4) to link the protein fouling formation with the hydrodynamic condition in the milk channel; and (5) to propose mechanisms for protein fouling mitigation in the different conditions tested.

2. Materials and Methods

2.1. Materials

The milk used in this study was a commercial homogenized and pasteurized skim milk Beatrice (Parmalat, Victoriaville, QC, Canada). Sodium sulphate (Na_2SO_4, ACS grade) and potassium chloride (KCl, ACS grade) were obtained from BDH (VWR International, Mississauga, ON, Canada). Chemicals for cleaning of the electrodialysis (ED) system, hydrochloric acid (HCl) and sodium hydroxide (NaOH), were purchased from Fisher Scientific (Thermo Fisher Scientific, Montreal, QC, Canada). The average concentrations of the milk were determined using a Delta Lactoscope FTIR dairy analyzer (Delta

Instruments, Drachten, Netherlands). The milk composition is presented in Table 1. This composition is consistent with data in the literature [18].

Table 1. Average composition of milk.

Fat PLS [1], % *w/w*	Protein, % *w/w*	Lactose, % *w/w*	Solids, % *w/w*	SNF [2], % *w/w*	Casein, g/L	NPN/CU [3], mg/100g
0.13 ± 0.01	3.37 ± 0.05	4.73 ± 0.05	9.14 ± 0.09	8.34 ± 0.09	27.07 ± 0.36	15.60 ± 1.72

[1] Phospholipids. [2] Solids not fat (SNF=Total solids – Fat). [3] Non Protein Nitrogen and Calculated Urea.

2.2. Methods

2.2.1. Electrodialysis Cell

The electroacidification cell was a Microflow-type cell (Electro-Cell AB, Täby, Sweden) consisting of four compartments separated by one Neosepta cationic membrane (CMX-fg) and two Neosepta bipolar membranes (BP-1E) (Astom, Tokyo, Japan) (Figure 1). The membranes tested had an effective surface area of 16 cm^2. The anode was a plate dimensionally stable electrode (DSA-O_2, Ti/IrO_2 coating) and the cathode a 316-stainless-steel electrode. This arrangement defines three closed loops containing equal volumes (300 mL) of the 20 g/L Na_2SO_4 electrolyte solution, 2 g/L KCl aqueous solution and the milk solution. The flow rates were equal to the flow rate tested for each solution used in the experiment. Each closed loop was connected to a separate external plastic tank, allowing a continuous recirculation. The ED system was not equipped to maintain a constant temperature, but since this parameter underwent low variations (between 25 and 37 °C) similar for each mode of current, the temperature was not further discussed.

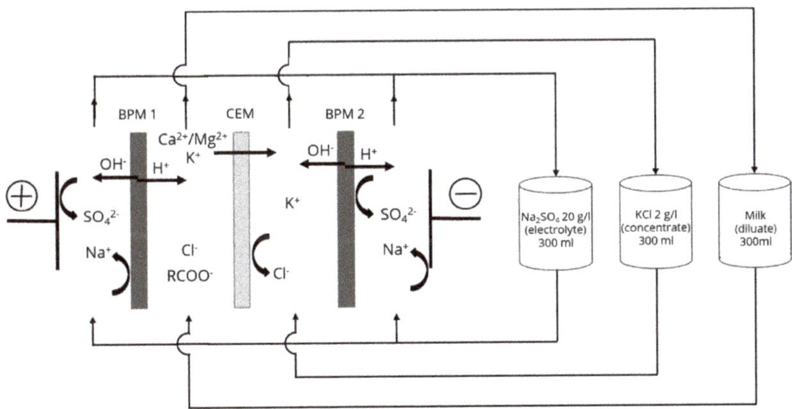

Figure 1. Electrodialysis cell configuration of electrodialysis with bipolar membranes (EDBM) process.

2.2.2. Protocol

EDBM was carried out using a constant current intensity of 50 mA (corresponding to a current density of 3.13 mA/cm^2) generated by using a Xantrex power supply, model HPD 60-5SX (Xantrex Technology Inc., Burnaby, BC, Canada). ED experiments were performed using CC or PEF with 10 s pulse and 50 s pause durations. In addition, for each mode of current, five different flow rates were investigated (400, 600, 800, 1000, and 1200 mL/min) corresponding to Reynolds numbers of 162, 242, 323, 404, and 485 to test the combined effect of current mode and flow rate on protein fouling during EDBM of skim milk. Three replicates of each condition were performed in this experiment. PEF was generated by a modified Bio-Rad Pulsewave 760 generator (Bio-Rad laboratories, Richmond,

BC, Canada). In this study, the conventional ED experiment under CC regime was considered as the control. Conventional ED experiment was stopped after 20 min whereas for PEF the duration of experiment was 120 min in order to compare the different current mode with respect to the same amount of charges transported.

2.2.3. Solution Conductivity

An YSI conductivity meter, model 3100 was used with an YSI immersion probe, model 3252, cell constant K = 0.1 cm^{-1} (Yellow Springs Instrument Co., Yellow Springs, OH, USA) to measure the conductivity of the milk and KCl solutions.

2.2.4. Membrane Thickness and Electrical Conductivity

The membranes were soaked in a 0.5 M NaCl solution for 30 min before and after each ED experiment for their characterization. The membrane thickness was measured using Marathon electronic digital micrometer, (Marathon watch company LTD, Richmond Hill, ON, Canada). The micrometer was equipped with a 10-mm-diameter flat contact point. The membrane thickness value was measured at six different locations on the effective membrane surface and then averaged [19]. The electrical conductivity of the membrane was calculated from the measured thickness and its electrical resistance, obtained from the membrane conductance. The conductance was measured using an YSI conductivity meter, model 3100 (Yellow Springs Instrument Co., Yellow Springs, OH, USA) equipped with a specially designed clip as described by Cifuentes-Araya [19]. A 0.5 M NaCl solution was used as a reference solution. Six conductance measurements of the reference solution and of the membrane in the reference solution were taken. The membrane conductance in the reference solution was taken on the effective membrane surface. Membrane conductivity (k) was calculated according to Equation (1) [20]. In Equation (1) k is the membrane electrical conductivity (in S/cm), L the thickness of the membrane (in cm) and A the electrode area (1 cm^2):

$$k = \frac{L}{R_m A} \quad (1)$$

The membrane electrical resistance (R_m) was calculated according to Equation (2). In Equation (2) G_m is a membrane conductance:

$$R_m = \frac{1}{G_m} \quad (2)$$

Membrane resistance (R_m) was obtained by subtracting the solution resistance (R_s) to the membrane resistance soaking in the solutions (R_{s+m}).

2.2.5. pH of the Diluate and Concentrate

The pH of milk was measured with a VWR Symphony SP70P pH-meter (VWR International, Montreal, QC, Canada). The pH of KCl solution was measured with an Thermo Scientific Orion Star A221 pH-meter (Thermo Fisher Scientific, Montreal, QC, Canada).

2.2.6. Fouling Weight

Fouling was collected from the BPM surface, and then freeze-dried in a Labconco Lyophyliser, model Freezone 4.5 (Labconco Corporation, Kansas City, MO, USA). After weighing the dried powders, they were kept at −20 °C until further analyses.

2.2.7. Membrane Surface Photographs

Digital photographs of the BPM cationic interface in contact with milk were taken after each EDBM experiment.

2.2.8. Number of Charges Transported

The number of charges transported was calculated according to Equation (3). In Equation (3) Q is the number of charges transported (in C), I is the current intensity (in A), t is the duration of the experiment (in s).

$$Q = It \tag{3}$$

When using PEF, I was constant during the pulse period (the same as in the CC mode), and zero during the pauses.

2.2.9. Energy Consumptions

The energy consumptions (in Wh) was calculated according to Equation (4). In equation (4), I is the current intensity (in A), $U(t)$ the voltage (in V) as a known from experiment function of time [19]. The time taken into account for the PEF mode of current was the effective time during the pulse periods.

$$EC = I \int U(t)dt \tag{4}$$

2.2.10. Reynolds Numbers

Reynolds number was calculated according to Equation (5) in order to evaluate the flow motion in a system. In Equation (5), ρ is the density of the fluid (in kg/m^3), v is the average flow velocity (in m/s), h is the intermembrane distance (in m), μ is the dynamic viscosity of the fluid (in Pa s) [21].

$$\mathrm{Re} = \frac{\rho v h}{\mu} \tag{5}$$

Density ρ as well as dynamic viscosity μ of skim milk are known parameters, which were taken equal to 1035 kg/m^3 and 0.0015 Pa s respectively.

2.2.11. Statistical Analyses

All data were subjected to a two-way or three-way analysis of variance (ANOVA) using SigmaPlot software (SigmaPlot version 12.5, Chicago, IL, USA). Fisher's Least Significant Difference (LSD) test was used to determine the effect of each factor under study on fouling kinetics.

3. Results and Discussion

3.1. Evolution of PH

3.1.1. In Skim Milk

It appeared from the statistical analysis, that the mode of current supply and flow rate had no significant effect ($p > 0.05$) on the variation of pH of the skim milk during EDBM. The regression curve calculated for the pH of milk as a function of number of charges transported (Figure 2) showed a decrease in pH of 0.23 ± 0.06 per C for all cases considered (3.4% decrease). The decrease in pH of milk can be explained by addition of H^+ which occurs by splitting of water molecules at the BPM cationic interface during electrochemical acidification of milk. The slow acidification of milk can be explained by its buffer capacity (i.e., proteins, weak acids) and the release of phosphate anions from the casein micelles, which neutralize the acidification effect of the H^+ addition [22]. Indeed, with decrease of pH, protein-bounded calcium (or magnesium) phosphates (or citrates) compounds convert to the soluble ionic form and remain in the whey fraction of milk [23]. Numerous papers proved the fact that the decrease in pH of milk leads to dissolving colloidal calcium phosphate and small amounts of magnesium [24–26] and causes the dissociation of casein from micelles [27]. Milk pH decreased in a similar way to the one observed previously by Bazinet et al. [28] and Masson et al. [29] but with a different cell design (effective surface area of 100 cm^2).

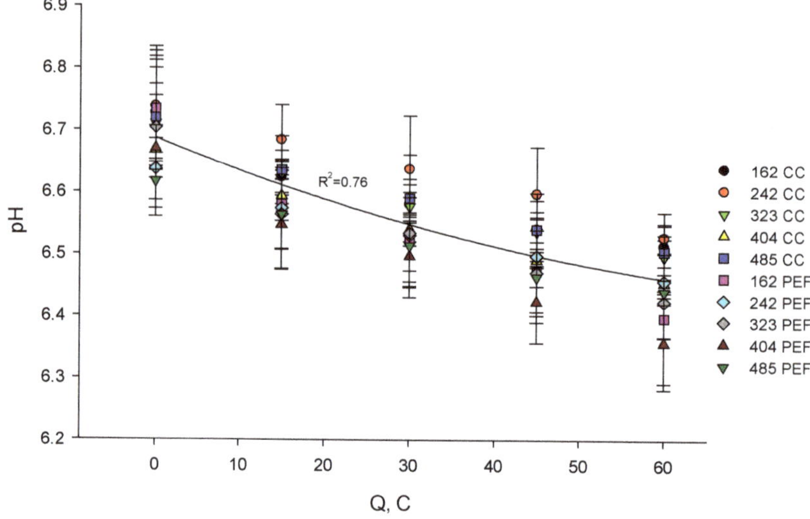

Figure 2. Evolution of milk pH as a function of the number of charges transported (in Coulomb) during EDBM carried out in different conditions of flow rate (corresponding to indicated Reynolds numbers) and current mode. Reynolds numbers of 162, 242, 323, 404 and 485 correspond to flow rates of 400, 600, 800, 1000 and 1200 mL/min respectively. CC—constant current and PEF—pulsed electric field.

3.1.2. In KCl Solution

The analysis of variance showed that the mode of current ($p < 0.001$) and flow rate ($p = 0.02$) have a significant effect on the variation of pH of the KCl solution during EDBM while the coupled effect of flow rate/mode of current has no effect ($p = 0.92$). Coupled effect represents the combined effects of both factors (flow rate and current mode) on the dependent measure (pH of KCl). Based on the p value it can be concluded that the current mode has the main effect on the pH of KCl. The conclusion that flow rate is a significant factor and coupled effect of flow rate/mode of current is not significant was mainly due to the high standard deviation of experimental data under CC mode and highest flow rate (which corresponds to a Reynolds number of 485), for other conditions the standard deviations were significantly small. Considering that, there was a major effect of current mode but the flow rate applied, and the resulting Reynolds number, for a mode of current did not influence on the variation of pH of the KCl solution. The curve of pH evolution calculated for the KCl solution as a function of number of charge transported (Figure 3) showed an increase in pH, which is different depending of the mode of current (CC vs PEF). Hence, whatever the flow rate of KCl solution, in the case of CC, the pH increased from 5.8 to 10.4 (79.3% increase), while it increased to 9.9 (70.7% increase) in the case of PEF. It is probably connected with the fact that during the pause lapse under PEF occurs a better mixing of the solution and equilibration allowing phosphate from micelle to better dissolve. The lower final pH of KCl solution under the PEF can also be connected with a leakage of OH^- through CMX membrane during a pause lapse [30] and decreasing of CP phenomenon and consequently with a decrease in water splitting [14]. Indeed, during the pause lapse of PEF, the ion concentration at the membrane interface can be partially restored, reducing the CP phenomena and membrane fouling during the subsequent pulse [31]. The difference in pH between CC and PEF appeared at the beginning of the process (up to 15 C) and then this difference remained constant until the end of experiment. Rapid increase of pH of the KCl solution at the beginning of EDBM occurred as the hydroxide electrogeneration progresses due to the water splitting at the BPM. Furthermore, the variation of pH is more important in the KCl solution that in the milk due to the different buffer capacities of the two solutions. The KCl has no buffer capacity while milk according to its composition has a high buffer capacity; phosphate, citrate, lactate, carbonate, acetate, and propionate ions are mainly responsible for the buffer capacity of milk [32].

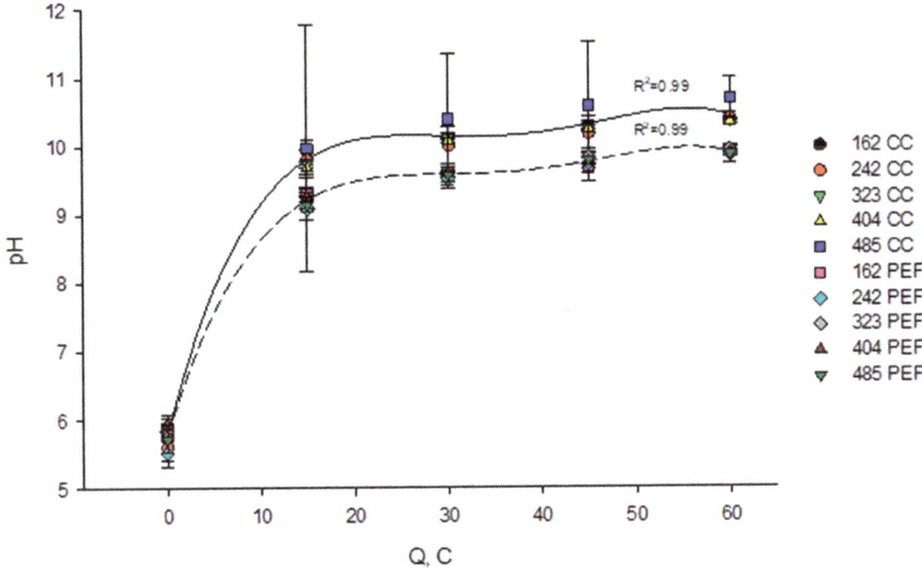

Figure 3. Evolution of KCl solution pH as a function of the number of charges transported (in Coulomb) during EDBM carried out in different flow rates (corresponding to indicated Reynolds numbers) and current modes (PEF in dashed line and CC in solid line). Reynolds numbers of 162, 242, 323, 404 and 485 correspond to flow rates of 400, 600, 800, 1000 and 1200 mL/min respectively. CC—constant current and PEF—pulsed electric field.

3.2. Evolution of Conductivity

3.2.1. In Skim Milk

It appeared from the statistical analysis, that the mode of current has a significant effect ($p < 0.001$) on the variation of skim milk conductivity during EDBM while the flow rate and the coupled effect of flow rate/mode of current have no effect ($p > 0.05$). The regression curves calculated for the milk conductivity as a function of number of charge transported (Figure 4) showed an increase in conductivity, different according to the mode of current (CC vs PEF). Hence, whatever the flow rate of the milk solution, in the case of CC, the electrical conductivity increased from 3100 to 3800 µS/cm (22.5% increase), while it increased to 4300 µS/cm (38.7% increase) in the case of PEF. The difference in electrical conductivity between both modes of current appeared at the beginning of the process (up to 15 C) and then it was quite constant until the end of the experiments. The conductivity changes of skim milk at a given pH is the result of H^+ generation, change in global soluble protein charge, dissolving of calcium, phosphate and magnesium from casein micelles, as well as potassium, sodium, chloride, hydrogen ions, and citrate already present in the soluble phase [33]. Indeed, according to [25,31] Ca^{2+} and Mg^{2+} are bound to the phosphoserine groups of different individual caseins of milk and remain as colloidal phosphocalcic (phosphomagnesic) bridges inside the casein micelles and are released as pH decreases which influences the resulting milk conductivity. Generally, electrical conductivity of milk was influenced by two competing phenomena: desalination and acidification. Cations migrate from the desalination channel by the action of CEM, which diminished the conductivity of solution. On the other hand, the hydrogen ions generated on the cationic interface of BPM and breakage of casein micelles contribute to the conductivity of milk. The increase of milk conductivity was probably connected to the release of calcium ions from casein micelle, which acts opposite to the demineralization effect and the rate of calcium release is higher than its migration through the CEM since many potassium ions are present in the solution. Indeed, it was reported by Bazinet et al. [34] that potassium is the first ion to migrate during EDBM, due to its higher electrical mobility, concentration and also the predominant one to leave the skim milk solution until a critical concentration of about 20% of its initial concentration was reached. A faster increase in electrical conductivity is observed at the beginning of EDBM under the PEF current mode. This would be linked to the fact that, under CC, electrogenerated H^+ would interact preferentially with phosphate ions released from casein micelle and consequently, these protons, would not contribute to the increase in milk conductivity. Indeed, previous works have already showed a decrease in milk conductivity during EDBM due to the demineralization effect of CEM [25,30,31].

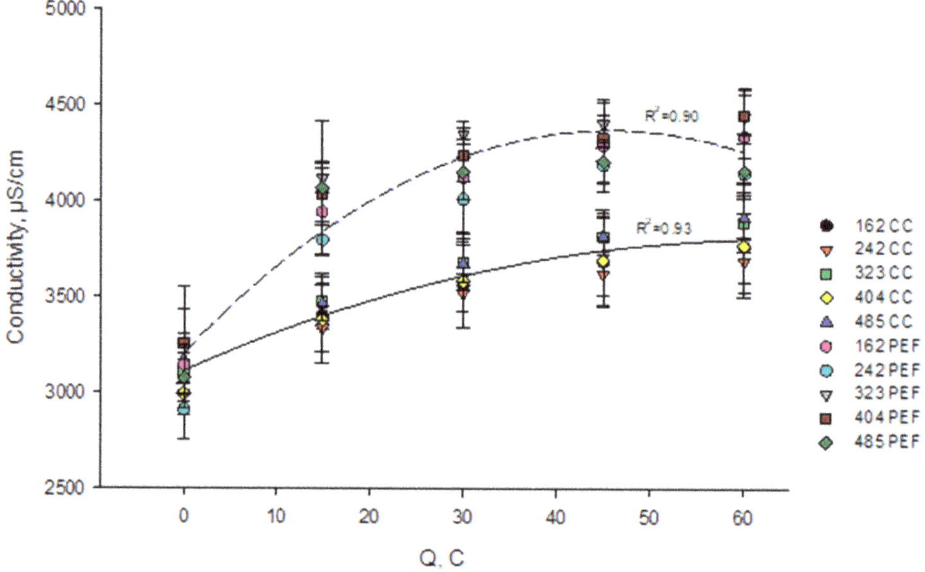

Figure 4. Evolution of milk conductivity as a function of a number of charges transported (in Coulomb) during EDBM carried out in different conditions of flow rate (corresponding to indicated Reynolds numbers) and current mode (PEF in dashed line and CC in solid line). Reynolds numbers of 162, 242, 323, 404 and 485 correspond to flow rates of 400, 600, 800, 1000 and 1200 mL/min respectively. CC—constant current and PEF—pulsed electric field.

3.2.2. In KCl Solution

The analysis of variance showed that the mode of current has a significant effect ($p < 0.001$) on the variation of KCl solution conductivity during EDBM while the flow rate and the coupled effect of flow rate/mode of current has no effect ($p > 0.05$). Hence, the electrical conductivity of the KCl solution increased for both current modes but differently and can be approximated linearly (Figure 5). Indeed, the variations of conductivity between the beginning and the end of EDBM, whatever the flow rate of solution, were of 251.5 mS/cm (corresponding to a 10.2% increase) and 211.0 mS/cm (8.4% increase) for CC and PEF respectively. Although there was a statistical difference, which was relatively small, it corresponds to less than 1.7% of the initial conductivity. Consequently, the evolution of KCl conductivity can be considered as the same and that whatever the flow rate and the current mode. Unlike the electrical conductivity of milk, for the KCl solution both process electrically-driven calcium, magnesium, and potassium ion migration and electrogeneration of hydroxide ions on the BPM surface acted together and lead to such an increase in conductivity during EDBM. Indeed, Lin Teng Shee et al. [35] have reported the fact that the H^+ and OH^- produced during the process of solution demineralization using BPMs contribute more to the conductivity than other ions. The same effect of increase of KCl conductivity was observed by Kravtsov et al. [36] for acid whey demineralization using EDBM. In the case of CC mode, a faster increase of the KCl solution conductivity was observed compared to a PEF current mode. It was connected with a more intensive water splitting phenomenon under CC current mode as we mentioned previously for pH evolution of the KCl solution.

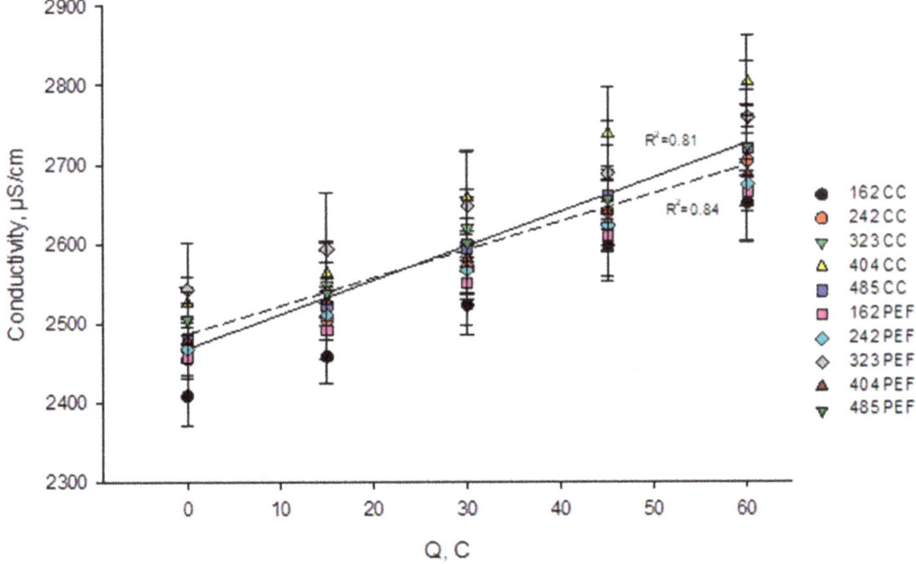

Figure 5. Evolution of KCl compartment conductivity as a function of the number of charges transported (in Coulomb) during EDBM carried out in different conditions of flow rate (corresponding to indicated Reynolds numbers) and current mode (pulsed electric field (PEF) in dashed line and continuous current (CC) in solid line). Reynolds numbers of 162, 242, 323, 404, and 485 correspond to flow rates of 400, 600, 800, 1000, and 1200 mL/min respectively. CC—constant current and PEF—pulsed electric field.

3.3. Membrane Parameters

According to the statistical analysis of the difference between membrane conductivity before and after experiments ($p < 0.001$, $p < 0.001$ and $p = 0.003$ for CEM, BPM1, and BPM2 respectively), all the membranes showed a decrease in conductivity after treatment (Table 2). It also appeared from the analysis of variance that flow rate has no significant effect ($p > 0.05$) on the conductivity variation of membranes after EDBM. For the CEM and BPM1 the mode of current has no significant effect on the conductivity ($p > 0.05$) while in the case of BPM2 this factor has a significant effect ($p = 0.04$) on the conductivity variations. The conductivity of CEM decreased probably due to the substitution of relatively mobile Na^+ by Ca^{2+} from milk [37]. Electrical conductivity of the first BPM decreased due to the presence of protein fouling on the cationic interface. This is known fact that the fouling and scaling formation decrease the conductivity and permselectivity of membranes [8,9]. Despite the fact that ANOVA showed that there was a significant difference of BPM2 conductivity after treatment in general and for different modes of current in particular we can observe that all the values are in the range of standard deviation. Indeed, an average conductivity of BPM2 before treatment is 7.45 ± 0.17 mS/cm, after is 7.15 ± 0.26 mS/cm for all the conditions averaged. If we compare different modes of current, we can notice that under CC the average conductivity of BPM2 is 7.30 ± 0.26 mS/cm, under PEF is 7.01 ± 0.24 mS/cm. Thus, we can conclude that the difference of conductivity of BPM is not significant before and after EDBM treatment.

Concerning the thickness of membranes, it was concluded that there was no significant difference in CMX thickness before and after experiments ($p > 0.05$) (Table 2). The averaged thickness of the CMX membranes was 0.149±0.003 mm. According to the ANOVA, the thickness of both BPMs significantly changed after all experiments ($p < 0.001$, $p = 0.002$ for BPM1 and BPM2 respectively) but the values did not depend on the flow rate ($p > 0.05$) and current mode ($p > 0.05$) during EDBM. Despite the fact that the changes of BPMs thicknesses are significant after EDBM we can notice that these values are in the range of standard deviation; BPM1 thickness before treatment is 0.246 ± 0.002 mm and 0.242 ± 0.002 mm after and for BPM2 0.245 ± 0.002 mm and 0.241 ± 0.002 mm respectively. Since that there is no real difference in thickness of the BPMs before and after EDBM treatment. The thickness of BPM in contact with milk did not change after treatment despite the presence of protein fouling for some cases due to the fact that fouling was removed from the membrane surface before thickness measurements.

3.4. Membrane Surface Integrity and Quantification of Protein Fouling

Concerning the CEM for both interfaces considered, as well as for the anionic side of the BPM, no fouling was observed whatever the conditions of flow rate and current mode applied. In contrast, the photographs of the cationic interface of the BPMs, in contact with the milk solution, showed that there was a major impact of current mode on the presence or not of a protein fouling at this interface while flow rate was less impacting (Figure 6).

Table 2. Membrane properties (thickness and conductivity) before and after EDBM carried out in different conditions of flow rate (corresponding to indicated Reynolds numbers) and current mode. Reynolds numbers of 162, 242, 323, 404 and 485 correspond to flow rates of 400, 600, 800, 1000 and 1200 mL/min respectively. CC—constant current, PEF—pulsed electric field, CEM—cation exchange membrane and BPM—bipolar membrane.

Current Mode			**CC**					**PEF**				
Reynolds Number			**162**	**242**	**323**	**404**	**485**	**162**	**242**	**323**	**404**	**485**
Conductivity (mS/cm)	CEM	Before	8.35 ± 0.20 a*	8.68 ± 0.30 a	8.49 ± 0.37 a	8.33 ± 0.41 a	8.53 ± 0.22 a	8.33 ± 0.33 a	8.35 ± 0.67 a	8.34 ± 0.69 a	8.02 ± 0.45 a	8.54 ± 0.58 a
		After	7.01 ± 0.17 b	7.41 ± 0.13 b	7.14 ± 0.27 b	7.16 ± 0.30 b	7.30 ± 0.51 b	7.25 ± 0.34 b	6.99 ± 0.45 b	7.04 ± 0.69 b	7.33 ± 0.20 b	7.30 ± 0.23 b
	BPM1	Before	7.59 ± 0.36 a	7.35 ± 0.40 a	7.44 ± 0.46 a	7.40 ± 0.18 a	6.84 ± 1.09 a	7.36 ± 0.21 a	7.14 ± 0.38 a	7.32 ± 0.14 a	7.82 ± 0.15 a	7.53 ± 0.17 a
		After	6.67 ± 0.46 b	6.55 ± 0.34 b	6.74 ± 0.57 b	6.70 ± 0.14 b	6.53 ± 0.30 b	6.58 ± 0.15 b	6.05 ± 0.65 b	6.35 ± 0.41 b	6.75 ± 0.22 b	6.87 ± 0.44 b
	BPM2	Before	7.39 ± 0.31 a	7.35 ± 0.23 a	7.57 ± 0.64 a	7.66 ± 0.46 a	7.54 ± 0.34 a	7.28 ± 0.18 a	7.39 ± 0.33 a	7.17 ± 0.08 a	7.48 ± 0.32 a	7.69 ± 0.07 a
		After	7.11 ± 0.46 bA	7.04 ± 0.39 bA	7.65 ± 0.48 bA	7.51 ± 0.46 bA	7.18 ± 0.09 bA	6.90 ± 0.06 bB	6.79 ± 0.27 bB	6.99 ± 0.34 bB	7.29 ± 0.45 bB	7.07 ± 0.47 bB
Thickness (mm)	CEM	Before	0.151 ± 0.005 a	0.149 ± 0.004 a	0.151 ± 0.006 a	0.144 ± 0.005 a	0.153 ± 0.004 a	0.151 ± 0.005 a	0.151 ± 0.004 a	0.148 ± 0.005 a	0.152 ± 0.006 a	0.143 ± 0.005 a
		After	0.150 ± 0.004 a	0.150 ± 0.004 a	0.148 ± 0.005 a	0.146 ± 0.010 a	0.149 ± 0.003 a	0.149 ± 0.002 a	0.148 ± 0.006 a	0.144 ± 0.004 a	0.155 ± 0.004 a	0.145 ± 0.005 a
	BPM1	Before	0.244 ± 0.004 a	0.244 ± 0.011 a	0.246 ± 0.002 a	0.247 ± 0.003 a	0.246 ± 0.003 a	0.247 ± 0.002 a	0.242 ± 0.006 a	0.246 ± 0.005 a	0.251 ± 0.001 a	0.249 ± 0.005 a
		After	0.240 ± 0.003 b	0.243 ± 0.004 b	0.244 ± 0.003 b	0.238 ± 0.004 b	0.240 ± 0.003 b	0.242 ± 0.003 b	0.243 ± 0.001 b	0.242 ± 0.002 b	0.246 ± 0.003 b	0.245 ± 0.001 b
	BPM2	Before	0.248 ± 0.004 a	0.245 ± 0.005 a	0.247 ± 0.007 a	0.247 ± 0.007 a	0.247 ± 0.005 a	0.246 ± 0.002 a	0.244 ± 0.004 a	0.241 ± 0.002 a	0.243 ± 0.004 a	0.247 ± 0.004 a
		After	0.241 ± 0.004 b	0.241 ± 0.003 b	0.245 ± 0.006 b	0.238 ± 0.004 b	0.245 ± 0.003 b	0.241 ± 0.002 b	0.238 ± 0.010 b	0.241 ± 0.001 b	0.243 ± 0.003 b	0.242 ± 0.002 b

* Data with different letters (a, b or A, B) are significantly different; lowercase letters indicate differences between conductivities and thicknesses before and after EDBM treatment for the same membrane; uppercase letters indicate differences between modes of current for the same membranes.

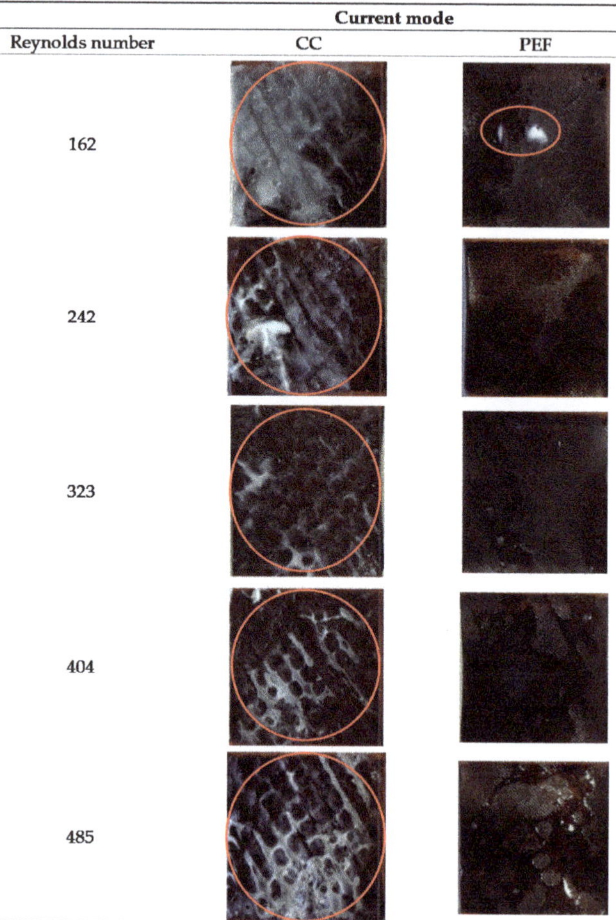

Figure 6. Photographs of the bipolar membrane (BPM) cationic interface in contact with milk after EDBM carried out in different flow rates (corresponding to indicated Reynolds numbers) and current modes (CC and PEF). The red circles indicate a fouling.

Considering the weight of protein fouling recovered on each cationic interface after EDBM treatment, the ANOVA showed that there is an effect of current mode ($p < 0.001$), flow rate ($p = 0.005$) as well as coupled effect mode of current/flow rate ($p = 0.006$). Indeed, with PEF mode of current the weight of protein fouling was quite negligible in comparison with CC with values, all flow rate conditions averaged, of 0.07 ± 0.08 mg/cm^2 and 5.56 ± 2.40 mg/cm^2 respectively. Confirmation of the positive impact of using PEF current regimes on preventing of protein fouling and scaling formation can be found in many papers [11]. In the work of Ruiz et al. [11] it was also demonstrated that PEF with 10–40 s pulse/pause combination whatever the conditions allowed to completely eliminate protein fouling from the AEM during demineralization by conventional ED of a casein model solution. The protein fouling formation under CC mode was observed because of a local pH change at the BPM interface, due to overlimiting conditions, since only small changes in the pH of the skim milk were observed during the treatment; the pH of the milk solution bulk was always over the isoelectric point (pH 4.6). Positive effects of using PEF for fouling minimization were explained by the fact that the PEF produces perturbations in electrophoretic movement of the substances forming the screening film on the surface of membrane [14]. These perturbations increase the mixing of solution within the boundary layer and hinder the formation of deposit at the membrane surface. The impact of flow rate has already been demonstrated by Bazinet et al. [7] on milk acidification but at only two different flow rate conditions (757 mL/min and 4542 mL/min) and with a different cell design (effective surface area of 100 cm^2). However, the impact of flow rate coupled with PEF has never been demonstrated before. Concerning flow rate, it has no impact during application of PEF with very low or no quantity of protein fouling while during CC, the weight of fouling was different according to the flow rate applied. Indeed, for CC, there was an almost linear decrease of the weight of protein fouling as a function of Reynolds numbers from 162 to 323 (which corresponds to a flow rate from 400 to 800 mL/min for the ED system used) since it decreased from 9.18 to 3.36 mg/cm^2 (Figure 7). A further increase in flow rate up to the Reynolds numbers of 485 did not lead to significant changes in the amount of fouling formed on the BPM surface.

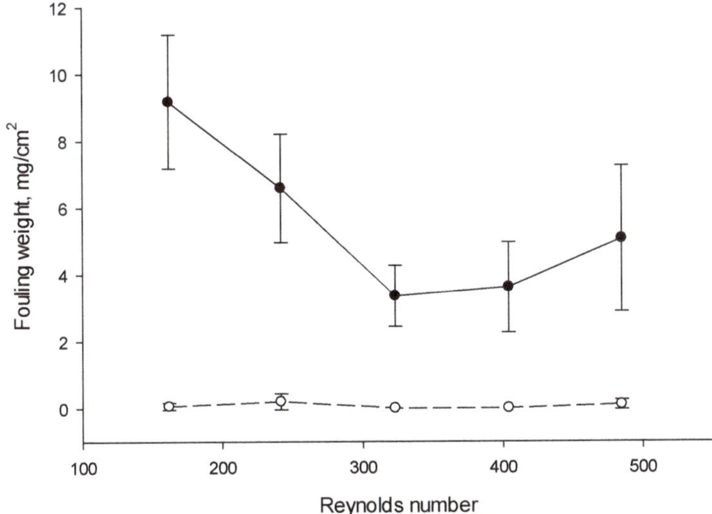

Figure 7. Weight of protein fouling (in mg/cm2) recovered on the BPM cationic interface in contact with milk after EDBM depending on the Reynolds numbers under CC and PEF current mode (PEF in dashed line and CC in solid line). Reynolds numbers of 162, 242, 323, 404 and 485 correspond to flow rates of 400, 600, 800, 1000 and 1200 mL/min respectively. CC—constant current and PEF—pulsed electric field.

According to literature data [38], all the Reynolds numbers reached in the present work correspond to a steady laminar fluid flow in the channel, while the turbulent fluid flow regime begins when the Reynolds numbers reach amount of 1000. Consequently, the decrease in the amount of protein fouling on the BPM surface with an increase of the Reynolds number up to value of 323 was observed due to the creation of unfavorable hydrodynamic conditions for the fouling attachment and growth. The similar effect for scaling formation was observed by Mikhaylin et al. [17] where authors showed that the use of higher flow rates during BMEA of skim milk coupled with ultrafiltration (UF) module leads to more than 38% decrease in scaling in comparison to the conventional EDBM-UF treatment. However, they did not mention the impact on protein fouling since they used a UF membrane to prevent such a fouling inside the EDBM cell. The positive coupled effect of PEF and flow rate on the fouling formation is due to the fact that during pauses the excess of H^+ ions at the cationic interface of the BPM, initiating the surface fouling, is dissipated due to the recirculation of solution and since no more H^+ are generated, avoiding consequently the fouling formation and accumulation while in the same time the solution flow flushes the potentially accumulated protein from the membrane surface.

3.5. Energy Consumptions

It appeared from the statistical analysis, that the mode of current has a significant effect ($p < 0.001$) on the energy consumptions (EC) during EDBM while the Reynolds number and the coupled effect Reynolds number/mode of current had no effect ($p > 0.05$). The curves calculated for the EC as a function of Reynolds number (Figure 8) showed quite similar values according to standard deviations, but different for the mode of current considered (CC vs PEF). Hence, whatever the Reynolds number, in the case of CC, the value of EC, all Reynolds numbers conditions averaged, was equal to 0.091 ± 0.001 Wh, while it is equal to 0.087 ± 0.001 Wh in the case of PEF. This difference corresponds to a 5% less energy consumption under PEF, which is related to the almost complete absence of protein fouling for the PEF conditions. Slight decrease of EC was also observed by Ruiz et al. [11] during ED of casein solution between conventional ED and PEF with 10–40 s pulses (2.770 Wh and 2.750 Wh respectively under 30 mA/cm^2) due to the elimination of protein fouling on AEM.

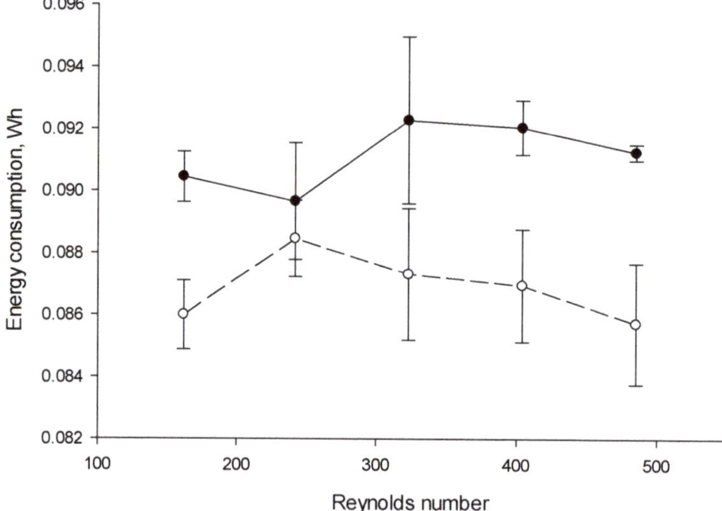

Figure 8. Energy consumptions (in Wh) during EDBM depending on the Reynolds numbers under CC and PEF current mode (PEF in dashed line and CC in solid line). Reynolds numbers of 162, 242, 323, 404, and 485 correspond to flow rates of 400, 600, 800, 1000 and 1200 mL/min respectively. CC—constant current and PEF—pulsed electric field.

4. Conclusions

In this work the influence of solution flow rate (and consequently Reynolds number) and PEF on the protein fouling formation at the cationic interface of BPM (where H^+ are electrogenerated) and on membrane properties during EDBM of skim milk were demonstrated. It was observed that the major impact on the presence or not of a protein fouling at the BPM cationic interface is the mode of current used while flow rate was less impacting. It can be concluded from obtained results that the application of a PEF mode of current prevented the fouling formation almost completely during EDBM, regardless of the flow rate due to the decrease of CP and less intense H^+ generation. Flow rate did not have a major impact during application of PEF with very low amount or complete absence of protein fouling while during CC, the weight of fouling was different according to the flow rate applied. The fouling formation under CC was due to the local pH changes at the BPM interface. When a CC mode was applied, a Reynolds number of 323 and over would be the optimal regime in these experimental conditions to minimize the protein fouling of the BPM during EDBM allowing to wash off part of the protein sediment from the membrane surface. CEM showed a decrease in conductivity after EDBM treatment due to the replacement of mobile single charged ions by double charged ions from milk. BPM which was in contact with milk solution also showed a decrease in conductivity due to the protein fouling formation, while conductivity of BPM2 did not change after all treatments. It was observed that EDBM process did not influence the membrane thickness. Using PEF leads to a 5% less EC in comparison with CC regime regardless of Reynolds number considered which is connected with an absence of protein fouling on the BPM membrane surface.

The next steps, currently underway, are to optimize PEF by testing different pulse–pause duration to find the optimal condition of this current regime and to model the EDBM process for a better understanding of protein fouling kinetics.

Author Contributions: Conceptualization, V.S.N. and L.B.; methodology, V.S.N. and L.B.; software, V.S.N.; validation, V.S.N., T.R.G., V.N., and L.B.; formal analysis, V.S.N.; investigation, V.S.N.; resources, V.S.N., T.R.G., V.N., and L.B.; data curation, V.S.N. and T.R.G.; writing—original draft preparation, V.S.N.; writing—review and editing, V.S.N., T.R.G., V.N., and L.B.; visualization, V.N. and L.B.; supervision, L.B.; project administration, L.B.; funding acquisition, V.N. and L.B. All authors have read and agreed to the published version of the manuscript.

Funding: This work was funded by the Natural Sciences and Engineering Research Council of Canada (NSERC) and the Russian Foundation for Basic Research (RFBR).

Acknowledgments: The Natural Sciences and Engineering Research Council of Canada (NSERC) financial support is acknowledged. This work was supported by the NSERC Discovery Grants Program (Grant SD 210829409 to Laurent Bazinet). This work was also financially supported by the Russian Foundation for Basic Research (RFBR), project number 19-38-90256. The authors thank Jacinthe Thibodeau, research professional at Laval University for her technical assistance and help at the lab.

Conflicts of Interest: The authors declare no conflict of interest.

References

1. Dominguez-Salas, P.; Galie, A.; Omore, A.; Omosa, E.; Ouma, E. Contributions of Milk Production to Food and Nutrition Security. In *Encyclopedia of Food Security and Sustainability*; Ferranti, P., Berry, E.M., Anderson, J.R., Eds.; Elsevier Science: Amsterdam, The Netherlands, 2019; pp. 278–291.
2. Gorbatova, K.K.; Gunkova, P.I. *Chemistry and Physics of Milk and Dairy Products*; GIORDR: Saint Petersburg, Russia, 2012; pp. 17–77. (In Russian)
3. Mier, M.P.; Ibanez, R.; Ortiz, I. Influence of process variables on the production of bovine milk casein by electrodialysis with bipolar membranes. *Biochem. Eng. J.* **2008**, *40*, 304–311. [CrossRef]
4. Audic, J.-L.; Chaufer, B.; Daufin, G. Non-food applications of milk components and dairy co-products: A review. *Le Lait.* **2003**, *83*, 417–438. [CrossRef]
5. Van der Horst, H.C.; Timmer, J.M.K.; Robbertsen, T.; Leenders, J. Use of nanofiltration for concentration and demineralization in the dairy industry: Model for mass transport. *J. Membr. Sci.* **1995**, *104*, 205–218. [CrossRef]
6. Houldsworth, D.W. Demineralization of whey by means of ion-exchange and electrodialysis. *J. Soc. Dairy Technol.* **1980**, *33*, 45–51. [CrossRef]
7. Bazinet, L.; Lamarche, F.; Ippersiel, D.; Amiot, J. Bipolar membrane electroacidification to produce bovine milk casein isolate. *J. Agric. Food Chem.* **1999**, *47*, 5291–5296. [CrossRef]
8. Thompson, D.W.; Tremblay, A.Y. Fouling in steady and unsteady state electrodialysis. *Desalination* **1983**, *47*, 181–188. [CrossRef]
9. Bleha, M.; Tishchenko, G.; Šumberová, V.; Kůdela, V. Characteristic of the critical state of membranes in ED-desalination of milk whey. *Desalination* **1992**, *86*, 173–186. [CrossRef]
10. Mikhaylin, S.; Bazinet, L. Fouling on ion-exchange membranes: Classification, characterization and strategies of prevention and control. *Adv. Colloid Interface Sci.* **2016**, *229*, 34–56. [CrossRef]
11. Ruiz, B.; Sistat, P.; Huguet, P.; Pourcelly, G.; Araya-Farias, M.; Bazinet, L. Application of relaxation periods during electrodialysis of a casein solution: Impact on anion-exchange membrane fouling. *J. Membr. Sci.* **2007**, *287*, 41–50. [CrossRef]

12. Dufton, G.; Mikhaylin, S.; Gaaloul, S.; Bazinet, L. Positive Impact of Pulsed Electric Field on Lactic Acid Removal, Demineralization and Membrane Scaling during Acid Whey Electrodialysis. *Int. J. Mol. Sci.* **2019**, *20*, 797. [CrossRef]
13. Mikhaylin, S.; Nikonenko, V.; Pismenskaya, N.; Pourcelly, G.; Choi, S.; Kwon, H.J.; Han, J.; Bazinet, L. How physico-chemical and surface properties of cation-exchange membrane affect membrane scaling and electroconvective vortices: Influence on performance of electrodialysis with pulsed electric field. *Desalination* **2016**, *393*, 102–114. [CrossRef]
14. Nikonenko, V.V.; Pismenskaya, N.D.; Belova, E.I.; Sistat, P.; Huguet, P.; Pourcelly, G.; Larchet, C. Intensive current transfer in membrane systems: Modelling, mechanisms and application in electrodialysis. *Adv. Colloid Interface Sci.* **2010**, *160*, 101–123. [CrossRef] [PubMed]
15. Sistat, P.; Huguet, P.; Ruiz, B.; Pourcelly, G.; Mareev, S.A.; Nikonenko, V.V. Effect of pulsed electric field on electrodialysis of a NaCl solution in sub-limiting current regime. *Electrochim. Acta* **2015**, *164*, 267–280. [CrossRef]
16. Mikhaylin, S.; Sion, A.-V. Improvement of a sustainable hybrid technology for caseins isoelectric precipitation (electrodialysis with bipolar membrane/ultrafiltration) by mitigation of scaling on cation-exchange membrane. *Innov. Food Sci. Emerg. Technol.* **2016**, *33*, 571–579. [CrossRef]
17. Lee, H.-J.; Sarfert, F.; Strathmann, H.; Moon, S.-H. Designing of an electrodialysis desalination plant. *Desalination* **2002**, *142*, 267–286. [CrossRef]
18. Muehlhoff, E.; Bennett, A.; McMahon, D. *Milk and Dairy Products in Human Nutrition*; FAO: Rome, Italy, 2013; pp. 41–90.
19. Cifuentes-Araya, N.; Pourcelly, G.; Bazinet, L. Impact of pulsed electric field on electrodialysis process performance and membrane fouling during consecutive demineralization of a model salt solution containing a high magnesium/calcium ratio. *J. Colloid Interface Sci.* **2011**, *361*, 79–89. [CrossRef]
20. Langevin, M.-E.; Bazinet, L. Ion-exchange membrane fouling by peptides: A phenomenon governed by electrostatic interactions. *J. Membr. Sci.* **2011**, *369*, 359–366. [CrossRef]
21. Tanaka, Y. *Ion Exchange Membranes: Fundamentals and Applications*, 1st ed.; Elsevier Science: Amsterdam, The Netherlands, 2007; pp. 205–244.
22. Buchanan, J.H.; Peterson, E.E. Buffers of milk and buffer value. *J. Dairy Sci.* **1927**, *10*, 224–331. [CrossRef]
23. Lucey, J.A.; Horne, D.S. Milk Salts: Technological Significance. In *Advanced Dairy Chemistry*; Springer: New York, NY, USA, 2009; pp. 351–389. [CrossRef]
24. Le Graet, Y.; Brulé, G. Les équilibres minéraux du lait: Influence du pH et de la force ionique. *Lait* **1993**, *73*, 51–60. [CrossRef]
25. Law, A.J.R.; Leaver, J. Effects of acidification and storage of milk on dissociation of bovine casein micelles. *J. Agric. Food Chem.* **1998**, *46*, 5008–5016. [CrossRef]
26. Attia, H.; Kherouatou, N.; Ayadi, J. Acidification chimique directe du lait: Correlation entre la mobilité du matériel micellaire et les micro et macrostructures des laits acidifiés. *Sci. Aliment.* **2000**, *20*, 289–307. [CrossRef]
27. Van Hooydonk, A.C.M.; Hagerdoorn, H.G.; Boerrigter, I.J. pH-induced physicochemical changes of casein micelles in milk and their effect on renneting. I. Effect of acidification on physichochemical properties. *Neth. Milk Dairy J.* **1986**, *40*, 281–296.
28. Bazinet, L.; Lamarche, F.; Ippersiel, D.; Gendron, C.; Mahdavi, B.; Amiot, J. Comparison of electrochemical and chemical acidification of skim milk. *J. Food Sci.* **2000**, *65*, 1303–1307. [CrossRef]
29. Masson, F.-A.; Mikhaylin, S.; Bazinet, L. Production of calcium- and magnesium-enriched caseins and caseinates by an ecofriendly technology. *J. Dairy Sci.* **2018**, *101*, 1–11. [CrossRef] [PubMed]
30. Bazinet, L.; Montpetit, D.; Ippersiel, D.; Amiot, J.; Lamarche, F. Identification of Skim Milk Electroacidification Fouling: A Microscopic Approach. *J. Colloid Interface Sci.* **2001**, *237*, 62–69. [CrossRef]
31. Lemay, N.; Mikhaylin, S.; Bazinet, L. Voltage spike and electroconvective vortices generation during electrodialysis under pulsed electric field: Impact on demineralization process efficiency and energy consumption. *Innov. Food Sci. Emerg.* **2019**, *52*, 221–231. [CrossRef]
32. Salaun, M.S.; Mietton, B.; Gaucheron, F. Buffering capacity of dairy products. *Int. Dairy J.* **2005**, *15*, 95–109. [CrossRef]
33. Bazinet, L.; Pouliot, Y.; Castaigne, F. Relative contributions of charged species to conductivity changes in skim milk during electrochemical acidification. *J. Memb. Sci.* **2010**, *352*, 32–40. [CrossRef]

34. Bazinet, L.; Ippersiel, D.; Gendron, C.; Beaudry, J.; Mahdavi, B.; Amiot, J.; Lamarche, F. Cationic balance in skim milk during bipolar membrane electroacidification. *J. Memb. Sci.* **2000**, *173*, 201–209. [CrossRef]
35. Lin Teng Shee, F.; Arul, J.; Brunet, S.; Bazinet, L. Performing a three-step process for conversion of chitosan to its oligomers using a unique bipolar membrane electrodialysis system. *J. Agric. Food Chem.* **2008**, *56*, 10019–10026. [CrossRef]
36. Kravtsov, V.; Kulikova, I.; Mikhaylin, S.; Bazinet, L. Alkalinization of acid whey by means of electrodialysis with bipolar membranes and analysis of induced membrane fouling. *J. Food Eng.* **2020**, *277*, 109891–109900. [CrossRef]
37. Bazinet, L.; Montpetit, D.; Ippersiel, D.; Mahdavi, B.; Amiot, J.; Lamarche, F. Neutralization of hydroxide generated during skim milk electroacidification and its effect on bipolar and cationic membrane integrity. *J. Memb. Sci.* **2003**, *216*, 229–239. [CrossRef]
38. Campione, A.; Gurreri, L.; Ciofalo, M.; Micale, G.; Tamburini, A.; Cipollina, A. Electrodialysis for water desalination: A critical assessment of recent developments on process fundamentals, models and applications. *Desalination* **2018**, *434*, 121–160. [CrossRef]

Article

The Development of Electroconvection at the Surface of a Heterogeneous Cation-Exchange Membrane Modified with Perfluorosulfonic Acid Polymer Film Containing Titanium Oxide

Violetta Gil *, Mikhail Porozhnyy, Olesya Rybalkina, Dmitrii Butylskii and Natalia Pismenskaya

Department of Physical Chemistry, Kuban State University, 149 Stavropolskaya st., 350040 Krasnodar, Russia; porozhnyj@mail.ru (M.P.); olesia93rus@mail.ru (O.R.); dmitrybutylsky@mail.ru (D.B.); n_pismen@mail.ru (N.P.)
* Correspondence: violetta_gil@mail.ru

Received: 12 May 2020; Accepted: 15 June 2020; Published: 17 June 2020

Abstract: One way to enhance mass transfer and reduce fouling in wastewater electrodialysis is stimulation of electroconvective mixing of the solution adjoining membranes by modifying their surfaces. Several samples were prepared by casting the perfluorosulfonic acid (PFSA) polymer film doped with TiO_2 nanoparticles onto the surface of the heterogeneous cation-exchange membrane MK-40. It is found that changes in surface characteristics conditioned by such modification lead to an increase in the limiting current density due to the stimulation of electroconvection, which develops according to the mechanism of electroosmosis of the first kind. The greatest increase in the current compared to the pristine membrane can be obtained by modification with the film being 20 μm thick and containing 3 wt% of TiO_2. The sample containing 6 wt% of TiO_2 provides higher mass transfer in overlimiting current modes due to the development of nonequilibrium electroconvection. A 1.5-fold increase in the thickness of the modifying film reduces the positive effect of introducing TiO_2 nanoparticles due to (1) partial shielding of the nanoparticles on the surface of the modified membrane; (2) a decrease in the tangential component of the electric force, which affects the development of electroconvection.

Keywords: fouling; ion-exchange membrane; surface modification; electroconvection; voltammetry; chronopotentiometry; impedance spectroscopy

---

1. Introduction

The broader use of electro-membrane technologies for separation, concentration, and isolation of valuable substances from wastewater and liquid media of the food industry is constrained by fouling which is an accumulation of mineral and organic pollutants on the surface and in the volume of ion-exchange membranes [1–4]. Fouling leads to an undesired decrease in useful mass transfer and increase in energy consumption [5]. It reduces the lifespan of membranes and requires spending on their regular cleaning [6]. The result is a significant increase in the cost of the final product. One way to prevent fouling is to stimulate electroconvective mixing of the solution [7–9] by targeted selection of ion-exchange membranes, modifying their surface [9] and/or using pulsed current modes [7–9].

Nowadays, due to the works of Dukhin, Mishchuk [10,11], Rubinstein, Zaltzman [12,13], Nikonenko, Urtenov [14,15], and a number of other researchers [16–18] there are certain concepts of the mechanisms of electroconvection (EC) and those factors that can induce electroconvective vortices. The details of these ideas are described in the review [19]. There are two main mechanisms for the development of EC. Bulk EC-I is due to the effect of an electric field on the residual space charge in a quasi-electroneutral electrolyte solution with an uneven concentration distribution. EC-II is caused by the electroosmotic slip of the space charge region (SCR). This SCR is formed in a depleted solution at the boundary with the membrane. A high degree of hydrophobicity of the membrane surface facilitates the fluid slip and stimulates EC-II [20–22].

Both EC-I and EC-II can be of two kinds currently known. (Quasi) equilibrium EC develops at low currents/voltages ($i \leq i_{\text{lim}}$; i_{lim} is the limiting current). In this case, the electrical double layer (EDL) at the membrane/depleted solution boundary is (quasi) equilibrium. The structure of the diffuse part of the EDL retains the Boltzmann distribution of ions which is shifted by an imposed external field [23]. In the case of EC-I, this kind of electroconvection is denoted by the term "electroosmosis of the first kind" [10,11]. Equilibrium EC can develop only at the electrically inhomogeneous and/or curved surface, which stimulates an appearance of the tangential component of an electric force. Its development is facilitated by a high surface charge [24]. In this case, small stable electroconvection vortices arise in the solution adjacent to a membrane. These vortices deliver a more concentrated solution to the depleted diffusion layer. This results in the "negative" conductance of the membrane system, when its resistance near the limiting current becomes lower than at $i << i_{\text{lim}}$, and also in the growth of the experimentally determined limiting current compared to that calculated by the convective-diffusion model [24,25].

In overlimiting current modes ($i \geq i_{\text{lim}}$), a nonequilibrium EC [13], also called "electroosmosis of the second kind", develops [11]. An indispensable condition for the development of a nonequilibrium EC is the presence of an extended SCR (which is outside the quasi-equilibrium EDL) and fluctuations in the concentration, electric fields, or the fluid velocity field, which, as a rule, occur in intensive current modes. The reason for the hydrodynamic instability after reaching a certain threshold value of the potential drop is the appearance of a positive feedback between the fluctuation of the local tangential force and the liquid electroosmotic slip velocity [12,26]. As a result, electroconvection vortices become larger than those in the case of equilibrium EC [27]. Their instability with time is the reason for the noticeable oscillations of the current (potential drops) observed in intensive current modes on the current-voltage characteristics (CVCs) and chronopotentiograms (ChPs) [9,28]. Partial removal of the diffusion limitations of salt delivery to the receiving side of the membrane is expressed in a reduction in length and an increase in the slope of the CVC plateau [9,28,29].

Thus, the development of EC is facilitated by electrical [9,28–30] and geometric [31,32] heterogeneity of the surface, a high degree of its hydrophobicity [24,25], and also a high electric charge [24]. The specified surface characteristics are achieved by profiling the surface of the membranes [31,32], varying the dispersion of the ion-exchange resin particles in heterogeneous membranes [33,34], and by modifying the membrane surface with various polyelectrolytes [29,30,35], in particular, conductive perfluorosulfonic acid (PFSA) polymer films [9,20,24,28]. Doping such a film with carbon nanotubes enhances electroconvection due to an increase in hydrophobicity and an increase in surface roughness [20]. It is very promising to increase the surface charge by doping the PFSA film with nanoparticles of metal oxides (such as TiO_2, ZrO_2, SiO_2). Currently, these dopants are mainly introduced into the pores of membranes and are widely used to increase their conductivity at low relative humidity in the fuel cells [36–39] and improving membrane permselectivity [40,41]. Recent studies revealed that the deposition of TiO_2-doped films on the surface increases the fouling resistance of the ultrafiltration, osmosis, and ion-exchange membranes [42–44]. For example, the authors of [44] modified the surface of the anion-exchange membrane by poly (sodium 4-styrene sulfonate) (PSS) with titanium dioxide nanoparticles. They found an increase in the surface charge and its hydrophilicity, as well as the self-cleaning ability of this surface due to photocatalytic decomposition caused by ultraviolet irradiation. Apparently, the effect of such surface modification on the development of electroconvection has not yet been investigated.

The aim of this work is to study the development of electroconvection as a function of the titanium oxide mass fraction in a perfluorosulfonic acid polymer film that modifies the surface of a heterogeneous cation-exchange membrane.

We hope that in the future, such a modification method can be used to create a new generation of ion-exchange membranes that are resistant to sedimentation and fouling during electrodialysis processing of liquid media of the food industry and natural waters.

2. Materials and Methods

2.1. Membranes

Commercial heterogeneous cation-exchange MK-40 membrane (Shchekinoazot, Pervomayskiy, Tula Region, Russia) is used as the substrate for preparing the modified membrane samples. The MK-40 membrane is produced by hot pressing of a mixture of polyethylene (inert binder) and powdered KU-2 ion-exchange resin (styrene-divinylbenzene sulfonated copolymer) and contain fixed sulfonic acid functional groups. Most of the surface (more than 80%) of the MK-40 membrane is non-conductive [45]. Grains of KU-2 ion-exchange resin, which form crests on the polyethylene surface, provide selective conductivity of the membrane.

The modified membranes were prepared by forming a thin homogeneous film on the MK-40 membrane surface by casting an LF-4SK solution (Plastpolymer, Saint Petersburg, Russia), containing a certain amount of titanium oxide particles. The LF-4SK is a solution of tetrafluoroethylene copolymer with sulfonated perfluorovinylether in isopropyl alcohol. Solvent evaporation gives a film identical to the MF-4SK membrane, a Russian analogue of the Nafion membrane. To prepare the modifier, the required amount of titanium butoxide (EKOS-1, Moscow, Russia) was added to the LF-4SK solution, then the modifier was placed in an ultrasonic bath PSB-2835-05 (Ultrasonic Equipment Center PSB-Gals, Moscow, Russia) to disperse the formed particles.

Modification of the membranes was carried out according to a method similar to that described in [46]. Before applying the modifier, the surface of the MK-40 substrate-membrane was roughened using a fine-grained abrasive sandpaper (P600). Application of the modifier solution on the membrane surface was carried out via drop casting. Preparing the MK-40_{21} sample, an LF-4SK solution containing no titanium oxide particles was used. The MK-40_{21} membrane was used as a reference sample to detect the effect of titanium oxide particles. The characteristics of the commercial and modified membranes are given in Table 1.

Table 1. Studied membranes and some characteristics.

Sample	Modifying Film Thickness[1], µm	Mass Fraction TiO_2 in the Film, wt%	Contact Angles, θ, Degrees (Swollen Membrane)
MK-40	–	–	55 ± 2
MK-40_{21}	21 ± 2	–	64 ± 3
MK-$40_{21+3\%}$	21 ± 2	3	61 ± 3
MK-$40_{21+6\%}$	21 ± 2	6	57 ± 2
MK-$40_{33+3\%}$	33 ± 2	3	60 ± 3
MK-$40_{33+6\%}$	33 ± 2	6	56 ± 2

[1] in swollen state

All membrane samples underwent standard salt pretreatment [47] before the measurements.

2.2. Methods

2.2.1. Voltammetry, Chronopotentiometry, and Impedance Spectroscopy

The electrochemical characteristics, the current-voltage curves, chronopotentiograms, and impedance spectra of the studied membranes were obtained using the experimental setup shown in Figure 1, its detailed description is presented in Supplementary Materials. The setup includes a laboratory four-compartment flow electrodialysis cell, described in detail in [45]. The compartments of the cell were formed by the studied cation-exchange membrane (C*) and auxiliary anion-exchange MA-41 (A) and cation-exchange MK-40 (C) membranes. Current was supplied to the cell and the potential drop was measured using an electrochemical complex Autolab PGSTAT-100 (EcoChemi, Utrecht, The Netherlands). The experimental procedure for obtaining ChP and CVC is detailed in [45].

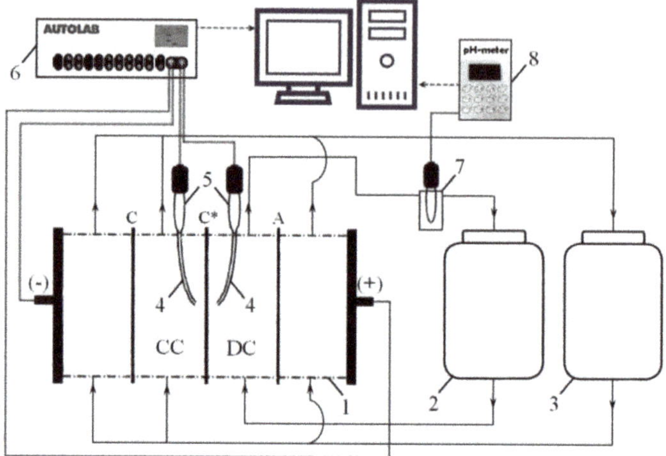

Figure 1. Principal scheme of the experimental setup: electrodialysis cell (1) consisting of one desalination compartment (DC), one concentration compartment (CC) and two electrode compartments; tanks with solutions (2, 3); Luggin's capillaries (4), connected with silver chloride electrodes (5); electrochemical complex Autolab PGSTAT-100 (6); flow pass cell with pH combination electrode (7), connected to pH-meter Expert 001 (8).

To measure the electrochemical impedance spectra, the closed silver chloride measuring electrodes (immersed in the saturated KCl solution) used in the ChP and CVC measurements were replaced by open silver chloride electrodes (immersed in 0.02 M NaCl solution). It takes about 2 h to get one spectrum. Before recording the complex impedance, the membrane was preliminarily held for 40 min at a given direct current density. Then alternating current was applied. The time required to achieve a quasistationary state at a given frequency was determined automatically by the Autolab PGSTAT-100 complex. Impedance spectrum recording starts at a low frequency equal to 0.003 Hz and ends at a frequency equal to 5×10^5 Hz. From the values of the real (ReZ) and imaginary (ImZ) components of the impedance of the membrane system obtained at a given frequency and at a given direct current density, the corresponding values (ReZ and ImZ) obtained at zero direct current ($i = 0$ A) were subtracted. This data processing allows one to exclude from consideration (1) the contribution of the measuring electrode system impedance to the studied membrane system impedance, (2) the ohmic resistance of the membrane and the response of the membrane system to the alternating current bias (the amplitude of alternating current is 0.25 mA). In preliminary experiments, it was shown that in the studied direct current range, polarization phenomena do not significantly affect the impedance of the measuring electrodes. The advantage of such processing of the impedance spectra is the reduction of errors, which occur as the result of electrochemical cell reassembling to determine the impedance of the measuring electrode system.

As a rule, the high-frequency arc [48,49] (in the range from 10^3 to 1.3×10^5 Hz) dominates the impedance spectrum obtained at zero direct current and presented in Nyquist plot. It characterizes the membrane and interphase boundaries [50,51]. The intercept by this arc on the ReZ axis corresponds to the ohmic resistance of the membranes R^{Ω} [48,49,52,53], and the frequency at the maximum of the ImZ, f_{max}, allows one to estimate the effective electric capacitance C of the membrane/solution interface by the formula [50]:

$$C = \frac{1}{2\pi f_{max}(-\mathrm{Im}Z_{max})} \tag{1}$$

where -ImZ_{max} is the maximum value of the imaginary impedance component at the high-frequency arc. This capacitance includes the capacitances of the EDLs on the membrane/solution boundaries, as well as the capacity caused by the asymmetry of the depleted and enriched diffusion boundary layers DBLs, which occurs when a direct electric current flows.

If the applied direct current is greater than zero, taking into account the experimental data processing mentioned above, the intercept by the high-frequency arc on the ReZ axis gives the ohmic resistance of the diffusion layers. The appearance of the Gerischer arc in the medium-frequency range of the spectrum (10–10^3 Hz) indicates the presence of an intense chemical reaction at the membrane/solution interface, namely, the generation of H^+ and OH^- ions, catalyzed by fixed groups of membranes [54]. The low-frequency (0.003–10 Hz) arc of the spectrum (finite-length Warburg impedance) characterizes the diffusion and electroconvective transfer of ions in DBL adjacent to a membrane. The DBL thickness can be found from difference in values of real component of the low-frequency arc for the lowest and the highest frequencies [55]. The ratio of the lengths of the intercepts by each of these arcs on the ReZ axis gives information on the contribution of each of the phenomena (the formation of concentration profiles near the membrane surface; the appearance and transport of new charge carriers at the membrane/solution interface; a decrease in the diffusion layer thickness due to electroconvection) to electrochemical characteristics of the studied membrane systems [35].

The experiments are conducted at a constant temperature of 25 ± 1 °C.

2.2.2. Diffusion Permeability and Conductivity Measurements

The concentration dependence of membrane conductivity is measured by the differential method using a clip-type cell [56,57] and a MOTECH MT4080 immittance meter at an AC frequency of 1 kHz. The measurements were carried out in NaCl solutions with different concentrations (0.1, 0.25, 0.5, 0.75, 1 M).

The experimental investigation of the concentration dependence of diffusion permeability was conducted at the same NaCl concentrations. The measurements were carried out using two-compartment cell and conductivity meter Expert 002 (Econix-Expert, Ltd., Moscow, Russia) [58].

The thickness of studied membranes was measured using a digital micrometer Schut Filetta (Schut Geometrical Metrology, Groningen, The Netherlands).

2.2.3. Contact Angles Measurements

The contact angles (θ) of the membranes under study were measured using the sessile drop technique [20] described in detail in Supplementary Materials. The studied membrane was in a swollen state, it was removed from the 0.02 M NaCl solution immediately before the measurements. The contact angles reported in this paper were registered 20 s after the applying a drop of distilled water on the membrane surface.

2.2.4. Surface Roughness Measurements

The surface roughness of the studied membranes was determined by using an optical microscope SOPTOP CX40M (Ningbo Sunny Instruments Co., Ltd., Yuyao, China). The surface of a membrane is shaped of so-called "hills" and "valleys", corresponding to the highest and lowest areas of the membrane relief. The difference between these two extreme forms of the membrane surface shape, parameter b (μm), characterizes the roughness and is measured by moving the objective and focusing on them, the focal length of the system being the same.

3. Results and Discussion

3.1. Substrate and Modified Membranes Characterization

The results of the contact angle measurements are presented in Table 1. Table 2 shows the values of the effective electric double layer capacity at the membrane/depleted solution interface, as well as the ohmic resistance of the membranes found from the analysis of the high-frequency arc of the electrochemical impedance spectrum.

As expected, perfluorosulfonic acid polymer casting onto the surface of the heterogeneous substrate MK-40 membrane leads to an increase in hydrophobicity of the surface that results in the higher value of contact angle (sample MK-40_{21}). This is attributed the less roughened surface of the modified membranes. Indeed, according to Wenzel [59], the effect of the surface roughness is to magnify its wetting properties: if the surface is hydrophilic, it will be more hydrophilic, the rougher the surface. Applying the PFSA layer results in flattening the surface of the modified membrane compared to the substrate membrane, which leads to lower hydrophilicity of the modified membrane surface and, consequently, the higher contact angle.

Adding the titanium dioxide nanoparticles into the modifying film (samples MK-$40_{21+3\%}$, MK-$40_{21+6\%}$, MK-$40_{33+3\%}$, and MK-$40_{33+6\%}$) results in increasing hydrophilicity of the membrane surface with increasing TiO_2 mass fraction due to hydrophilic properties of the particles [60] and as the result of the increase in its roughness [61].

Figure 2 shows that applying the PFSA layer results in flattening the surface of the modified membrane compared to the substrate membrane. Adding the titanium dioxide nanoparticles into the modifying film results in decreasing the flattening effect. The greater the fraction of TiO_2, the lesser the effect of the modifying film. However, all the modified membranes have the less roughened surface compared to the substrate membrane.

The application of a modifying film practically does not change the ohmic resistance of the studied membranes, but noticeably affects the effective electric capacitance of their interphase boundary (Table 2). This capacity increases slightly after depositing the PFSA polymer film (sample MK-40_{21}) and increases noticeably after inserting TiO_2 particles into this film (samples MK-$40_{21+3\%}$ and MK-$40_{21+6\%}$). Note that this increase is observed if the thickness of the modifying film remains almost constant. In the case where the thickness of the PFSA polymer film increases by 1.5 times (samples MK-$40_{33+3\%}$ and MK-$40_{33+6\%}$), the addition of TiO_2 nanoparticles to it practically does not lead to changes in the effective capacitance of the interphase boundary. It is known from studies in the field of electrode kinetics that the capacitance of the interphase boundary increases with increasing electric charge of the dense part of the electric double layer [50,62]. Apparently, coating the entire surface of the MK-40_{21} sample with an ion-conducting film leads to an increase in its electric charge in comparison with the substrate membrane (MK-40 sample), about 80% of which is occupied by inert polyethylene [9,28,46]. The introduction of 3 wt% particles into this film (sample MK-$40_{21+3\%}$) increases the charge of the membrane/solution interface. A further increase in the percentage of nanoparticles in the modifying film leads to an increase in the roughness of the modifying film/depleted solution interface, as well as a reorientation of the side chains of PFSA polymer with sulfone fixed groups towards charged nanoparticles. The result of this reorientation is "repulsion" to the surface of the main chains of the modifying material that do not have an electric charge. Similar phenomena are discussed in the works [63,64]. Apparently, in the case of using thicker films (samples MK-$40_{33+3\%}$ and MK-$40_{33+6\%}$), the reorientation of the side chains becomes crucial. Therefore, the introduction of nanoparticles does not lead to noticeable changes in the effective capacitance of the interphase boundary. To make sure our assumptions are correct, in the future we plan to conduct a direct measurement of the charge of the samples under study.

Table 2. Characteristic points of the high-frequency region of the studied membranes impedance spectra and the effective capacities of the electric double layer and the ohmic resistances of the membranes found from these points.

Sample	f_{max}, Hz	$-ImZ_{max}$, Ohm	R^{Ω}, Ohm	C, µF
MK-40	13,318	22.2	55 ± 2	0.5 ± 0.2
MK-40_{21}	10,809	21.7	54 ± 2	0.7 ± 0.2
MK-$40_{21+3\%}$	8506	19.0	50 ± 2	1.0 ± 0.2
MK-$40_{21+6\%}$	8506	20.4	49 ± 2	0.9 ± 0.2
MK-$40_{33+3\%}$	10,809	20.2	51 ± 2	0.7 ± 0.2
MK-$40_{33+6\%}$	10,809	21.8	52 ± 2	0.7 ± 0.2

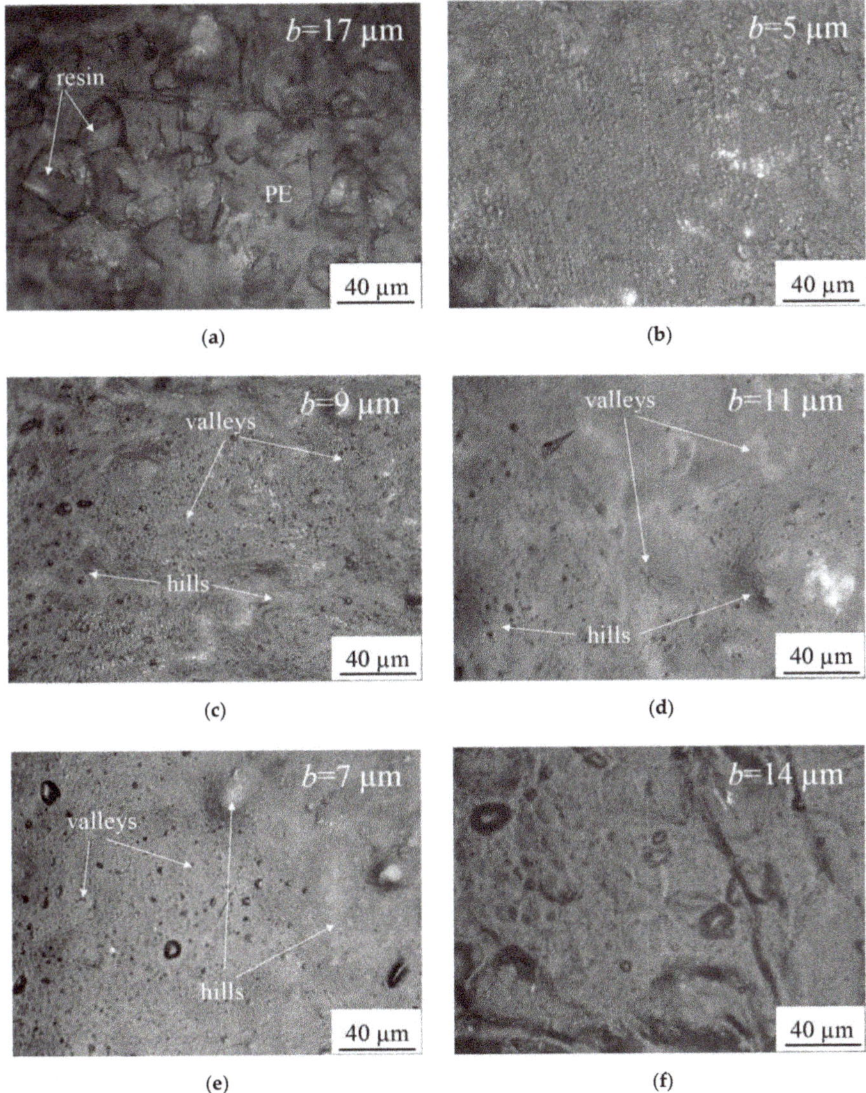

Figure 2. Optical images of the studied membranes: MK-40 (**a**), MK-40_{21} (**b**), MK-$40_{21+3\%}$ (**c**), MK-$40_{21+6\%}$ (**d**), MK-$40_{33+3\%}$ (**e**), MK-$40_{33+6\%}$ (**f**). Parameter b characterizes the roughness of the surface and shows the difference between the highest and lowest areas of the membrane relief.

An important parameter characterizing the membrane permselectivity is the counterion transport number, t_i^*. In the case of the studied cation-exchange membranes and sodium chloride solution, the value of t_+^* shows the portion of the electric current transferred by the Na^+ ion in conditions of the absence of diffusion. To evaluate the t_+^* value, it is possible to use the following relationship, [65]:

$$t_-^* = \frac{1}{2} - \sqrt{\frac{1}{4} - \frac{P^* F^2 c}{2RT\kappa^*}}, \qquad t_+^* = 1 - t_-^* \tag{2}$$

where t_-^* is the transport number of coion; P^* and κ^* are the diffusion permeability and the electrical conductivity of the membrane, respectively; c is the concentration of the electrolyte solution; F, R, and T have their usual meanings. Using the experimentally determined values of electrical conductivity and diffusion permeability the counterion transport numbers in the studied membranes are calculated and presented in Figure 3 as function of NaCl solution concentration. It follows from this dependence that the modification of the samples practically does not change the counterion transport numbers in comparison with the substrate membrane. Values of t_i^* remain high throughout the studied range of NaCl solution concentrations. The concentration dependencies of the conterion (Na^+) transport number are approximated and extrapolated to the region of lower concentration (0.02 M), which corresponds to the measurements of the electrochemical characteristics (CVC, ChP, and impedance spectra). As a result, the values of the conterion transport number in the studied membranes at this concentration are virtually the same and equal to 1.

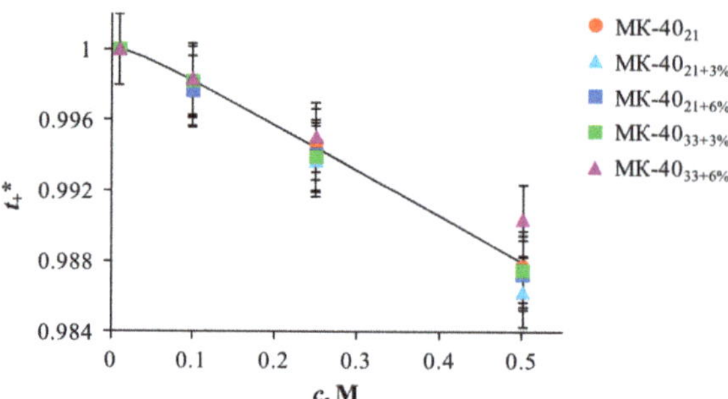

Figure 3. Concentration dependences of the conterion (Na^+) transport number in the studied membranes.

3.2. *Electroconvection and Water Splitting at the Surface of the Substrate Membrane and Modified Membranes*

3.2.1. Current-Voltage Curves

Figure 4 shows the ratio of the current density to its theoretical limiting value, $i/i_{\rm lim}^{\rm th}$ as well as the difference in the pH between the outlet (pH_{out}) and inlet (pH_{in}) solution passing through the desalination compartment (DC) of the electrodialysis cell, ΔpH ($\Delta pH = pH_{out} - pH_{in}$), as functions of reduced potential drop, $\Delta\varphi'$. This is a potential drop between the Luggin's capillaries, from which the ohmic component is subtracted. The exact definition of $\Delta\varphi'$ is given in the Supplementary Materials. The value $i_{\rm lim}^{\rm th}$ was calculated using the Leveque equation [66], which was derived in the framework of the convective-diffusion model under the assumption that the membrane has a smooth, homogeneous ion-conducting surface. In the conditions of our experiments, this calculation gives $i_{\rm lim}^{\rm th} = 2.01$ mA cm^{-2} (see Supplementary Materials).

Three sections can be distinguished on the CVC [12,67]: an initial section (I), a slope plateau (II), and a section of a sharp growth of current density (III). It is believed [13,25,35], that the increase in the experimental current over the theoretical value $i_{\rm lim}^{\rm th}$ is determined by the EC, which develops according to the mechanism of electroosmosis of the first kind. The slope of the plateau characterizes the increase in the depleted diffusion layer conductivity as compared to that achieved at $i_{\rm lim}^{\rm th}$ due to the development of equilibrium EC [12,19]. The plateau length (determined by the potential drop, at which section II of the CVC turns into section III) corresponds to the threshold value of the potential drop at which a nonequilibrium EC occurs [12,19]. These characteristic parameters found by graphical treatment of the CVC (Figure 4) are summarized in Table 3.

Before discussing the obtained results, we recall that the modification of the membranes did not lead to any noticeable change in their selectivity (Figure 3), which could affect the values of the limiting currents.

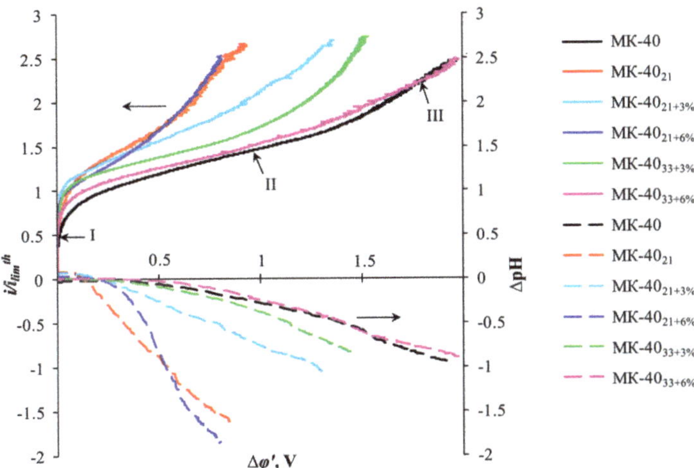

Figure 4. Current-voltage characteristics (solid lines) and the difference in pH between the outlet and inlet solution passing through the desalination compartment of the electrodialysis cell (dashed lines) of the pristine and modified membranes.

Table 3. Some parameters of current-voltage curves, current-voltage characteristics (CVCs).

Sample	i_{lim}^{exp}, mA cm^{-2}	$i_{lim}^{exp}/i_{lim}^{th}$	Plateau Length, V	Plateau Slope, mS cm^{-2}
MK-40	1.57	0.78	1.43	0.61
MK-40$_{21}$	1.81	0.90	0.60	1.58
MK-40$_{21+3\%}$	2.19	1.09	0.85	1.08
MK-40$_{21+6\%}$	1.97	0.98	0.50	1.33
MK-40$_{33+3\%}$	2.01	1.00	1.03	0.78
MK-40$_{33+6\%}$	1.71	0.85	1.25	0.62

From these data it follows that in the case of the MK-40 substrate membrane the value $i_{lim}^{exp}/i_{lim}^{th}$ is less than unity. This result is expected, because only 20% of the surface of this membrane conducts electric current. The increase in current above the value $i_{lim}^{exp}/i_{lim}^{th}$ = 0.20 (which would correspond to the fraction of the conductive surface) is due to the tangential transport of salt from non-conductive regions to conductive ones, as well as due to electroconvection. The latter is stimulated by the "funnel effect" described by Rubinstein et al. [68], that is, curvature of streamlines and their crowding at inclusions of ion-exchange resin particles in polyethylene on the MK-40 surface [20] (Figure 5a). The value of the potential drop (plateau length), at which the development of a nonequilibrium EC begins, is close to the theoretically predicted one in the articles by Rubinstein and Zaltsman [12].

As in previous works [20,28], the deposition of the PFSA polymer film is accompanied by an increase in the $i_{lim}^{exp}/i_{lim}^{th}$ ratio, a decrease in the plateau length, and an increase in the conductivity of the depleted diffusion layer. This is facilitated by a decrease in surface hydrophilicity (Table 1), as well as by an increase in ion-conducting gates through which streamlines enter the ion-exchange membrane. Inserting 3 wt% TiO_2 into the PFSA polymer film (sample MK-40$_{21+3\%}$) stimulates the development of all types of EC. Indeed, the $i_{lim}^{exp}/i_{lim}^{th}$ ratio, which is an indicator of the contribution of electroosmosis of the first kind to the increase in near-surface salt concentration, increases (Table 3). The plateau length, that is, the threshold potential drop at which a nonequilibrium EC occurs, is reduced in comparison with the MK-40 and MK-40$_{21}$ samples. Apparently, this is facilitated by an increase in the electric charge while maintaining a rather high degree of hydrophobicity of the MK-40$_{21+3\%}$ sample surface (see Section 3.1).

Note that even a slight increase in surface hydrophilicity while maintaining almost the same value of the effective capacitance of the interphase boundary of the modifying film/depleted solution (see Section 3.1) led to a decrease in the $i_{\text{lim}}^{\text{exp}}/i_{\text{lim}}^{\text{th}}$ value in the case of MK-$40_{21+6\%}$ sample as compared to MK-$40_{21+3\%}$. Nevertheless, in the case of MK-$40_{21+6\%}$ sample, the plateau length turned out to be minimal, and the plateau inclination turned out to be maximum among all the studied membranes. It can be assumed that the ratio of surface characteristics achieved in this sample turned out to be the most optimal for facilitating the development of nonequilibrium EC. Which of these characteristics became decisive for the stimulation of this type of EC is to be determined in subsequent experiments.

MK-$40_{33+3\%}$ and MK-$40_{33+6\%}$ samples, which were modified with a thicker PFSA polymer film, demonstrated a more modest ability to stimulate EC in comparison with not only MK-$40_{21+3\%}$ and MK-$40_{21+6\%}$ samples, but also with a sample that does not contain TiO_2 particles. The reasons for such a behavior of the MK-$40_{33+3\%}$ and MK-$40_{33+6\%}$ samples were already discussed in Section 3.1. In addition, it can be assumed that at a thickness of the modifying film of about 20 μm (MK-40_{21}, MK-$40_{21+3\%}$, MK-$40_{21+6\%}$ samples), the streamlines remain sufficiently curved to stimulate the increase in the tangential component of the electric force at the modifying film/depleted diffusion layer (Figure 5b). A 1.5-fold increase in the thickness of this film (MK-$40_{33+3\%}$ and MK-$40_{33+6\%}$ samples) apparently weakens the curvature of streamlines in the near-membrane layer of the solution, thereby reducing the tangential component of the electric force (Figure 5c).

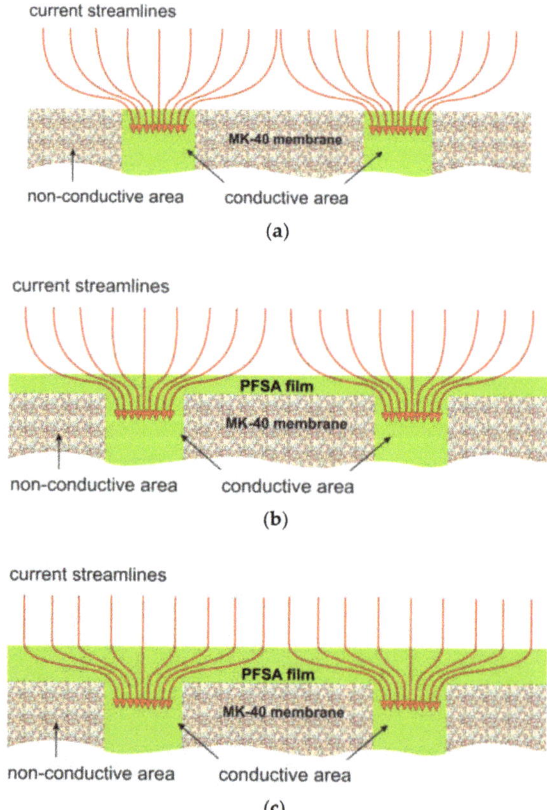

Figure 5. Schematic representation of streamlines near the surface of the MK-40 (**a**) membrane and that, modified by PFSA polymer film of various thicknesses: about 20 μm (**b**) and about 30 μm (**c**).

3.2.2. Chronopotentiometric Curves

Chronopotentiometry data provide additional details on the effect of modification on the degree of the membrane surface heterogeneity and on the mechanisms of EC development in the studied membrane systems.

Figure 6 shows the chronopotentiometric curves, ChPs, of the studied membranes obtained at $i/i_{\text{lim}}^{\text{th}} = 2$. When plotting ChP, the reduced potential drop, $\Delta\varphi'$, [69] is used (see the Supplementary Materials). Table 4 summarizes the transition times, τ_{exp}, for various $i/i_{\text{lim}}^{\text{th}}$. They were determined by the inflection points on the initial sections of the ChPs (indicated by points in Figure 6). According to the classical Sand theory [70] and later studies [71], the inflection point on the ChP corresponds to the state of the system where the electrolyte concentration at the membrane surface decreases to minimum values, and the new mechanism of electric charge transfer begins to play a dominant role. These can be large electroconvection vortices that develop according to the mechanism of nonequilibrium electroconvection and deliver a more concentrated electrolyte solution to the membrane surface, and/or the water splitting, which is a source of new charge carriers (H^+/OH^-) ions [72]. A number of recent studies have shown that equilibrium electroconvection, which develops according to the mechanism of electroosmosis of the first kind, contributes to an increase in the transition times [25,35]. Small EC vortices formed by this mechanism cannot completely obviate the diffusion limitations of electrolyte delivery, which increase with increasing time from the moment the current is turned on. However, the delivery of a slightly more concentrated solution from the depleted diffusion layer shifts the moment of reaching the minimum surface concentration to greater times.

From the obtained data (Figure 6, Table 4) it follows that any modification of the MK-40 substrate membrane contributes to an increase in the transition times. The greatest differences in the behavior of the modified samples are observed in the current range $1.25 < i/i_{\text{lim}}^{\text{th}} < 2.5$. Moreover, the best ability to stimulate equilibrium EC, which is expressed in the prolongation of transition times, demonstrate MK-40$_{21+3\%}$ and MK-40$_{33+3\%}$ samples. These results are consistent with voltammetry data (Section 3.2.1). Apparently, at too high currents, the values of the transition times are reached so quickly that the equilibrium EC does not have time to fully develop.

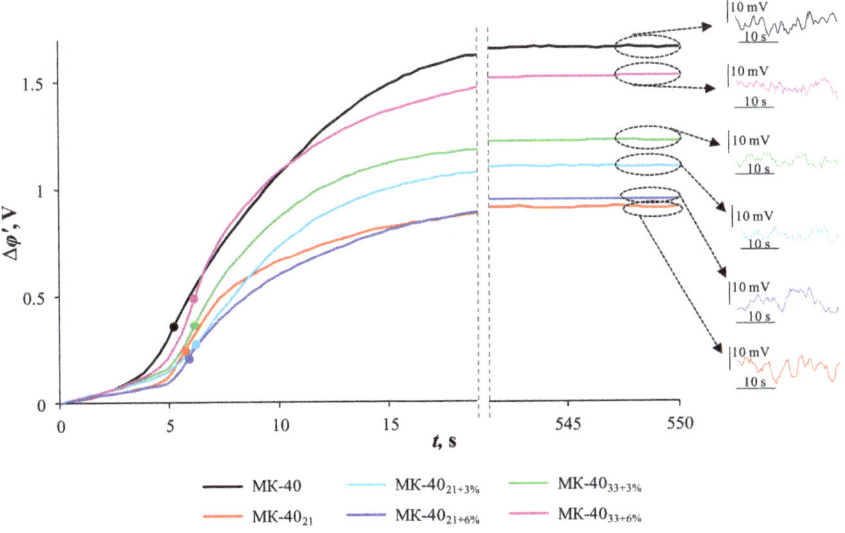

Figure 6. Chronopotentiometric curves of the studied membranes measured at $i/i_{\text{lim}}^{\text{th}} = 2$. The circles show the inflection points related to τ_{exp}.

Table 4. Values of the experimental transition time, τ_{exp}, at various i/i_{lim}^{th} for the studied membranes.

i/i_{lim}^{th}	MK-40	MK-40_{21}	MK-$40_{21+3\%}$	MK-$40_{21+6\%}$	MK-$40_{33+3\%}$	MK-$40_{33+6\%}$
1.25	15.2 ± 0.2	18.6 ± 0.2	18.4 ± 0.2	17.5 ± 0.2	17.7 ± 0.2	17.6 ± 0.2
1.5	9.2 ± 0.2	10.5 ± 0.2	12.1 ± 0.2	9.9 ± 0.2	11.8 ± 0.2	11.7 ± 0.2
2.0	5.2 ± 0.2	5.7 ± 0.2	6.2 ± 0.2	5.9 ± 0.2	6.1 ± 0.2	6.1 ± 0.2
2.5	2.9 ± 0.2	3.6 ± 0.2	3.8 ± 0.2	3.7 ± 0.2	3.8 ± 0.2	3.8 ± 0.2

As already mentioned in Section 1, in the case of a nonequilibrium EC, large vortices form clusters. Under conditions of forced convection of the solution, these clusters move along the surface of the membranes under the measuring capillaries, which is expressed in the oscillations of the recorded potential drop. We registered such oscillations for all studied membranes (Figure 6). The largest amplitude of oscillations, which indicates the development of larger EC cluster structures, is demonstrated by MK-40_{21} and MK-$40_{21+6\%}$ samples. This result also confirms the conclusions made by us after analyzing the CVCs.

It should be noted that the chronopotentiometry data allow one to assess the degree of membrane surface electrical heterogeneity: the higher this heterogeneity, the faster the potential drop increases in the first few seconds after turning on the current [71]. From the obtained data (Figure 6), it follows that the surface of the MK-40_{21} and MK-$40_{21+6\%}$ samples is less electrically heterogeneous in comparison with other studied membranes. This experimental fact allows one to conclude that the high ability of the MK-$40_{21+6\%}$ sample to stimulate EC is mainly caused by an increase in the electric charge of the modifying film/solution interface rather than by an increase in the surface roughness after introducing 6 %wt TiO_2 into the PFSA polymer film.

3.2.3. Impedance Spectra under Direct Current Condition and the pH Difference at the Outlet and Inlet of the Desalination Channel of the Electrodialysis Cell

Figures 7 and 8 show the electrochemical impedance spectra obtained for the studied membranes at 1.5 and 1.8 i/i_{lim}^{th}. All these spectra consist of two arcs. The first high-frequency arc is characterized by frequencies at the point of maximum along the ordinate axis from 8.5 to 17.5 kHz. Similar characteristic frequencies for this arc (5–11 kHz) are demonstrated by the homogeneous cation-exchange membrane CMX (Astom, Tokyo, Japan), studied under similar conditions [73].

The length of the high-frequency arc chord corresponds to the resistance of the membrane system, R_I (in our case, mainly a depleted diffusion layer), which depends on the degree of polarization caused by direct current [48,49]. From the obtained data, it follows that the maximum value of such resistance at a given direct current is achieved in the case of a substrate membrane MK-40. Any modification of its surface leads to a decrease in R_I. The values decrease in the same series in which the ability of membranes to stimulate EC increases: MK-40 > MK-$40_{33+6\%}$ ≥ MK-$40_{33+3\%}$ > MK-$40_{21+3\%}$ > MK-$40_{21+6\%}$ ≈ MK-40_{21}. The appearance of the Warburg impedance arc is caused by the time delay of changes in the electrolyte concentration in the quasi-electroneutral layer of the solution adjacent to the membrane in response to the alternating current bias [49,74,75]. According to the classical assumptions [49,75], the reason for this delay (and the delay caused by it in changes in the potential drop) is the finite rate of the electrolyte diffusion. According to modern concepts, this rate can change in the presence of electroconvective vortices at the membrane surface. Their appearance is accompanied by distortion in the regular shape of the arc. The more intensive electroconvective mixing, the stronger the spread of points in this region of the spectrum. The characteristic value of the frequency at the maximum point of the Warburg arc is 10 mHz [35,73]. Our results confirm that the second arc corresponds to Warburg impedance. Moreover, the effect of electroconvection on this arc (expressed in an increasing spread of points), increases with increasing density of the overlimiting current. This effect is characteristic for all the studied samples. Apparently, the electroconvective mixing provides a sufficiently high salt concentration at the membrane/solution interface. As a result, even at sufficiently high current densities (Figures 7 and 8), the Gerischer arc, which appears in the case of a homogeneous CMX membrane with the same sulfonic fixed groups at $i/i_{\lim}^{\mathrm{th}}$ = 1.6, cannot be identified on the impedance spectra. As mentioned earlier, the Gerischer impedance [54] corresponds to the resistance caused by a chemical reaction (in our case, water splitting) at the membrane surface/depleted diffusion layer interface. In the case of the CMX membrane, the characteristic frequencies corresponding to the maximum points of this arc have values of about 4 Hz. In the considered systems, these values are situated in the region of the spectrum where the second arc is bordering the high-frequency arc. Note that this region is characterized by the absence of spread of points and does not represent a straight line passing at a 45° angle to the ReZ axis (such characteristics should have the initial part of the finite-length Warburg impedance [50]). Apparently, the deviation of this part of the spectrum from the classical form is caused by the influence of water splitting. An indirect evidence of this reaction, at least at high current densities, is the fact that the ReZ takes a negative value, recorded in the case of MK-40 ($i/i_{\lim}^{\mathrm{th}}$ = 1.8) in the frequency range 35–500 kHz. A similar shape of spectra was observed in the case of water splitting for many membrane systems [35,73]. However, the rate of this reaction in studied systems is so small that the Gerischer arch does not appear.

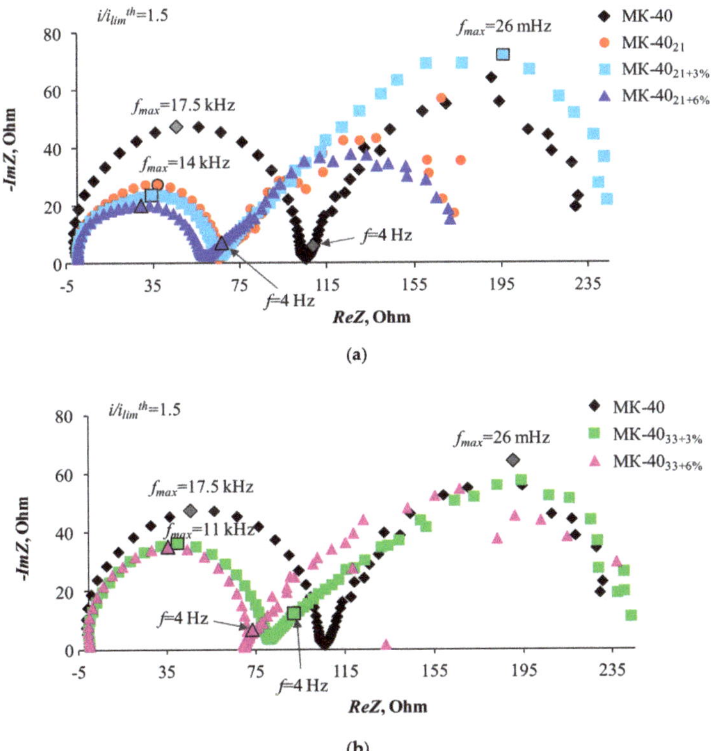

Figure 7. Electrochemical impedance spectra of the MK-40 as well as MK-40$_{21}$, MK-40$_{21+3\%}$, MK-40$_{21+6\%}$ (**a**) and MK-40$_{33+3\%}$, MK-40$_{33+6\%}$ (**b**) samples at $i/i_{\lim}^{\mathrm{th}}$ = 1.5.

Nevertheless, this water splitting at the cation-exchange membrane/solution interface, affects the pH difference at the outlet and inlet of the desalination channel, which is formed by the studied cation-exchange membranes and the MA-41 anion exchange membrane. This follows from Figure 4 that shows the difference in the pH between the outlet and the inlet solution passing through the desalination compartment of the electrodialysis cell, ΔpH, as functions of reduced potential drop, $\Delta\varphi'$.

The MA-41 membrane contains a certain amount of secondary and tertiary amines, whose catalytic activity with respect to the water dissociation reaction is quite high [76,77]. The protons generated on the surface of this membrane acidify the desalinated solution. As a result, ΔpH of the solution is less than zero. In the case where the water splitting also occurs on the surface of the cation-exchange membrane, hydroxyl ions enter the solution. As a result, ΔpH of the solution should tend to zero or even positive values. The closer the ΔpH of the solution to zero, the more intensive the water splitting on the surface of the cation-exchange membrane. Indeed, in intensive current modes, the water splitting rate on the MK-$40_{33+6\%}$ surface is close to that observed on the MK-40 substrate membrane. The lowest water splitting rate occurs in the case of MK-40_{21} and MK-$40_{21+6\%}$ samples. As already mentioned, the cause of the decrease in water splitting is electroconvection, which shifts intensive water splitting to the range of higher potential drops.

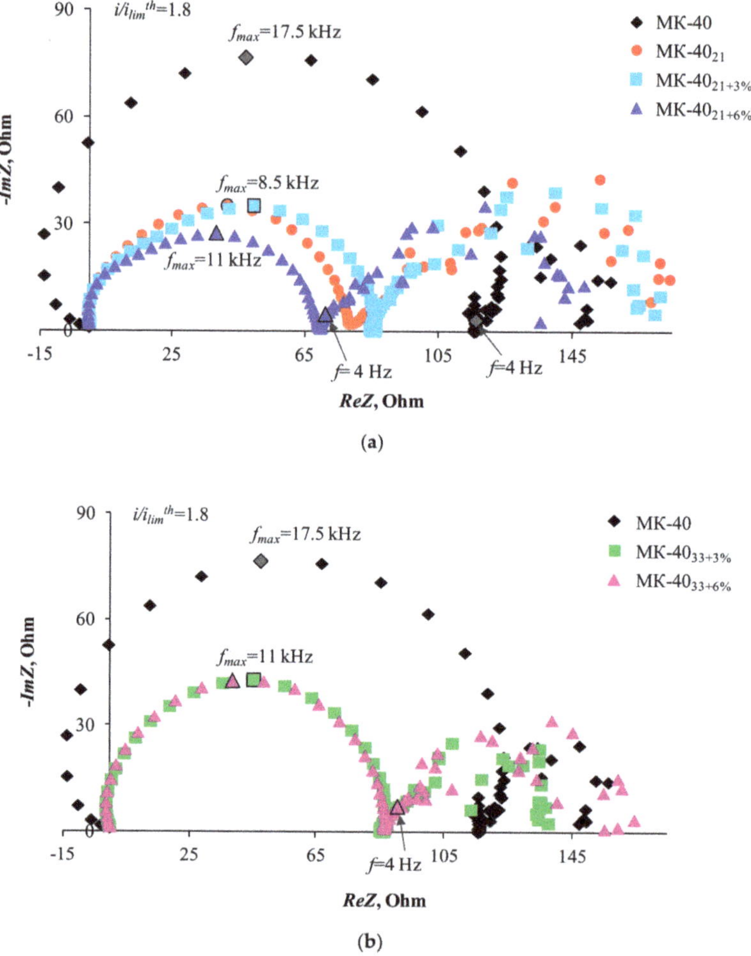

Figure 8. Electrochemical impedance spectra of the MK-40 as well as MK-40_{21}, MK-$40_{21+3\%}$, MK-$40_{21+6\%}$ (**a**) and MK-$40_{33+3\%}$, MK-$40_{33+6\%}$ (**b**) samples at $i/i_{\mathrm{lim}}^{\mathrm{th}} = 1.8$.

4. Conclusions

The surface of the MK-40 heterogeneous cation-exchange membrane was modified with a Nafion-type PFSA polymer film containing TiO_2 nanoparticles.

It was shown that such modification does not affect the selectivity of the obtained samples. All of them, like the MK-40 substrate membrane, are characterized by counterion transport numbers (Na^+) close to 1. Introduction of TiO_2 nanoparticles (from 3 to 6 wt%) into the modifying film leads to an increase in surface charge while maintaining a sufficiently low degree of its hydrophilicity.

These changes in surface characteristics result in an increase of the experimental limiting current density on 9–39% due to the stimulation of electroconvection, developing by the mechanism of electroosmosis of the first kind. The greatest increase in current as compared to the substrate membrane can be obtained at a particle concentration of 3 wt%. A sample containing 6 wt% provides greater mass transfer in overlimiting current modes due to the development of nonequilibrium electroconvection. Electroconvective mixing of the solution leads to an increase in the near-surface salt concentration as compared to the substrate membrane. As a result, water splitting on the surface of modified samples exerts an extremely insignificant effect on their mass transfer characteristics in intensive current modes.

The above results are achieved with a thickness of the modifying film equal to 20 μm. A 1.5-fold increase in the thickness of this film reduces the positive effect of the introduction of TiO_2 nanoparticles due to (1) partial shielding of the nanoparticles on the surface of the modified membrane; (2) a decrease in the tangential component of the electric force, which affects the development of electroconvection.

The tested relatively simple method of surface modification of commercial membranes can be used to increase mass transfer and increase their resistivity to sedimentation and fouling in the processes of electrodialysis processing of natural water and liquid media of the food industry.

Supplementary Materials: Supplementary materials can be found at http://www.mdpi.com/2077-0375/10/6/125/s1. Figure S1: Principal scheme of the experimental setup: electrodialysis cell (1) consisting of one desalination compartment (DC), one concentration compartment (CC) and two electrode compartments; tanks with solutions (2, 3); Luggin's capillaries (4), connected with silver chloride electrodes (5); electrochemical complex Autolab PGSTAT-100 (6); flow pass cell with pH combination electrode (7), connected to pH-meter Expert 001 (8). Figure S2: The experimental setup for contact angles measurements: dispenser of the test liquid (1); closed optically transparent box (2); camera with sufficient magnification (3); drop of the test liquid (4); test sample (5); filter paper (6); layer of the test liquid (7); sample stand (8).

Author Contributions: Conceptualization, N.P. and V.G.; Methodology, N.P., V.G. and M.P.; Formal Analysis, N.P., V.G. and M.P.; Investigation, M.P., O.R. and D.B.; Writing—Original Draft Preparation, V.G. and M.P.; Writing—Review and Editing, V.G. and N.P.; Experimentation, M.P., O.R. and D.B.; Supervision, N.P.; Project Administration, V.G.; Funding Acquisition, V.G. All authors approved the final article. All authors have read and agreed to the published version of the manuscript.

Funding: This study was realized with the financial support of the Russian Science Foundation, Grant no. 19-79-00347.

Acknowledgments: We are grateful to V. Konshin for his valuable advice about the methods of membrane modification.

Conflicts of Interest: The authors declare no conflict of interest.

References

1. Barros, K.S.; Scarazzato, T.; Pérez-Herranz, V.; Espinosa, D.C.R. Treatment of Cyanide-Free Wastewater from Brass Electrodeposition with EDTA by Electrodialysis: Evaluation of Underlimiting and Overlimiting Operations. *Membranes (Basel)* **2020**, *10*, 69. [CrossRef] [PubMed]
2. Shi, L.; Hu, Z.; Simplicio, W.S.; Qiu, S.; Xiao, L.; Harhen, B.; Zhan, X. Antibiotics in nutrient recovery from pig manure via electrodialysis reversal: Sorption and migration associated with membrane fouling. *J. Memb. Sci.* **2020**. [CrossRef]
3. Mikhaylin, S.; Bazinet, L. Fouling on ion-exchange membranes: Classification, characterization and strategies of prevention and control. *Adv. Colloid Interface Sci.* **2016**, *229*, 34–56. [CrossRef] [PubMed]
4. Ghalloussi, R.; Chaabane, L.; Larchet, C.; Dammak, L.; Grande, D. Structural and physicochemical investigation of ageing of ion-exchange membranes in electrodialysis for food industry. *Sep. Purif. Technol.* **2014**, *123*, 229–234. [CrossRef]

5. Shi, L.; Xie, S.; Hu, Z.; Wu, G.; Morrison, L.; Croot, P.; Hu, H.; Zhan, X. Nutrient recovery from pig manure digestate using electrodialysis reversal: Membrane fouling and feasibility of long-term operation. *J. Memb. Sci.* **2019**. [CrossRef]
6. Garcia-Vasquez, W.; Dammak, L.; Larchet, C.; Nikonenko, V.; Grande, D. Effects of acid–base cleaning procedure on structure and properties of anion-exchange membranes used in electrodialysis. *J. Memb. Sci.* **2016**, *507*, 12–23. [CrossRef]
7. Mikhaylin, S.; Nikonenko, V.; Pourcelly, G.; Bazinet, L. Intensification of demineralization process and decrease in scaling by application of pulsed electric field with short pulse/pause conditions. *J. Memb. Sci.* **2014**, *468*, 389–399. [CrossRef]
8. Dufton, G.; Mikhaylin, S.; Gaaloul, S.; Bazinet, L. Positive Impact of Pulsed Electric Field on Lactic Acid Removal, Demineralization and Membrane Scaling during Acid Whey Electrodialysis. *Int. J. Mol. Sci.* **2019**, *20*, 797. [CrossRef]
9. Andreeva, M.A.; Gil, V.V.; Pismenskaya, N.D.; Dammak, L.; Kononenko, N.A.; Larchet, C.; Grande, D.; Nikonenko, V.V. Mitigation of membrane scaling in electrodialysis by electroconvection enhancement, pH adjustment and pulsed electric field application. *J. Memb. Sci.* **2018**, *549*, 129–140. [CrossRef]
10. Dukhin, S.S.; Mishchuk, N.A.; Takhistov, P.V. Electroosmosis of the second kind and unrestricted current increase in the mixed monolayer of an ion-exchanger. *Colloid J. USSR* **1989**, *51*, 540–542.
11. Mishchuk, N.A. Concentration polarization of interface and non-linear electrokinetic phenomena. *Adv. Colloid Interface Sci.* **2010**, *160*, 16–39. [CrossRef] [PubMed]
12. Rubinstein, I.; Zaltzman, B. Electro-osmotically induced convection at a permselective membrane. *Phys. Rev. E* **2000**, *62*, 2238–2251. [CrossRef] [PubMed]
13. Rubinstein, I.; Zaltzman, B. Equilibrium electroconvective instability. *Phys. Rev. Lett.* **2015**. [CrossRef] [PubMed]
14. Uzdenova, A.M.; Kovalenko, A.V.; Urtenov, M.K.; Nikonenko, V.V. Effect of electroconvection during pulsed electric field electrodialysis. Numerical experiments. *Electrochem. Commun.* **2015**, *51*, 1–5. [CrossRef]
15. Urtenov, M.K.; Uzdenova, A.M.; Kovalenko, A.V.; Nikonenko, V.V.; Pismenskaya, N.D.; Vasil'eva, V.I.; Sistat, P.; Pourcelly, G. Basic mathematical model of overlimiting transfer enhanced by electroconvection in flow-through electrodialysis membrane cells. *J. Memb. Sci.* **2013**, *447*, 190–202. [CrossRef]
16. Pham, V.S.; Li, Z.; Lim, K.M.; White, J.K.; Han, J. Direct numerical simulation of electroconvective instability and hysteretic current-voltage response of a permselective membrane. *Phys. Rev. E* **2012**, *86*, 046310. [CrossRef]
17. Demekhin, E.A.; Nikitin, N.V.; Shelistov, V.S. Direct numerical simulation of electrokinetic instability and transition to chaotic motion. *Phys. Fluids* **2013**. [CrossRef]
18. De Valença, J.C.; Wagterveld, R.M.; Lammertink, R.G.H.; Tsai, P.A. Dynamics of microvortices induced by ion concentration polarization. *Phys. Rev. E—Stat. Nonlinear Soft Matter Phys.* **2015**. [CrossRef]
19. Nikonenko, V.V.; Mareev, S.A.; Pis'menskaya, N.D.; Uzdenova, A.M.; Kovalenko, A.V.; Urtenov, M.K.; Pourcelly, G. Effect of electroconvection and its use in intensifying the mass transfer in electrodialysis (Review). *Russ. J. Electrochem.* **2017**, *53*, 1122–1144. [CrossRef]
20. Belashova, E.D.; Melnik, N.A.; Pismenskaya, N.D.; Shevtsova, K.A.; Nebavsky, A.V.; Lebedev, K.A.; Nikonenko, V.V. Overlimiting mass transfer through cation-exchange membranes modified by Nafion film and carbon nanotubes. *Electrochim. Acta* **2012**. [CrossRef]
21. Shelistov, V.S.; Demekhin, E.A.; Ganchenko, G.S. Electrokinetic instability near charge-selective hydrophobic surfaces. *Phys. Rev. E* **2014**, *90*, 013001. [CrossRef] [PubMed]
22. Vinogradova, O.I.; Dubov, A.L. Superhydrophobic Textures for Microfluidics. *Mendeleev Commun.* **2012**, *22*, 229–236. [CrossRef]
23. Levich, V.G.; Tobias, C.W. Physicochemical Hydrodynamics. *J. Electrochem. Soc.* **1963**, *110*, 251C. [CrossRef]
24. Nebavskaya, K.A.; Sarapulova, V.V.; Sabbatovskiy, K.G.; Sobolev, V.D.; Pismenskaya, N.D.; Sistat, P.; Cretin, M.; Nikonenko, V.V. Impact of ion exchange membrane surface charge and hydrophobicity on electroconvection at underlimiting and overlimiting currents. *J. Memb. Sci.* **2017**, *523*, 36–44. [CrossRef]
25. Korzhova, E.; Pismenskaya, N.; Lopatin, D.; Baranov, O.; Dammak, L.; Nikonenko, V. Effect of surface hydrophobization on chronopotentiometric behavior of an AMX anion-exchange membrane at overlimiting currents. *J. Memb. Sci.* **2016**, *500*, 161–170. [CrossRef]
26. Karatay, E.; Andersen, M.B.; Wessling, M.; Mani, A. Coupling between Buoyancy Forces and Electroconvective Instability near Ion-Selective Surfaces. *Phys. Rev. Lett.* **2016**. [CrossRef]

27. Rubinstein, S.M.; Manukyan, G.; Staicu, A.; Rubinstein, I.; Zaltzman, B.; Lammertink, R.G.H.; Mugele, F.; Wessling, M. Direct observation of a nonequilibrium electro-osmotic instability. *Phys. Rev. Lett.* **2008**, *101*. [CrossRef]
28. Gil, V.V.; Andreeva, M.A.; Jansezian, L.; Han, J.; Pismenskaya, N.D.; Nikonenko, V.V.; Larchet, C.; Dammak, L. Impact of heterogeneous cation-exchange membrane surface modification on chronopotentiometric and current–voltage characteristics in NaCl, CaCl 2 and MgCl 2 solutions. *Electrochim. Acta* **2018**, *281*, 472–485. [CrossRef]
29. Roghmans, F.; Evdochenko, E.; Stockmeier, F.; Schneider, S.; Smailji, A.; Tiwari, R.; Mikosch, A.; Karatay, E.; Kühne, A.; Walther, A.; et al. 2D Patterned Ion-Exchange Membranes Induce Electroconvection. *Adv. Mater. Interfaces* **2019**, *6*, 1801309. [CrossRef]
30. Davidson, S.M.; Wessling, M.; Mani, A. On the Dynamical Regimes of Pattern-Accelerated Electroconvection. *Sci. Rep.* **2016**, *6*, 22505. [CrossRef]
31. Benneker, A.M.; Gumuscu, B.; Derckx, E.G.H.; Lammertink, R.G.H.; Eijkel, J.C.T.; Wood, J.A. Enhanced ion transport using geometrically structured charge selective interfaces. *Lab Chip* **2018**, *18*, 1652–1660. [CrossRef] [PubMed]
32. Pawlowski, S.; Crespo, J.G.; Velizarov, S. Profiled ion exchange membranes: A comprehensible review. *Int. J. Mol. Sci.* **2019**, *20*, 165. [CrossRef] [PubMed]
33. Vasil'eva, V.I.; Akberova, E.M.; Zabolotsky, V.I.; Novak, L.; Kostylev, D.V. Effect of Dispersity of a Sulfonated Cation-Exchanger on the Current—Voltage Characteristics of Heterogeneous Membranes Ralex CM Pes. *Pet. Chem.* **2018**. [CrossRef]
34. Akberova, E.M.; Vasil'eva, V.I.; Zabolotsky, V.I.; Novak, L. Effect of the sulfocation-exchanger dispersity on the surface morphology, microrelief of heterogeneous membranes and development of electroconvection in intense current modes. *J. Memb. Sci.* **2018**. [CrossRef]
35. Pismenskaya, N.D.; Pokhidnia, E.V.; Pourcelly, G.; Nikonenko, V.V. Can the electrochemical performance of heterogeneous ion-exchange membranes be better than that of homogeneous membranes? *J. Memb. Sci.* **2018**, *566*, 54–68. [CrossRef]
36. Laberty-Robert, C.; Vallé, K.; Pereira, F.; Sanchez, C. Design and properties of functional hybrid organic–inorganic membranes for fuel cells. *Chem. Soc. Rev.* **2011**, *40*, 961. [CrossRef]
37. Jalani, N.H.; Dunn, K.; Datta, R. Synthesis and characterization of Nafion®-MO2 (M = Zr, Si, Ti) nanocomposite membranes for higher temperature PEM fuel cells. *Electrochim. Acta* **2005**, *51*, 553–560. [CrossRef]
38. Adjemian, K.T.; Lee, S.J.; Srinivasan, S.; Benziger, J.; Bocarsly, A.B. Silicon Oxide Nafion Composite Membranes for Proton-Exchange Membrane Fuel Cell Operation at 80–140°C. *J. Electrochem. Soc.* **2002**. [CrossRef]
39. Miyake, N.; Wainright, J.S.; Savinell, R.F. Evaluation of a Sol-Gel Derived Nafion/Silica Hybrid Membrane for Proton Electrolyte Membrane Fuel Cell Applications: I. Proton Conductivity and Water Content. *J. Electrochem. Soc.* **2001**. [CrossRef]
40. Safronova, E.Y.; Prikhno, I.A.; Yurkov, G.Y.; Yaroslavtsev, A.B. Nanocomposite membrane materials based on nafion and cesium acid salt of phosphotungstic heteropolyacid. *Chem. Eng. Trans.* **2015**, *43*, 679–684. [CrossRef]
41. Nemati, M.; Hosseini, S.M.; Bagheripour, E.; Madaeni, S.S. Electrodialysis heterogeneous anion exchange membranes filled with TiO2 nanoparticles: Membranes' fabrication and characterization. *J. Membr. Sci. Res.* **2015**, *1*, 135–140. [CrossRef]
42. Li, X.; Li, J.; Van Der Bruggen, B.; Sun, X.; Shen, J.; Han, W.; Wang, L. Fouling behavior of polyethersulfone ultrafiltration membranes functionalized with sol-gel formed ZnO nanoparticles. *RSC Adv.* **2015**. [CrossRef]
43. Tiraferri, A.; Kang, Y.; Giannelis, E.P.; Elimelech, M. Highly Hydrophilic Thin-Film Composite Forward Osmosis Membranes Functionalized with Surface-Tailored Nanoparticles. *ACS Appl. Mater. Interfaces* **2012**, *4*, 5044–5053. [CrossRef] [PubMed]
44. Wang, X.; Chen, G.Q.; Zhang, W.; Deng, H. Surface-modified anion exchange membranes with self-cleaning ability and enhanced antifouling properties. *J. Taiwan Inst. Chem. Eng.* **2019**. [CrossRef]
45. Volodina, E.; Pismenskaya, N.; Nikonenko, V.; Larchet, C.; Pourcelly, G. Ion transfer across ion-exchange membranes with homogeneous and heterogeneous surfaces. *J. Colloid Interface Sci.* **2005**, *285*, 247–258. [CrossRef]
46. Pismenskaya, N.; Melnik, N.; Nevakshenova, E.; Nebavskaya, K.; Nikonenko, V. Enhancing Ion Transfer in Overlimiting Electrodialysis of Dilute Solutions by Modifying the Surface of Heterogeneous Ion-Exchange Membranes. *Int. J. Chem. Eng.* **2012**, *2012*, 1–11. [CrossRef]

47. Berezina, N.P.; Kononenko, N.A.; Dyomina, O.A.; Gnusin, N.P. Characterization of ion-exchange membrane materials: Properties vs structure. *Adv. Colloid Interface Sci.* **2008**, *139*, 3–28. [CrossRef]
48. Moya, A.A. Electrochemical Impedance of Ion-Exchange Membranes with Interfacial Charge Transfer Resistances. *J. Phys. Chem. C* **2016**. [CrossRef]
49. Moya, A.A.; Moleón, J.A. Study of the electrical properties of bi-layer ion-exchange membrane systems. *J. Electroanal. Chem.* **2010**. [CrossRef]
50. Barsoukov, E.; Macdonald, J.R. Chapter 4 Applications of Impedance Spectroscopy. In *Impedance Spectroscopy: Theory, Experiment, and Applications*; Barsoukov, E., Macdonald, J.R., Eds.; John Wiley & Sons, Inc.: Hoboken, NJ, USA, 2005; ISBN 9780471716242.
51. Nikonenko, V.V.; Kozmai, A.E. Electrical equivalent circuit of an ion-exchange membrane system. *Electrochim. Acta* **2011**. [CrossRef]
52. Vorotyntsev, M.A.; Badiali, J.P.; Inzelt, G. Electrochemical impedance spectroscopy of thin films with two mobile charge carriers: Effects of the interfacial charging. *J. Electroanal. Chem.* **1999**. [CrossRef]
53. Vorotyntsev, M.A. Impedance of thin films with two mobile charge carriers. Interfacial exchange of both species with adjacent media. Effect of the double layer charges. *Electrochim. Acta* **2002**, *47*, 2071–2079. [CrossRef]
54. Kniaginicheva, E.; Pismenskaya, N.; Melnikov, S.; Belashova, E.; Sistat, P.; Cretin, M.; Nikonenko, V. Water splitting at an anion-exchange membrane as studied by impedance spectroscopy. *J. Memb. Sci.* **2015**, *496*, 78–83. [CrossRef]
55. Kozmai, A.E.; Nikonenko, V.V.; Pismenskaya, N.D.; Mareev, S.A.; Belova, E.I.; Sistat, P. Use of electrochemical impedance spectroscopy for determining the diffusion layer thickness at the surface of ion-exchange membranes. *Pet. Chem.* **2012**, *52*, 614–624. [CrossRef]
56. Lteif, R.; Dammak, L.; Larchet, C.; Auclair, B. Membrane electric conductivity: A study of the effect of the concentration and nature of the electrolyte and of the structure of the membrane. *Eur. Polym. J.* **1999**, *35*, 1187–1195. [CrossRef]
57. Karpenko, L.V.; Demina, O.A.; Dvorkina, G.A.; Parshikov, S.B.; Larchet, C.; Auclair, B.; Berezina, N.P. Comparative study of methods used for the determination of electroconductivity of ion-exchange membranes. *Russ. J. Electrochem.* **2001**, *37*, 328–335. [CrossRef]
58. Pismenskaya, N.D.; Nevakshenova, E.E.; Nikonenko, V.V. Using a Single Set of Structural and Kinetic Parameters of the Microheterogeneous Model to Describe the Sorption and Kinetic Properties of Ion-Exchange Membranes. *Pet. Chem.* **2018**, *58*, 465–473. [CrossRef]
59. Wenzel, R.N. Resistance of solid surfaces to wetting by water. *Ind. Eng. Chem.* **1936**, *28*, 988–994. [CrossRef]
60. Rahimpour, A.; Jahanshahi, M.; Rajaeian, B.; Rahimnejad, M. TiO2 entrapped nano-composite PVDF/SPES membranes: Preparation, characterization, antifouling and antibacterial properties. *Desalination* **2011**. [CrossRef]
61. Erbil, H.Y. The debate on the dependence of apparent contact angles on drop contact area or three-phase contact line: A review. *Surf. Sci. Rep.* **2014**, *69*, 325–365. [CrossRef]
62. Grafov, B.M.; Ukshe, E.A. *Electrochemical AC Circuits*; Nauka: Moscow, Russia, 1973.
63. Goswami, S.; Klaus, S.; Benziger, J. Wetting and absorption of water drops on nafion films. *Langmuir* **2008**. [CrossRef] [PubMed]
64. Bass, M.; Berman, A.; Singh, A.; Konovalov, O.; Freger, V. Surface structure of nafion in vapor and liquid. *J. Phys. Chem. B* **2010**. [CrossRef]
65. Larchet, C.; Dammak, L.; Auclair, B.; Parchikov, S.; Nikonenko, V. A simplified procedure for ion-exchange membrane characterisation. *New J. Chem.* **2004**. [CrossRef]
66. Newman, J.S. *Electrochemical Systems, Prentice-Hall International Series in the Physical and Chemical Engineering Sciences*; Prentice-Hall: Englewood Clis, NJ, USA, 1972; ISBN 978-0-13-248922-5.
67. Choi, J.H.; Lee, H.J.; Moon, S.H. Effects of electrolytes on the transport phenomena in a cation-exchange membrane. *J. Colloid Interface Sci.* **2001**. [CrossRef] [PubMed]
68. Rubinstein, I.; Zaltzman, B.; Pundik, T. Ion-exchange funneling in thin-film coating modification of heterogeneous electrodialysis membranes. *Phys. Rev. E* **2002**, *65*, 041507. [CrossRef] [PubMed]
69. Rösler, H.-W.; Maletzki, F.; Staude, E. Ion transfer across electrodialysis membranes in the overlimiting current range: Chronopotentiometric studies. *J. Memb. Sci.* **1992**, *72*, 171–179. [CrossRef]
70. Sand, H.J.S., III. On the concentration at the electrodes in a solution, with special reference to the liberation of hydrogen by electrolysis of a mixture of copper sulphate and sulphuric acid. *Lond. Edinb. Dublin Philos. Mag. J. Sci.* **1901**, *1*, 45–79. [CrossRef]

71. Mareev, S.A.; Nichka, V.S.; Butylskii, D.Y.; Urtenov, M.K.; Pismenskaya, N.D.; Apel, P.Y.; Nikonenko, V.V. Chronopotentiometric Response of an Electrically Heterogeneous Permselective Surface: 3D Modeling of Transition Time and Experiment. *J. Phys. Chem. C* **2016**, *120*, 13113–13119. [CrossRef]
72. Sistat, P.; Pourcelly, G. Chronopotentiometric response of an ion-exchange membrane in the underlimiting current-range. Transport phenomena within the diffusion layers. *J. Memb. Sci.* **1997**, *123*, 121–131. [CrossRef]
73. Rybalkina, O.A.; Tsygurina, K.A.; Melnikova, E.D.; Pourcelly, G.; Nikonenko, V.V.; Pismenskaya, N.D. Catalytic effect of ammonia-containing species on water splitting during electrodialysis with ion-exchange membranes. *Electrochim. Acta* **2019**. [CrossRef]
74. Sistat, P.; Kozmai, A.; Pismenskaya, N.; Larchet, C.; Pourcelly, G.; Nikonenko, V. Low-frequency impedance of an ion-exchange membrane system. *Electrochim. Acta* **2008**, *53*, 6380–6390. [CrossRef]
75. Barbero, G. Warburg's impedance revisited. *Phys. Chem. Chem. Phys.* **2016**. [CrossRef] [PubMed]
76. Simons, R. Strong electric field effects on proton transfer between membrane-bound amines and water. *Nature* **1979**, *280*, 824–826. [CrossRef]
77. Zabolotskii, V.I.; Shel'deshov, N.V.; Gnusin, N.P. Dissociation of Water Molecules in Systems with Ion-exchange Membranes. *Russ. Chem. Rev.* **1988**, *57*, 801–808. [CrossRef]

Article

Preliminary Study on Enzymatic-Based Cleaning of Cation-Exchange Membranes Used in Electrodialysis System in Red Wine Production

Myriam Bdiri [1,*], Asma Bensghaier [1], Lobna Chaabane [1], Anton Kozmai [2], Lassaad Baklouti [3] and Christian Larchet [1]

1 Institut de Chimie et des Matériaux Paris-Est (ICMPE), Université Paris-Est, UMR 7182 CNRS, 2 Rue Henri Dunant, 94320 Thiais, France
2 Membrane Institute, Kuban State University, 149 Stavropolskaya Street, Krasnodar 350040, Russia
3 Department of Chemistry Sciences and Arts College, Al-Rass Province Qassim University, BP35, (KSA) Ar Rass 58876, Saudi Arabia
* Correspondence: myriam.bdiri@u-pec.fr; Tel.: +33-(0)1-4978-1173

Received: 7 July 2019; Accepted: 21 August 2019; Published: 3 September 2019

Abstract: The use of enzymatic agents as biological solutions for cleaning ion-exchange membranes fouled by organic compounds during electrodialysis (ED) treatments in the food industry could be an interesting alternative to chemical cleanings implemented at an industrial scale. This paper is focused on testing the cleaning efficiency of three enzyme classes (β-glucanase, protease, and polyphenol oxidase) chosen for their specific actions on polysaccharides, proteins, and phenolic compounds, respectively, fouled on a homogeneous cation-exchange membrane (referred CMX-Sb) used for tartaric stabilization of red wine by ED in industry. First, enzymatic cleaning tests were performed using each enzyme solution separately with two different concentrations (0.1 and 1.0 g/L) at different incubation temperatures (30, 35, 40, 45, and 50 °C). The evolution of membrane parameters (electrical conductivity, ion-exchange capacity, and contact angle) was determined to estimate the efficiency of the membrane's principal action as well as its side activities. Based on these tests, we determined the optimal operating conditions for optimal recovery of the studied characteristics. Then, cleaning with three successive enzyme solutions or the use of two enzymes simultaneously in an enzyme mixture were tested taking into account the optimal conditions of their enzymatic activity (concentration, temperatures, and pH). This study led to significant results, indicating effective external and internal cleaning by the studied enzymes (a recovery of at least 25% of the electrical conductivity, 14% of the ion-exchange capacity, and 12% of the contact angle), and demonstrated the presence of possible enzyme combinations for the enhancement of the global cleaning efficiency or reducing cleaning durations. These results prove, for the first time, the applicability of enzymatic cleanings to membranes, the inertia of their action towards polymer matrix to the extent that the choice of enzymes is specific to the fouling substrates.

Keywords: ion-exchange membrane; tartaric stabilization of wine; enzymatic cleaning; organic fouling

---

1. Introduction

Organic Fouling of ion-exchange membranes (IEMs) represents an important issue for industrials and researchers in all fields of membrane and electromembrane systems. This phenomenon concerns the deposition or the adsorption of undesirable organic matters contained in effluents and beverages treated on the surface or inside the membrane matrix. In IEMs the mechanisms involved in fouling are principally caused by electrostatic interactions between fixed functional sites of the polymer, as well as charged and ionisable molecules in solutions [1]. Interactions are also caused by hydrogen bonds between linked water in matrix and heteroatoms (O, N, or S) of organic foulants, and hydrophobic interactions (π–π or n–π stacking) due to the stacking of aromatic rings in both treated solutions and IEM polymers [2,3]. Organic fouling alters the membranes' physicochemical characteristics by increasing their electrical resistance and changing their surface properties such as hydrophobicity [4–6], and requires the implementation of regular cleaning cycles that increase the process cost in industry.

Chemical cleaning procedure is one of the most common and easy practices for the cleaning of IEMs in electrodialysis (ED) and electromembrane processes in general, especially the clean-in-place (CIP) operations that are currently implemented at an industrial scale. CIP generally involves the use of alternating washing sequences by acid and alkali solutions with or without additional surfactants [7], which generates prolonged and repetitive contacts of polymer materials of IEMs with solutions of low and high pH. Alkali cleaning solutions facilitate the removal of organic particles from the surface by the modification of the ionisable charges, making them more soluble and provoking the saponification of the lipids and acid solutions, which are also efficient for the removal of mineral particles [8]. However, recent studies have demonstrated that prolonged contact with acids and bases generates significant mechanical and structural alterations of the membrane polymer matrix marked by the apparition of structure defects as well as significant membrane swelling which lead to a drop in the IEMs selectivity and performance, and reduce their life duration [4,5,9–12].

It is important to consider the efficiency, processability, and cost of the cleaning methods, as well as the effects of chemical solutions on the membranes' performance and lifetime. It is consequently essential to develop more gentle cleaning strategies with minimum drawbacks on membrane materials.

The application of biological cleaning solutions that involve the use of enzymatic agents could be an advantageous alternative to the chemical solutions or could at least represent a new method that reduces the frequency of chemical cleanings. Enzymes have a three-dimensional protein structure and react very specifically on their substrate with catalytic properties [13], which makes their action specific to the nature of each foulant. To date, enzymes are not communally used for the cleaning of IEMs, and to our knowledge, no scientific research on the enzymatic cleaning of IEMs has been reported. However, enzymatic procedures have been studied and used at an industrial scale for the cleaning of porous membranes based on polymer materials fouled by organic compounds during filtration processes such as polyethersulfone (PES) ultrafiltration (UF) membranes fouled by proteins [14]. Other studies compared enzymatic cleaning to usual chemical cleaning (alkali/acid) of UF membranes fouled with proteins [15], or investigated the use of enzymes with chemical surfactants in alternating cleaning sequences of polysulfone UF membranes fouled with bovine serum albumin (BSA) and whey [16]. Enzyme activity was found to be targeted, efficient, and without negative effects on membrane materials, and to increase the cleaning efficiency when used prior to surfactant agents. Most studies involving porous membranes' cleaning with enzymes have been based on the use of hydrolases, a specific class of enzymes that catalyze hydrolysis reactions and allows degradation of organic matter in an aqueous medium. There are several kinds of hydrolases, and the most efficient in the case of organic fouling during beverage treatments or in the dairy industry are the proteases that hydrolyze peptidic bonds in proteins, the glucanases that hydrolyze osidic bonds in polysaccharides, etc. [13,17]. These actions facilitate the solubilization of organic matter or the degradation of complex structures that could accumulate in the pores (in the case of porous membranes) or in the structure defects and interstitial spaces in the case of dense membranes based on polymer matrix [5,10].

If one takes into account the protein nature of an enzyme, its activity could be highly affected, inhibited, or even irreversibly denaturized by the structure changes that could occur under different conditions, such as pH and temperature. The effectiveness of the enzyme's action also depends on its concentration in the medium in relation to the concentration of the substrates and its chemical structure [13]. Consequently, the use of enzymes in membrane cleaning operations requires prior optimization of the operating conditions in order to improve the washing efficiency [14,16,18]. The use of a moderate pH to comply with the conditions of enzymatic activity is therefore an advantage for ion-exchange membranes, a reason why these kinds of agents could be interesting as an alternative cleaning to the use of chemicals such as acids, bases, detergents, or oxidizing agents.

This work was performed to study the feasibility of enzymatic cleaning of homogeneous CMX-Sb fouled by red wine at advanced stages of tartaric stabilization by ED. Tartaric stabilization is used to reduce the concentration of potassium hydrogen tartrate (KHT) in wine to avoid its precipitation into the bottle and preserve wine quality. Conventional electrodialysis is the most commonly used configuration for this technique, it consists of an electrodialysis cell composed of a succession of chambers. Each ED chamber is formed by a feed compartment (diluate) corresponding to the treated wine, and a brine solution (concentrate) that usually corresponds to a solution of KCl, or KNO_3, etc., and each compartment is formed by a cation- (CEM) and an anion-exchange membrane (AEM). The ED cell is constituted of a succession of chambers formed by alternating the AEM and CEM [19].

In this study, three enzymes were used for their specific action on fouled organic compounds: Rohalase® BXL (β-glucanase for the hydrolysis of the peptide bonds of proteins), Corolase® 7089 (protease for the hydrolysis of high molecular protein into low molecular peptides), and Tyrosinase® (polyphenol oxidase for the oxidation of phenolic acids and polyphenolic substrates). Cleaning tests were performed on industrially fouled CMX-Sb, and the evolution of physicochemical parameters of the treated membranes was determined to estimate the cleaning efficiency. The principal objectives of the cleaning tests performed on the fouled CMX-Sb were: optimizing the operating conditions for a maximum efficiency for each enzyme, and applying the three enzymatic solutions separately, successively, and simultaneously in an enzyme mixture.

2. Materials and Methods

2.1. Membranes

Homogeneous cation-exchange membranes CMX-Sb (Astom, Japan) were used in this study, a batch of new membranes (CMX-Sb(n)) and a batch of industrially used membranes (CMX-Sb(u))—taken from an ED stack after ~2738 h of tartaric stabilization of red wine before replacement.

Used membranes are extensively used in industry and are consequently much damaged. It would be interesting to extend their life duration. CMX-Sb are comprised of a poly(styrene-co-divinylbenzene) (PS-DVB) copolymer with sulfonic groups and reinforced by a PVC cloth. As the same batches of CMX-Sb were previously characterized and studied for chemical cleaning or fouling mechanism identification [5,10], the characteristics determined are summarized in Table 1. It involves the ion-exchange capacity (IEC), thickness (T_m), membrane electric conductivity (κ_m), contact angle (θ), water content or water uptake (W_C), the volume faction of the inter-gel solution f_2, and tensile strength parameters: Young's modulus (E) and rupture point (Rp).

Table 1. Summary of the main characteristics of studied CMX-Sb.

Characteristics	CMX-Sb(n)	CMX-Sb(u)
IEC (mmol g^{-1})	2.47 ± 0.13	1.40 ± 0.06
T_m (μm)	177 ± 3	270 ± 4
κ_m (mS cm^{-1})	13.4 ± 0.6	6.2 ± 0.4
θ (°)	47 ± 1.3	49.3 ± 0.9
W_C (%)	26.1 ± 0.9	40.0 ± 2.1
f_2	0.11 ± 0.01	0.04 ± 0.01
E (MPa)	555 ± 11	264 ± 17
R_p (MPa)	29 ± 3	11 ± 1

IEC: Ion-Exchange Capacity. κ_m: Electrical conductivity. θ: Contact angle. Wc: Water content. f_2: Volume fraction of the inter-gel solution. E: Young's Modulus. / R_p: Rupture point.

In our study, we were mainly interested in following the evolution of the more significant parameters: κ_m, IEC, and θ after enzymatic cleanings. Table 1 shows that after prolonged industrial use, the values of both IEC and κ_m decreased by more than 50% [6,20] under the accumulation of colloidal particles of phenolic compounds that inhibit partially or totally the functional sites. The organic foulants of wine solutions are principally formed of relatively dense low-conducting organic colloidal particles in the inter-gel spaces [21] of the IEM polymer matrices where they replace the electroneutral solution and lead to the decrease in electric conductivity of the fouled membrane [10].

2.2. Enzymatic Agents

Three kinds of enzymes were chosen for the cleaning solutions. The choice of enzymes was made according to their properties to degrade the main molecules present in the wine and susceptible to foul the membranes. Rohalase® BXL (pH_{opt} = 4.5, $T_{storage} \leq 10$ °C) and Corolase® 7089 (pH_{opt} = 7, $T_{storage}$ = [0–6] °C) were purchased from from AB enzyme (Darmstadt, Germany) in liquid form (~5 g L^{-1}).

Tyrosinase® (pH_{opt} = 6.5, $T_{storage} = -20$ °C) was purchased from Sigma Aldrich (St. Louis, MO, USA) in lyophilized form.

Optimum pH for enzymatic activity is given by the suppliers, and it was recommended to work in a temperature interval varying between 30 and 50 °C. Temperature storage was respected to avoid the loss on enzyme activity.

Rohalase® BXL is a β-glucanase compound with high content of β-1.3- and β-1.6-glucanase which principally perform hydrolysis of the osidic bonds in polysaccharides. The enzyme contains high protease side activities that correspond to the hydrolysis of protein peptide bonds. It was obtained from specific cultures of a classically modified Trichoderma strain. Rohalase® BXL belongs to food and specialties enzymes' category which is used for winemaking. The enzyme splits colloids and β-glucane derived from botrytis infected grapes and improves the filterability of wines [22].

Corolase® 7089 is a neutral protease that contains exclusively endopeptidase activity responsible for the hydrolysis of high molecular protein into low molecular peptides. An endopeptidase is a form of protease that breaks the bonds inside the protein chain from several sources. It is obtained from Bacillus subtilis cultures. Corolase® 7089 can be used in the fermentation of a variety of alcoholic beverages such as beer, wine, and spirits [23].

Tyrosinase® principal activity corresponds to the catalysis of the mono and diphenolic compounds oxidation to corresponding quinones with the concomitant reduction of molecular oxygen to water. In addition, Tyrosinases oxidize phenolic acids and polyphenolic substrates such as catechins. It is obtained principally from mushrooms and bacteria [24,25]. This kind of tyrosinase has been chosen to act on the phenolic compounds that are abundant in red grapes, which represent the raw material for red wines and can contain on average 3–6 g of total polyphenols per liter [26].

2.3. Chemicals

Glacial acetic acid (>99.85%), sodium acetate (99 atom% ^{13}C), potassium phosphate dibasic, potassium phosphate monobasic for buffer solutions, and sodium hydroxide (anhydrous; >98%) were purchased from Sigma Aldrich (St. Louis, MO, USA). Hydrochloric acid 37% was purchased from VWR Chemicals (France) and crystalline powder of sodium hydroxide (>99%) from Alfa Aesar (Germany).

2.4. Protocols

New CMX-Sb were stabilized before characterizations and cleaning treatments according to the protocol described in the French standard NF X 45-200 [27] and used CMX-Sb were directly stored in 0.1 M NaCl conditioning solution to avoid the possible evolutions of these membranes.

First, for the optimization of enzyme concentrations, cleaning tests were performed for each enzyme with two different concentrations taken into account the masse of pure enzymes per liter of solvent (0.1 and 1.0 g L^{-1}) at five distinct incubation temperatures (30, 35, 40, 45, and 50 °C). The choice of preconized concentrations was based on two principal information: one given by our enzyme supplier and the other taken from our bibliographic study. The enzyme supplier (AB enzymes) gave a recommended concentrations of 0.5 g L^{-1} for Rohalase® and 0.2 g L^{-1} for Corolase® during at least 2 h for the cleaning of cross-membranes and filter cartridges. Usual concentrations used in previous works for the enzymatic cleaning of porous membranes are 0.1 [14,17] to 0.75 wt % [16].

The solvent of each enzyme corresponded to a specific buffer that maintains its pH for optimal activity (sodium acetate at pH 4.5 for Rohalase®, and potassium phosphate at pH 6.5 and 7 for Corolase® and Tyrosinase®, respectively). For each enzyme, 6 samples of 5 cm^2 were cut from the used CMX-Sb, and each sample was soaked in 10 mL of the diluted enzyme solution at specific concentrations (0.1 or 1.0 g L^{-1}) and temperatures (30, 35, 40, 45, or 50 °C) for 20 h in 50 mL closed flasks. Temperatures were maintained by soaking the flasks in a thermo-regulated bath. The duration of cleaning was chosen to be ten times longer than that recommended by AB enzymes for cleaning cross-membranes and filter cartridges, since CEMs have a high degree of use and are highly fouled.

After choosing an optimal concentration and incubation temperature for each enzyme, a membrane sample of 5 cm^2 was soaked in a 10 mL solution of each enzyme at its optimal pH successively during a 10 h period. The duration was half that implemented for each individual enzyme because the different cleaning modes were combined and to reduce the total time of cleaning from 60 to 30 h for each membrane. Before changing the solution, the sample was rinsed with 100 mL ultra-pure water (Milli-Q water).

To finish, a cleaning operation was performed on a 5 cm^2 membrane sample using a 10 mL mixture of two enzymes, Corolase® and Tyrosinase® because of their close values of optimal pH in a potassium phosphate buffer solution at 6.8 pH for 10 h at their optimal concentration and temperature.

Following all the cleaning operations, the IEC, κ_m, and θ were determined to estimate the efficiency of the membranes' parameter recovery.

2.5. *Analysis*

2.5.1. Ion-Exchange Capacity

The ion-exchange capacity (IEC) was determined by soaking the CMX-Sb sample in 25 mL cm^{-2} of 1 M HCl for 2 h, and then rinsing it with ultra-pure water (Milli-Q water). The sample was then immersed in a solution at 23 mL cm^{-2} of 0.1 M NaCl + 2 mL cm^{-2} of 0.1 M NaOH for at least 2 h. The hydroxide amount was immediately determined by titrating using a 0.01 M HCl. For CMX-Sb(u), the immersion times in the corresponding solutions were doubled to ensure the internal equilibrium of the fouled membranes. Ultimately, the dry mass was determined using a moisture analyzer HB43-S Mettler-Toledo in 40 min at 80 °C.

2.5.2. Electric Conductivity

Membrane electric conductivity (κ_m) was measured in a 1 M NaCl solution using a clip-type cell and a conductivity meter CDM92 (Radiometer) at 25 ± 0.5 °C maintained in a thermo-regulated water reservoir. Two measurements of the NaCl solution's electric resistance were needed, with (R_2) and without (R_1) the membrane. Thus, we used the following equation:

$$\kappa_m = \frac{T_m}{(R_2 - R_1)\,A}$$

where A is the area of the electrodes (1 cm^2).

2.5.3. Contact Angle

An FM40 Easy Drop (KRUSS, Germany) was used to determine the contact angle (θ) between a pure water drop and the membrane surface lightly wiped with a filter paper after rinsing with ultra-pure water (Milli-Q water).

3. Results and Discussion

3.1. Optimizing the Operating Conditions

In order to highlight the evolution of the studied parameters following the 20 h cleaning operations with each enzyme solution at different concentrations (0.1 and 1.0 g L^{-1}) and temperatures (30, 35, 40, 45, and 50 °C), and compared to the initial values for CMX-Sb(n), evolution of the $\kappa_m(t)/\kappa_m(n)$, IEC(t)/IEC(n), and θ(t)/θ(n) ratios are reported in Figures 1–3, respectively. Since the studies CEMs are not perfectly homogeneous due to their method of preparation and the heterogenic distribution of the organic fouling on their whole surface, their parameters could sensitively change from one sample to another. This is due to the fact that the ratios to the initial values were taken into account.

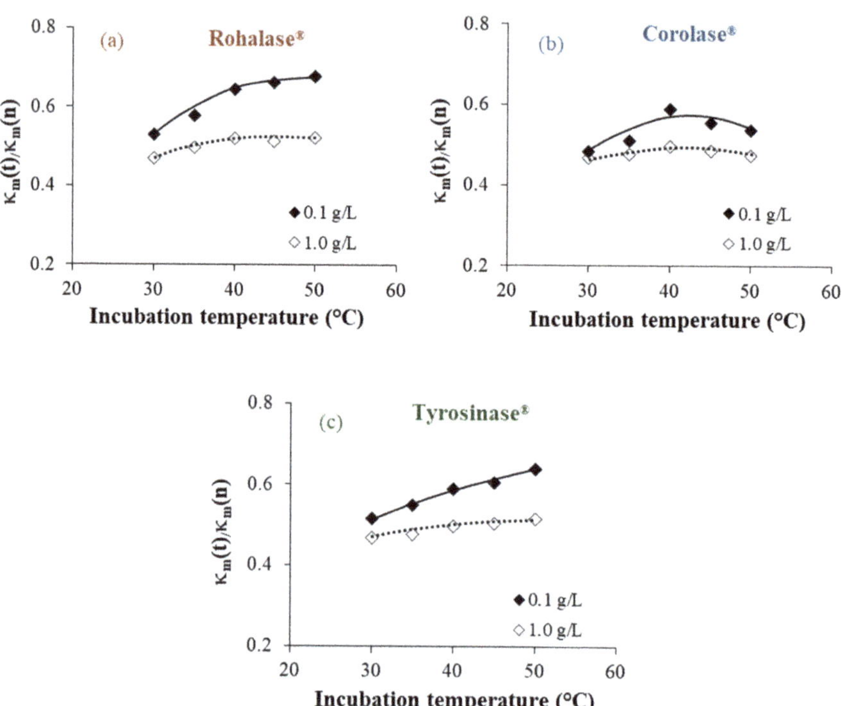

Figure 1. Dependence of $\kappa_m(t)$ for the treated CMX-Sb on incubation temperatures and concentrations of Rohalase® (**a**); Corolase® (**b**), and Tyrosinase® (**c**).

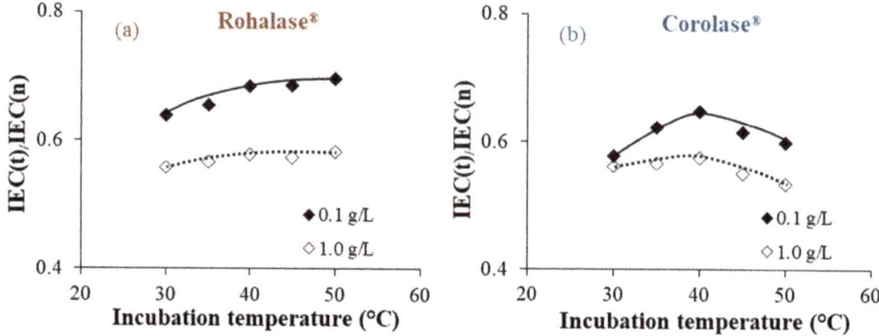

Figure 2. *Cont.*

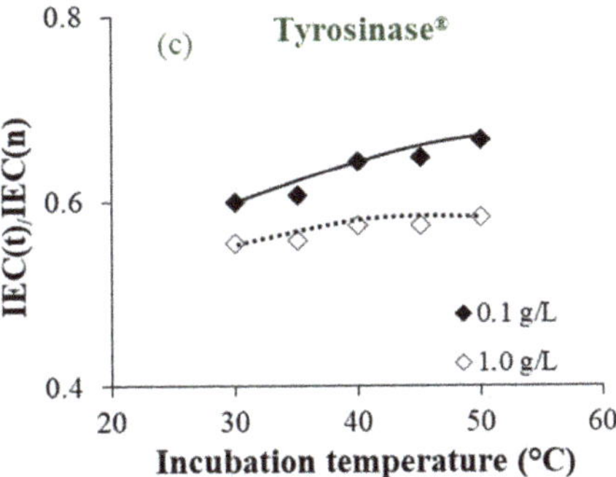

Figure 2. Dependence of the IEC(t) for the treated CMX-Sb on incubation temperatures and concentrations of Rohalase® (**a**); Corolase® (**b**), and Tyrosinase® (**c**).

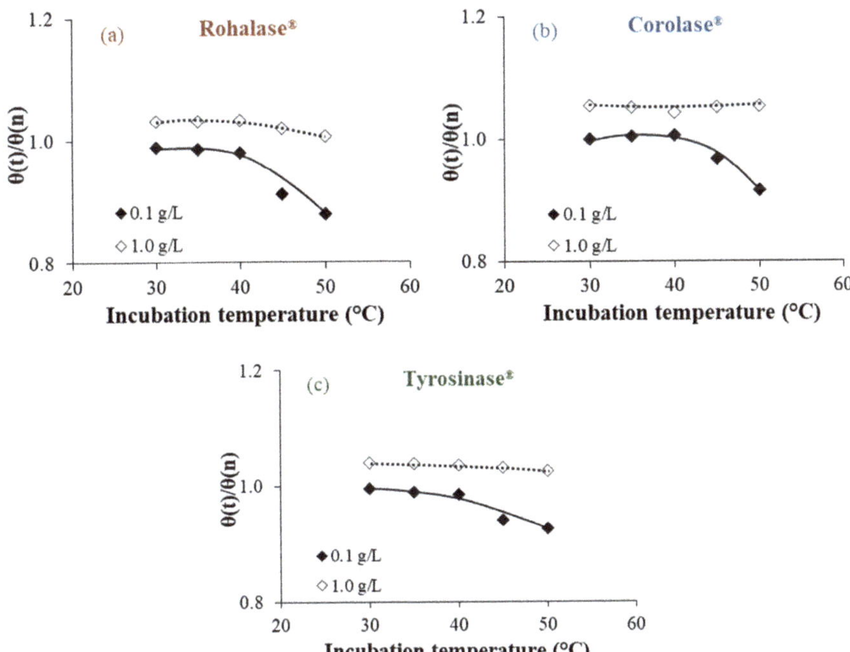

Figure 3. Dependence of the θ(t) for the treated CMX-Sb on incubation temperatures and concentrations of Rohalase® (**a**), Corolase® (**b**), and Tyrosinase® (**c**).

We observed that after 20 h of cleaning, the best results were obtained at lower enzyme concentrations (0.1 g L^{-1}) and at a common temperature of 50 °C. As enzymes already have catalytic activity at higher concentrations (1.0 g L^{-1}), it could be possible that enzymes are proteins that act as foulants. This hypothesis was suggested by Muñoz-Aguado et al. [16], where the same enzymatic cleaning was successively repeated during alternating enzyme-surfactant cleaning sequences on UF membranes fouled with BSA and whey.

Under these conditions, significant percentages of the studied parameters were recovered after the cleaning operations compared to the non-treated membranes. These were the percentages obtained after cleaning with each enzyme solution: Rohalase® (+45% in κ_m, +23% in IEC and −16% of the θ), Corolase® (+ 26% in κ_m, + 14% in IEC and −13% of the θ), and Tyrosinase® (+37% in κ_m, +18%

in IEC and −12% of the θ). The increase observed for the membranes' electric conductivity and the ion-exchange capacity following the cleaning procedures could be attributed to the efficiency of internal cleaning. Our previous studies, on the same CMX-Sb and their homologue anion-exchange membranes (AMX-Sb), show that internal organic fouling after prolonged use leads to partial inhibition of fixed functional sites due to electrostatic interactions with charged groups and to the accumulation of relatively dense low-conducting organic colloidal particles in the inter-gel spaces where they replace the electroneutral solution [5,10]. Despite the three-dimensional protein structures of enzymes, their relatively high molecular weight [13], and the dense structure of the polymer material of the CMX-Sb tested, the enzymes acted upon their specific substrates as shown in Figures 1–3, which confirm that the enzymes penetrated into the polymer matrix. It could be explained by the high degree of swelling of the CMX-Sb(u) (an increase by 50% in thickness and water content was observed for CMX-Sb(u) comparing to CMX-Sb(n) after prolonged industrial use as shown in Table 1) facilitated the penetration of molecules with high molecular weights such as polysaccharides and proteins during beverage treatments.

The decrease in contact angle indicated a decrease in the hydrophobicity of the membrane's surface after treatments due to the partial removal of hydrophobic organic foulants.

The membrane's thickness (T_m) and the tensile strength parameters (E and R_p) of the CMX-Sb were also determined after the different enzymatic cleanings, and no significant changes were noticed. It can be concluded that the activity of the tested enzymes had no negative effects on the mechanical properties of the membranes and that it did not provoke additional swelling of the polymer matrix. Consequently, enzymatic agents proved their efficiency for both external and internal cleaning of the studied CMX-Sb. It should be noted that the chosen enzymes are inert towards membrane matrices as their action is specific to their substrates.

Rohalase® principal activity as β-glucanase enzymes is the hydrolysis of polysaccharides. Furthermore, its activity was the most significant according to the higher recovery percentages observed. Polysaccharides, such as arabinanes, galactanes, arabino–galactanes, etc., with a molecular weight ranging from 40 to 250 kDa or β-glucanes that are linear molecules with molecular weight ranging from 25 and 270 kDa, are abundant raw materials for beverage preparation such as fruits (grape, apple, etc.) [28–30].

Tyrosinase® principal activity is to catalyze the oxidation of mono- and di-phenolic compounds to corresponding quinones with the concomitant reduction of molecular oxygen to water. One of the more studied side activities of bacterial Tyrosinase® is the oxidation of phenolic compounds such as phenolic acids (caffeic acid) and polyphenolic substrates such as catechins [25,31,32]. Phenolic compounds are amphipathic molecules that contain hydrophobic aromatic rings and hydrophilic phenolic hydroxyl groups [26,33]. They are also abundant in beverages (wines especially red wine, fruits and vegetables juices, and derivatives) and reach more than 6 g of total polyphenols per liter of solution. The amount of such foulant compounds vary from a kind of beverage to another but they are highly responsible for the organic fouling of IEM polymer materials even at low concentrations and after a short duration of contact with solutions. It was demonstrated by Sarapulova and al. [3] that during the first hours of contact with red wine (during an operation of fouling simulation on an anion-exchange membrane), strongly hydrated and relatively small organic molecules, such as anthocyanins, penetrate easily into the matrix. The latest components are colored pigments of general structure in C_{15} (C_6–C_3–C_6) and approximate molecular weight of 200 g mol^{-1} (corresponding to 0.2 kDa) that belong to the flavonoids class, a category of phenolic compounds [34,35]. Bigger molecules such as tannins and proteins can enter the membrane due to matrix swelling and the apparition of structure defects (mainly caused by alkali and acidic cleaning operations in industry [5,9,10]) [3]. This interpretation is also valid in the case of CMX-Sb. In fact, due to the nature of functional sites and their negative charge (sulfonic groups), the phenomenon of Donnan exclusion of hydroxyl groups occurs, which decreases the pH value of the internal charged solution in the membrane compared to the externally treated solution that has an acidic pH generally around 3.5 [1,3,36]. Internal CMX-Sb pH in this condition is slightly more acidic, which favors the presence of positively charged anthocyanins. These anthocyanins are rich in chromene cycles that exist in the form of positively charged flavylium cations and could have more tendency to interact with negatively charged functional groups of the matrix. These cycles are also present in condensed tannins (general structure of tannins in (C_{15})) and contribute to fouling [3,37]. Anthocyanins have a great tendency to interact with simple sugars and polysaccharides, or with homologue species such as other anthocyanins or phenolic acids, and be present in condensed forms or in aggregates [26,38]. Consequently, the action of polysaccharide hydrolysis and phenolic compound oxidation could be useful for the decomposition of such complex structures inside the membranes.

The cleaning strategy using the Corolase® enzyme resulted in less recovery compared to the other two enzymes. It is a protease that hydrolyzes proteins and reduces their structure. Even if the amount of proteins is less important than polysaccharides, simple sugars, and phenolic compounds in food beverages, the efficiency of protease proved the effective hydrolysis of proteins. Indeed, proteins are present in wines at a content of ~1.0 g L^{-1} and could be accumulated in swollen polymer matrices of IEMs in simple or condensed forms with tannins [26]. Rohalase® and Tyrosinase® enzymes have a side protease activity, another reason why the results obtained for these two enzymes are more important.

It might be possible to test enzymatic agents as regular cleaning solutions from the beginning of ED treatments during fouling simulation on ED cells in laboratories. It could be predicted that enzyme activity would be more efficient insofar as it could prevent the accumulation of organic foulants and their side reactions such as condensation, polymerization, etc. inside the membranes' matrix, and cleaning would be limited to the surface as membranes would not be subjected to the important swelling phenomenon at the first stage of fouling by organic matters.

3.2. Membrane Cleaning Using Three Enzymatic Solutions Successively and a Mixture of Two Compatible Enzymes

The previous optimization of enzyme concentrations and incubation temperatures determined the optimal operating conditions for each enzyme: 0.1 g L^{-1} at 50 °C for Rohalase®, Corolase®, and Tyrosinase®. After 30 h of successive cleaning (first Rohalase®, second Corolase®, and then Tyrosinase®) of a used CMX-Sb sample, a recovery of ~50% in κ_m, 33% in IEC, and 15% in θ was obtained. It should be taken into account that the duration of cleaning with each enzyme solution was two times less than the tests during operating conditions' optimization. It was reduced to 10 h by each enzyme instead of 20 h. It could be assumed that Rohalase® enzymes were efficient for polysaccharides hydrolysis, followed by proteins, and then effective oxidation of phenolic compounds by the catalytic effect of Tyrosinase®. This order led to a reduction in the size of foulants' complex structures such as tannin–polysaccharide aggregates [33] or the condensed forms of anthocyanins–polysaccharides under the principal action of Rohalase®. Then Tyrosinase® was able to oxidize their free states.

The order of sample soaking in the enzyme solutions was chosen arbitrarily. For our future studies on IEMs cleaning with enzymatic agents, the effect of the order will be studied to optimize the results obtained and to increase the efficiency of the cleaning sequences.

The mixture of enzymes (enzymes cocktail) for the last cleaning test contained both Corolase® (pH_{opt} = 7) and Tyrosinase® (pH_{opt} = 6.5) because of their close values of optimal pH in a potassium phosphate buffer solution at 6.8 pH. The considered concentration was 0.1 g L^{-1} for each enzyme at an incubation temperature of 50 °C and the treatment duration was chosen to be 10 h. This strategy led to 25% increase in κ_m, 18% increase in IEC, and 13% increase in θ. These results could give an idea about the important effect of Rohalase® for intern cleaning and the reduction of polysaccharides and their condensed forms with phenolic compounds (anthocyanins and tannins) on the CMX-Sb fouling.

Table 2 includes a summary of the recovery rates of the significant IEM parameters studied (κ_m, IEC, and θ) after the different enzymatic cleaning operations.

Table 2. Summary of the recovery rates of κ_m, IEC, and θ after the different enzymatic cleanings.

Cleaning Operations	Total Duration (h)	% κ_m (mS cm^{-1})	% IEC (mmol g^{-1})	% θ (°)
Rohalase® BXL	20	+45	+23	−16
Corolase® 7089	20	+26	+14	−13
Tyrosinase®	20	+37	+18	−12
Successive cleanings	30 (10 h/enzyme)	+50	+33	−15
Enzymes cocktail	10	+25	+18	−13

4. Conclusions

This preliminary study includes the enzymatic cleaning of CMX-Sb used for the treatment of red wine by ED in the food industry. Three different enzymes were used for the experiments; Rohalase® BX-BXL (a β-glucanase for the hydrolysis of osidic bonds in polysaccharide), Corolase® 7089 (a neutral protease for the hydrolyses of high molecular protein into low molecular peptides), and Tyrosinase® that principally catalyze the mono- and di-phenolic compounds' oxidation to corresponding quinones, and also the oxidation of phenolic acids and polyphenolic substrates.

The enzymes' activities were found to be more efficient at lower concentrations (0.1 g L^{-1}). It was assumed that at higher concentrations (1.0 g L^{-1}), the enzymes that are basically proteins could foul the polymer matrix.

We identify suitable enzymes for IEM cleaning through this study. However, it would be preferable to choose more specific enzymes taking into account the exact nature of foulants with an adequate evaluation of enzyme activity during the cleaning processes. For ED treatments of red wines, IEMs are in prolonged contact with phenolic compounds, proteins, polysaccharides, and simples sugars, etc. It seems that the best strategy for estimating the cleaning efficiency of enzymatic agents is to study the activity of each enzyme towards its substrate separately on fouled IEM samples after fouling simulations by each of the main foulants. Consequently, the amount of recovered foulants in the cleaning solution could be analyzed (by UV-vis detection, for example) without any interference with other components.

It will be also interesting to investigate enzymatic cleaning on anion-exchange membranes for comparative analysis.

Author Contributions: Conceptualization and methodology, M.B. and A.B.; validation, C.L. and L.C.; formal analysis and investigation, M.B.; data curation, M.B.; writing—original draft preparation, M.B.; writing—review and editing, A.K., L.B., and C.L.; supervision, C.L.; project administration, C.L.

Funding: This research received no external funding.

Acknowledgments: The authors gratefully thank the AB enzyme industry (Darmstadt, Germany) for providing free samples of Rohalase® BXL and Corolase® 7089 in liquid form in their specific buffers (~5 g L^{-1}).

Conflicts of Interest: The authors declare no conflict of interest.

References

1. Langevin, M.-E.; Bazinet, L. Ion-exchange membrane fouling by peptides: A phenomenon governed by electrostatic interactions. *J. Membr. Sci.* **2011**, *369*, 359–366. [CrossRef]
2. Mikhaylin, S.; Bazinet, L. Fouling on ion-exchange membranes: Classification, characterization and strategies of prevention and control. *Adv. Colloid Interface Sci.* **2016**, *229*, 34–56. [CrossRef] [PubMed]
3. Sarapulova, V.; Nevakshenova, E.; Nebavskaya, X.; Kozmai, A.; Aleshkina, D.; Pourcelly, G.; Nikonenko, V.; Pismenskaya, N. Characterization of bulk and surface properties of anion-exchange membranes in initial stages of fouling by red wine. *J. Membr. Sci.* **2018**, *559*, 170–182. [CrossRef]
4. Ghalloussi, R.; Garcia-Vasquez, W.; Chaabane, L.; Dammak, L.; Larchet, C.; Deabate, S.V.; Nevakshenova, E.; Nikonenko, V.; Grande, D. Ageing of ion-exchange membranes in electrodialysis: A structural and physicochemical investigation. *J. Membr. Sci.* **2013**, *436*, 68–78. [CrossRef]

5. Bdiri, M.; Dammak, L.; Chaabane, L.; Larchet, C.; Hellal, F.; Nikonenko, V.; Pismenskaya, N.D. Cleaning of cation-exchange membranes used in electrodialysis for food industry by chemical solutions. *Sep. Purif. Technol.* **2018**, *199*, 114–123. [CrossRef]
6. Ghalloussi, R.; Chaabane, L.; Larchet, C.; Dammak, L.; Grande, D. Structural and physicochemical investigation of ageing of ion-exchange membranes in electrodialysis for food industry. *Sep. Purif. Technol.* **2014**, *123*, 229–234. [CrossRef]
7. Šímová, H.; Kysela, V.; Černín, A. Demineralization of natural sweet whey by electrodialysis at pilot-plant scale. *Desalin. Water Treat* **2010**, *14*, 170–173. [CrossRef]
8. Regula, C.; Carretier, E.; Wyart, Y.; Gésan-Guiziou, G.; Vincent, A.; Boudot, D.; Moulin, P. Chemical cleaning/disinfection and ageing of organic UF membranes: A review. *Water Res.* **2014**, *56*, 325–365. [CrossRef]
9. Garcia-Vasquez, W.; Dammak, L.; Larchet, C.; Nikonenko, V.; Grande, D. Effects of acid–base cleaning procedure on structure and properties of anion-exchange membranes used in electrodialysis. *J. Membr. Sci.* **2016**, *507*, 12–23. [CrossRef]
10. Bdiri, M.; Dammak, L.; Larchet, C.; Hellal, F.; Porozhnyy, M.; Nevakshenova, E.; Pismenskaya, N.; Nikonenko, V. Characterization and cleaning of anion-exchange membranes used in electrodialysis of polyphenol-containing food industry solutions; comparison with cation-exchange membranes. *Sep. Purif. Technol.* **2019**, *210*, 636–650. [CrossRef]
11. Doi, S.; Yasukawa, M.; Kakihana, Y.; Higa, M. Alkali attack on anion exchange membranes with PVC backing and binder: Effect on performance and correlation between them. *J. Membr. Sci.* **2019**, *573*, 85–96. [CrossRef]
12. Doi, S.; Kinoshita, M.; Yasukawa, M.; Higa, M. Alkali Attack on Anion Exchange Membranes with PVC Backing and Binder: II Prediction of Electrical and Mechanical Performances from Simple Optical Analyses. *Membranes* **2018**, *8*, 133. [CrossRef] [PubMed]
13. Fersht, A. *Enzyme Structure and Mechanism*, 2nd ed.; W.H. Freeman: New York, NY, USA, 1985; ISBN 978-0-7167-1614-3.
14. Petrus, H.B.; Li, H.; Chen, V.; Norazman, N. Enzymatic cleaning of ultrafiltration membranes fouled by protein mixture solutions. *J. Membr. Sci.* **2008**, *325*, 783–792. [CrossRef]
15. te Poele, S.; van der Graaf, J. Enzymatic cleaning in ultrafiltration of wastewater treatment plant effluent. *Desalination* **2005**, *179*, 73–81. [CrossRef]
16. Muñoz-Aguado, M.J.; Wiley, D.; Fane, A.G. Enzymatic detergent cleaning of polysul-phone membrane fouled with BSA and whey. *J. Membr. Sci.* **1996**, *117*, 175–187. [CrossRef]
17. Rudolph, G.; Schagerlöf, H.; Morkeberg Krogh, K.B.; Jönsson, A.-S.; Lipnizki, F. Investigations of Alkaline and Enzymatic Membrane Cleaning of Ultrafiltration Membranes Fouled by Thermomechanical Pulping Process Water. *Membranes* **2018**, *8*, 91. [CrossRef] [PubMed]
18. Bilad, M.R.; Baten, M.; Pollet, A.; Courtin, C.; Wouters, J.; Verbiest, T.; Vankelecom, I.F.J. A novel in-situ Enzymatic Cleaning Method for Reducing Membrane Fouling in Membrane Bioreactors (MBRs). *Indones. J. Sci. Technol.* **2016**, *1*, 1–22. [CrossRef]
19. Gonçalves, F.; Fernandes, C.; Cameira dos Santos, P.; de Pinho, M.N. Wine tartaric stabilization by electrodialysis and its assessment by the saturation temperature. *J. Food Eng.* **2003**, *59*, 229–235. [CrossRef]
20. Garcia-Vasquez, W.; Dammak, L.; Larchet, C.; Nikonenko, V.; Pismenskaya, N.; Grande, D. Evolution of anion-exchange membrane properties in a full scale electrodialysis stack. *J. Membr. Sci.* **2013**, *446*, 255–265. [CrossRef]
21. Zabolotsky, V.I.; Nikonenko, V.V. Effect of structural membrane inhomogeneity on transport properties. *J. Membr. Sci.* **1993**, *79*, 181–198. [CrossRef]
22. AB Enzymes, Rohalase Rohament, Products Brochure. 2011. Available online: https://www.abenzymes.com/media/2375/abenzymes_brochure_grainprocessingproducts_jul2018.pdf (accessed on 18 September 2018).
23. AB Enzymes, Corolase 7089, Products Brochure. 2009. Available online: http://usc.ge/files/produqtebi/ludi/eng/4.%20Corolase_7089.pdf (accessed on 18 September 2018).
24. Sigma Aldrich, Tyrosinase, Products Catalog. Available online: https://www.sigmaaldrich.com/catalog/search?term=Tyrosine&interface=All&N=0&mode=partialmax&lang=en®ion=SX&focus=product&F=PR&ST=RS&N3=mode%20matchpartialmax&N5=All (accessed on 18 September 2018).
25. Faccio, G.; Kruus, K.; Saloheimo, M.; Thöny-Meyer, L. Bacterial tyrosinases and their applications. *Process Biochem.* **2012**, *47*, 1749–1760. [CrossRef]

26. Macheix, J.-J.; Fleuriet, A.; Jay-Allemand, C. *Les composés phénoliques des végétaux: un exemple de métabolites secondaires d'importance économique*; Presses Polytechniques et Universitaires Romandes: Lausanne, Switzerland, 2005; ISBN 978-2-88074-625-4.
27. *French Standard NF X 45-200*; AFNOR: La Plaine Saint-Denis, France, 1995.
28. Ulbricht, M.; Ansorge, W.; Danielzik, I.; König, M.; Schuster, O. Fouling in microfiltration of wine: The influence of the membrane polymer on adsorption of polyphenols and polysaccharides. *Sep. Purif. Technol.* **2009**, *68*, 335–342. [CrossRef]
29. Loubser, P. The Role of Polysaccharides in Vinification and their Contribution to Aspects such as Body, Better Mouthfeel and Stability. *Wineland Magazine*, 1 November 2001.
30. Knee, M. Polysaccharides and glycoproteins of apple fruit cell walls. *Phytochemistry* **1973**, *12*, 637–653. [CrossRef]
31. Fairhead, M.; Thöny-Meyer, L. Bacterial tyrosinases: old enzymes with new relevance to biotechnology. *New Biotechnol.* **2012**, *29*, 183–191. [CrossRef] [PubMed]
32. Le Roes-Hill, M.; Palmer, Z.; Rohland, J.; Kirby, B.M.; Burton, S.G. Partial purification and characterisation of two actinomycete tyrosinases and their application in cross-linking reactions. *J. Mol. Catal. B Enzym.* **2015**, *122*, 353–364. [CrossRef]
33. Riou, V.; Vernhet, A.; Doco, T.; Moutounet, M. Aggregation of grape seed tannins in model wine—effect of wine polysaccharides. *Food Hydrocoll.* **2002**, *16*, 17–23. [CrossRef]
34. Bell, E.A.; Charlwood, B.V.; Harborne, J.B. (Eds.) Plant Phenolics. In *Secondary Plant Products*; Encyclopedia of Plant Physiology; Springer: Berlin/Heidelberg, Germany, 1980; Volume 8, ISBN 978-3-642-67362-7.
35. Macheix, J.-J.; Fleuriet, A. *Fruit Phenolics*; CRC Press: Boca Raton, FL, USA, 1990; ISBN 978-0-8493-4968-3.
36. Sarapulova, V.; Nevakshenova, E.; Pismenskaya, N.; Dammak, L.; Nikonenko, V. Unusual concentration dependence of ion-exchange membrane conductivity in ampholyte-containing solutions: Effect of ampholyte nature. *J. Membr. Sci.* **2015**, *479*, 28–38. [CrossRef]
37. Ribéreau-Gayon, P.; Glories, Y.; Maujean, A.; Dubourdieu, D. Phenolic Compounds. In *Handbook of Enology*; John Wiley & Sons, Ltd.: Hoboken, NJ, USA, 2006; ISBN 978-0-470-01039-6.
38. Ibeas, V.; Correia, A.C.; Jordão, A.M. Wine tartrate stabilization by different levels of cation exchange resin treatments: impact on chemical composition, phenolic profile and organoleptic properties of red wines. *Food Res. Int.* **2015**, *69*, 364–372. [CrossRef]

MDPI
St. Alban-Anlage 66
4052 Basel
Switzerland
Tel. +41 61 683 77 34
Fax +41 61 302 89 18
www.mdpi.com

Membranes Editorial Office
E-mail: membranes@mdpi.com
www.mdpi.com/journal/membranes

www.ingramcontent.com/pod-product-compliance
Lightning Source LLC
LaVergne TN
LVHW071033170726
843515LV00006B/1272

* 9 7 8 3 0 3 6 5 2 7 2 4 6 *